T0256127

LOW-POWER CMOS VLSI CIRCUIT DESIGN

LOW-POWER CMOS VLSI CIRCUIT DESIGN

KAUSHIK ROY
Purdue University

SHARAT PRASAD
Cisco

A Wiley Interscience Publication

JOHN WILEY & SONS, INC.

New York / Chichester / Weinheim / Brisbane / Singapore / Toronto

Library of Congress Cataloging in Publication Data:

Roy, Kaushik.
    Low power CMOS VLSI circuit design / Kaushik Roy,
Sharat Prasad.
        p. cm.
    "A Wiley-Interscience publication."
    Includes index.
    ISBN 0-471-11488-X (alk. paper)
    1. Low voltage integrated circuits--Design and
construction. 2. Integrated circuits--Very large
scale integration--Design and construction. 3. Metal
oxide semiconductors, Complementary--Design and
construction. I. Prasad, Sharat. II. Title.
TK7874.66.R69   2000
621.39'5--dc21   98-19426

10 9 8 7 6 5 4 3 2 1

This book is dedicated to our parents
and grandparents

CONTENTS

4 SYNTHESIS FOR LOW POWER   **143**

PREFACE

The scaling of silicon technology has been ongoing for over forty years. We are on our way to commercializing devices having a minimum feature size of one-tenth of a micron. The push for miniaturization comes from the demand for higher functionality and higher performance at a lower cost. As a result, successively higher levels of integration have been driving up the power consumption of chips. Today heat removal and power distribution are at the forefront of the problems faced by chip designers.

In recent years portability has become important. Historically, portable applications were characterized by low throughput requirements such as for a wristwatch. This is no longer true. Among the new portable applications are hand-held multimedia terminals with video display and capture, audio reproduction and capture, voice recognition, and handwriting recognition capabilities. These capabilities call for a tremendous amount of computational capacity. This computational capacity has to be realized with very low power requirements in order for the battery to have a satisfactory life span. This book is an attempt to provide the reader with an in depth understanding of the sources of power dissipation in digital CMOS circuits and to provide techniques for the design of low-power circuit chips with high computational capacity.

Chapter 1 is an introduction to the field of low power CMOS VLSI circuit design.

In Chapter 2, the physics of power dissipation in MOSFET devices and circuits is presented. Expressions for the different components of power dissipation are derived. The emphasis is on submicron and deep-submicron devices. Issues addressed include the short channel effects, for example, the threshold voltage shift and ways of combating undesirable consequences, such as, increased leakage power. To conclude the chapter, limits on low power CMOS technology are derived and ways of approaching those limits are discussed.

Designing the multimillion transistor chips of today would be unthinkable without appropriate computer design tools. The very first computer design tool for low power design have to be power estimators. Power estimation is the subject of Chapter 3. Both average and maximum power estimates are considered. Conventional circuit simulation techniques as well as pattern independent probabilistic techniques are presented. The problem of power estimation in combinatorial logic circuits is difficult owing to spatial correlation among internal signals. The problem of power dissipation in sequential logic circuits is even more difficult owing to the presence of temporal correlation in addition to the spatial correlation.

Computer-aided design (CAD) tools are also widely used to automatically design circuits with gates in an ASIC library. A desired circuit is specified at the register transfer level (RTL) or even the behavioral level. Automatic tools are then used, quite often in an iterative loop, to generate a technology independent logic gate realization to map the circuit to a specific technology library, and finally to optimize the technology mapped circuit. Impressive progress has been made in making these tools account for power dissipation and produce more power-efficient designs. Tools for the automatic synthesis of low power CMOS circuits is the subject of Chapter 4. Besides the tools, the libraries themselves need to be optimized to reduce power dissipation. Towards this goal, new circuit styles have been developed and existing ones improved.

As lowering the supply voltage is a very effective way of lowering power dissipation, Chapter 5 is devoted to the design and test of low-voltage circuits. A large part of designing a low-voltage device is controlling leakage currents. Chapter 5 includes a detailed discussion on controlling leakage currents.

While some of the developments addressing circuit styles can be called incremental, this does not include adiabatic or charge recovery circuits. Adiabatic circuits are considered in Chapter 6. During an adiabatic process no loss or gain of heat occurs. A quasi-adiabatic process approximates this ideal. That very low energy computation could be possible using quasi-adiabatic circuits follows from the second law of thermodynamics. The law states that the entropy of a closed system remains unchanged or increases if the thermodynamic processes in the system proceed from one equilibrium state to another. In a computational process, only the steps that destroy information and increase disorder are subject to a lower limit on energy dissipation. Hence quasi-adiabatic circuits hold the promise of lower energy consumption than the limits that the law imposes on conventional non-adiabatic processes.

Chapter 7 focuses on the design of low power static random-access memory (SRAM). Common forms of storage cells for SRAM are presented.

Then common design and architectural techniques for minimizing power are described. The chapter concludes by presenting a model for quantifying the various components of power dissipation in SRAM.

Chapter 8 looks at how the power dissipation of a system comprised of software as well as hardware components can be reduced. In any system, all of its modules may not be in use at all times. System power management is a general term used to refer to a set of techniques that exploit this fact to minimize power consumption. Modules of a system that are temporarily not in use can have the supply to them turned off or the clock stopped.

In any system based on a stored-program processor, it is the program that directs much of the activity of the hardware. Consequently, besides the hardware, the program or the software can have a significant impact on the power dissipation of such a system. The power dissipation model of a typical application program is described. A variety of software power optimization techniques are then presented. The chapter also highlights the difference between power and energy consumption.

This book had its beginning while both authors were at the erstwhile VLSI Design Laboratory at Texas Instruments (TI), Incorporated. We thank our colleagues at TI, as well Purdue University, Samsung Telecommunications America and Cisco Systems for encouragement and support. The book is a result of nearly eight years of work at Texas Instruments and Purdue University. We would like to thank the graduate students at Purdue who helped form the foundations of many of the Chapters—Liqiong Wei, Dinesh Somasekhar, Khurram Muhammad, Zhanping Chen, Tan-Li Chou, Mark Johnson, Chuan-Yu Wang, Yibin Ye, Priya Patil, N. Sankarayya, Hendrawan Soeleman, Sheldon Zhang, Shiyou Zhao, Rongtian Zhang, Magnus Lundberg, and Ali Keshavarzi. We would also like to thank Rwitti Roy for spending hours reading and correcting the manuscript. We have tried to give appropriate credit to the sources that we have used. We offer apologies for any omissions. Any omission is unintentional and we take full blame for it. All the pieces of our work that we have chosen to include in this book have been published in refereed journals or conferences. We thank the staff of Wiley Interscience and our editor for having patience as we made progress through this book at a highly varying rate. Writing a book while holding down full-time positions invariably leads to trading off some leisure and family time for time at the keyboard. We thank members of our family for being supportive.

KAUSHIK ROY
SHARAT PRASAD

May 1999

CHAPTER 1

LOW-POWER CMOS VLSI DESIGN

1.1 INTRODUCTION

Arguably, invention of the transistor was a giant leap forward for low-power microelectronics that has remained unequaled to date, even by the virtual torrent of developments it forbore. Operation of a vacuum tube required several hundred volts of anode voltage and few watts of power. In comparison the transistor required only milliwatts of power. Since the invention of the transistor, decades ago, through the years leading to the 1990's, power dissipation, though not entirely ignored, was of little concern. The greater emphasis was on performance and miniaturization. Applications powered by a battery—pocket calculators, hearing aids, implantable pacemakers, portable military equipment used by individual soldier and, most importantly, wristwatches—drove low-power electronics. In all such applications, it is important to prolong the battery life as much as possible. And now, with the growing trend towards portable computing and wireless communication, power dissipation has become one of the most critical factors in the continued development of the microelectronics technology. There are two reasons for this:

I. To continue to improve the performance of the circuits and to integrate more functions into each chip, feature size has to continue to shrink. As a result the magnitude of power per unit area is growing and the accompanying problem of heat removal and cooling is worsening. Examples are the general-purpose microprocessors. Even with the scaling down of the supply voltage from 5 to 3.3 and then 3.3 to 2.5 V,

1

power dissipation has not come down. A plateau at about 30 W, possibly as a consequence of the escalating packaging and cooling costs for power densities of the order of 50 W/cm^2 [1], seems to have been reached. If this problem is not addressed, either the very large cost and volume of the cooling subsystem or curtailment in the functionality will have to be accepted.

II. Portable battery-powered applications of the past were characterized by low computational requirement. The last few years have seen the emergence of portable applications that require processing power up until now. Two vanguards of these new applications are the notebook computer and the digital personal communication services (PCSs). People are beginning to expect to have access to same computing power, information resources, and communication abilities when they are traveling as they do when they are at their desk. A representative of what the very near future holds is the portable multimedia terminal. Such terminals will accept voice input as well as hand-written (with a special pen on a touch-sensitive surface) input. Unfortunately, with the technology available today, effective speech or hand-writing recognition requires significant amounts of space and power. For example, a full board and 20 W of power are required to realize a 20,000-word dictation vocabulary [1]. Conventional nickel–cadmium battery technology only provides a 26 W of power for each pound of weight [2–4]. Once again, advances in the area of low-power microelectronics are required to make the vision of the inexpensive and portable multimedia terminal a reality.

As a result, today, in the late 1990s, it is widely accepted that power efficiency is a design goal at par in importance with miniaturization and performance. In spite of this acceptance, the practice of low-power design methodologies is being adopted at a slow pace due to the widespread changes called for by these methodologies. Minimizing power consumption calls for conscious effort at each abstraction level and at each phase of the design process.

1.2 SOURCES OF POWER DISSIPATION

There are three sources of power dissipation in a digital complementary metal–oxide–semiconductor (CMOS) circuit. The first source are the logic transitions. As the "nodes" in a digital CMOS circuit transition back and forth between the two logic levels, the parasitic capacitances are charged and discharged. Current flows through the channel resistance of the transistors, and electrical energy is converted into heat and dissipated away. As suggested by this informal description, this component of power dissipation is proportional to the supply voltage, node voltage swing, and the average switched capacitance per cycle. As the voltage swing in most cases is simply equal to the supply voltage, the dissipation due to transitions varies overall as the square of the supply voltage. Short-circuit currents that flow directly from supply to ground when the n-subnetwork and the p-subnetwork of a CMOS gate both conduct simultaneously are the second source of power dissipation. With input(s) to the gate stable at either logic level, only one of the two subnetworks conduct and no short-circuit currents flow. But when the output of a gate is changing in response to change in input(s), both subnetworks conduct simultaneously for a brief interval. The duration of the interval depends on the input and the output transition (rise or fall) times and so does the short-circuit dissipation. Both of the above sources of power dissipation in CMOS circuits are related to transitions at gate outputs and are therefore collectively referred to as dynamic dissipation. In contrast, the third and the last source of dissipation is the leakage current that flows when the input(s) to, and therefore the outputs of, a gate are not changing. This is called static dissipation. In current day technology the magnitude of leakage current is low and is usually neglected. But as the supply voltage is being scaled down to reduce dynamic power, MOS field-effect transistors (MOSFETs) with low threshold voltages have to be used. The lower the threshold voltage, the lower the degree to which MOSFETs in the logic gates are turned off and the higher is the standby leakage current.

1.3   DESIGNING FOR LOW POWER

The power dissipation attributable to the three sources described above can be influenced at different levels of the overall design process.

Since the dominant component of power dissipation in CMOS circuits (due to logic transitions) varies as the square of the supply voltage, significant savings in power dissipation can be obtained from operation at a reduced supply voltage. If the supply voltage is reduced while the threshold voltages stay the same, reduced noise margins result. To improve noise margins, the threshold voltages need to be made smaller too. However, the subthreshold leakage current increases exponentially when the threshold voltage is reduced. The higher static dissipation may offset the reduction in transitions component of the dissipation. Hence the devices need to be designed to have threshold voltages that maximize the net reduction in the dissipation.

The transitions component of the dissipation also depends on the frequency or the probability of occurrence of the transitions. If a high probability of transitions is assumed and correspondingly low supply and threshold voltages chosen, to reduce the transitions component of the power dissipation and provide acceptable noise margins, respectively, the increase in the static dissipation may be large. As the supply voltage is reduced, the power–delay product of CMOS circuits decreases and the delays increase monotonically. Hence, while it is desirable to use the lowest possible supply voltage, in practice only as low a supply voltage can be used as corresponds to a delay that can be compensated by other means, and steps can be taken to retain the system level throughput at the desired level.

One way of influencing the delay of a CMOS circuit is by changing the channel-width-to-channel-length ratio of the devices in the circuit. The power–delay product for an inverter driving another inverter through an interconnect of certain length varies with the width-to-length ratio of the devices. If the interconnect capacitance is not insignificant, the power–delay product initially decreases and then increases when the width-to-length ratio is increased and the supply voltage is reduced to keep the delay constant. Hence, there exists a combination of the supply voltage and the width-to-length ratio that is optimal from the power–delay product consideration.

The way to assure that the system level throughput does not degrade as supply voltage is reduced is by exploiting parallelism and pipelining. Hence as the supply voltage is reduced, the degree of parallelism or the number of stages of pipelining is increased to compensate for the increased delay. But then the latency increases. Overhead control circuitry also has to be added. As such circuitry itself consumes power, there exists a point beyond which power, rather than decreasing, increases. Even so, great reductions in power dissipation by factors as large as 10, have been shown to be obtainable theoretically [4] as well as in practice [5].

Phenomena that were nonexistent or insignificant earlier are important issues in the submicrometer regime. One of these, the hot-carrier effect, is exhibited when feature size is scaled down while keeping the supply voltage constant. High electric fields result and cause the devices to degrade: Threshold voltages increase, transconductance decreases, and subthreshold currents increase. One solution is to use the lightly doped drain (LDD) structure. As the LDD structure exhibits higher series parasitic resistance, an optimum supply voltage again exists. For larger values of supply voltage, the delay increases. The occurrence of velocity saturation in submicrometer devices makes delay relatively independent of supply voltage [6]. Hence, for not too large a delay penalty, reducing the supply voltage can reduce power dissipation.

As more circuit can be accommodated per unit area, off-chip input–output (I/O)-power may become the dominant power-consuming function [7] unless a significant amount of memory [usually static random-access memory (RAM) and in some cases dynamic RAM] and analog functions are integrated on-chip.

The analog functions may require only 5% of the total transistors but present a complex circuit design and technology challenge. Perhaps a novel substrate technology such as silicon on insulator (SOI) with its much better crosstalk characteristics [8] will enable easier integration of analog and digital circuits.

Circuit level choices also impact the power dissipation of CMOS circuits. Usually a number of approaches and topologies are available for implementing various logic and arithmetic functions. Choices between static versus dynamic style, pass-gate versus normal CMOS realization versus asynchronous circuits have to be made.

At the logic level, automatic tools can be used to locally transform the circuit and select realizations for its pieces from a precharacterized library so as to reduce transitions and parasitic capacitance at circuit nodes and therefore circuit power dissipation. At a higher level various structural choices exist for realizing any given logic functions; for example for an adder one can select one of ripple-carry, carry-look-ahead, or carry-select realizations.

In synchronous circuits, even when the outputs computed by a block of combinatorial logic are not required, the block keeps computing its outputs from observed inputs every clock cycle. In order to save power, entire execution units comprising of combinatorial logic and their state registers can be put in stand-by mode by disabling the clock and/or powering down the unit. Special circuitry is required to detect and power-down unused units and then power them up again when they need to be used.

The rate of increase in the total amount of memory per chip as well as rate of increase in the memory requirement of new applications has more than kept pace with the rate of reduction in power dissipation per bit of memory. As a result, in spite of the tremendous reductions in power dissipation obtained from each generation of memory to the next, in many applications, the major portion of the instantaneous peak-power dissipation occurs in the memory.

In case of dynamic RAM (DRAM) memory, the most effective way to reduce power of any memory size is to lower the voltage and increase the effective capacitance to maintain sufficient charge in the cell. The new array organizations introduced recently [9] present many possibilities of lowering the power. A far greater challenge at this point in time is to get to the next generation of memory chips with a capacity of 4 Gb. At the 1-Gb generation, there is no design margin remaining after the required bit area is subtracted from the available bit area. The implications are that it may not be possible to implement the capacitor-and-transistor cell for the 4 Gb memory in the conventional folded bit line architecture.

1.4 CONCLUSIONS

This chapter was an aerial view of the designing low-power CMOS circuits landscape. The motivations and challenges were emphasized. In the following chapters, we will return to these topics again and address them in detail.

REFERENCES

[1] J. M. C. Stork, "Technology Leverage for Ultra-Low Power Information Systems," *Proc. IEEE*, vol. 83, no. 4, 1995.

[2] T. Bell, "Incredible Shrinking Computers," *IEEE Spectrum*, pp. 37–43, May 1991.

[3] B. Eager, "Advances in Rechargeable Batteries Pace Portable Computer Growth," paper presented at the Silicon Valley Computer Conference, 1991, pp. 693–697.

[4] A. P. Chandrakashan, "Low Power CMOS Digital Design," *IEEE. J. Solid State Circuits*, vol. 27, no. 4, pp. 473–483, 1992.

[5] A. P. Chandrakashan et al., "A Low Power Chipset for Portable MultiMedia Applications," *Proc. IEEE ISSCC*, pp. 82–83, 1994.

[6] H. K. Bakoglu, *Circuits, Interconnections and Packaging for VLSI*, Addison-Wesley, Reading, MA, 1990.

[7] D. K. Liu and C. Svensson, "Power Consumption Estimation in CMOS VLSI Chips," *IEEE J. Solid State Circuits*, vol. 29, pp. 663–670, 1994.

[8] R. B. Merrill et al., "Effect of Substrate Material on Crosstalk in Mixed Analog/Digital Circuits," *IEDM Dig. Tech. Papers*, pp. 433–436, 1994.

[9] M. Takada, "Low Power Memory Design," IEDM Short Course Program, 1993.

CHAPTER 2

PHYSICS OF POWER DISSIPATION IN CMOS FET DEVICES

2.1 INTRODUCTION

This chapter begins with a study of the physics of small geometry MOSFET devices. Many phenomena that are absent in larger geometry devices occur in submicrometer devices and greatly affect performance and power consumption. These phenomena and ways of combating their undesirable effects are described. We begin with the metal–insulator–semiconductor (MIS) structure in Section 2.2.1. Analytical expressions for the threshold voltage of a MIS diode, the depth of the depletion region, the quanta of charges in the inversion layer, and the thickness of the inversion layer are derived. Section 2.2.2 considers the long-channel MOSFET. Impact of the body effect on the threshold voltage is analyzed. A model of the subthreshold behavior of MOSFETs is presented and used to estimate the subthreshold current. The important device characteristic called the *subthreshold swing* is introduced. Many phenomena, which are absent in larger geometry devices, occur in sub-micron devices and greatly affect several aspects of their performance including their power consumption. These phenomena and ways of combating their undesirable effects are described in this section. In Section 2.2.3, a model of the submicrometer MOSFETs based on *drain-induced barrier lowering* is used to study the shift in the threshold voltage due to the short-channel effect. Other submicrometer phenomena—*narrow-gate-width effects, substrate bias dependence,* and the *reverse short-channel effects*—are studied in following sections. Section 2.2.3.2 discusses the subsurface punchthrough and ways of preventing it.

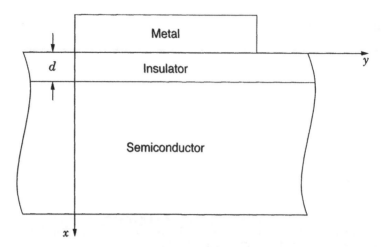

Figure 2.1 The MIS structure.

This study of the physics of MOSFET devices prepares us for the second section, which examines various components of overall power consumption in CMOS very large scale integrated (VLSI) circuit chips. The final section of the chapter derives limits on how far we may be able to lower the power consumption and discusses ways of approaching those limits.

2.2 PHYSICS OF POWER DISSIPATION IN MOSFET DEVICES

2.2.1 The MIS Structure

The stability and reliability of all semiconductor devices are intimately related to their surface conditions. Semiconductor surface conditions heavily influence even the basic working of some devices (e.g., the MOSFET). The MIS structure, besides being a device (a voltage variable capacitor and a diode), is an excellent tool for the study of semiconductor surfaces. In this section the ideal MIS diode will be discussed. Nonideal characteristics will be briefly introduced at the end of the section.

Figure 2.1 shows the MIS structure. A layer of thickness d of insulating material is sandwiched between a metal plate and the semiconductor substrate. For the present discussion, let the semiconductor be of p-type. A voltage V is applied between the metal plate and the substrate. Let us first consider the case when $V = 0$. As we are considering an *ideal* MIS diode, the energy difference ϕ_{ms} between the metal work function[1] ϕ_m and the semi-

---

[1]The work function is defined as the minimum energy necessary for a metal electron in a metal–vacuum system to escape into the vacuum from an initial energy at the Fermi level. In a metal–semiconductor system, the metal work function may still be used but only with the free-space permittivity ε_0 replaced by the permittivity ε_s of the semiconductor medium.

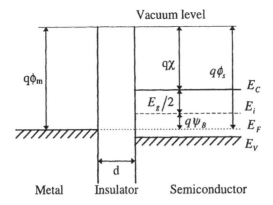

Figure 2.2   Energy bands in an unbiased MIS diode.

conductor work function is zero, that is,

$$\phi_{ms} \equiv \phi_m - \left(\chi + \frac{E_g}{2q} + \psi_B \right) = 0 \tag{2.1}$$

where χ is the semiconductor electron affinity,[2] E_g the band gap, ϕ_B the potential barrier[3] between the metal and the insulator, and ψ_B the potential difference between the Fermi level E_F and the intrinsic Fermi level E_i. Furthermore, in an ideal MIS diode the insulator has infinite resistance and does not have either mobile charge carriers or charge centers. As a result, the Fermi level in the metal lines up with the Fermi level in the semiconductor. The Fermi level in the metal itself is same throughout (consequence of the assumption of uniform doping). This is called the flat-band condition as in the energy band diagram, the energy levels E_c, E_v, and E_i appear as straight lines (Figure 2.2).

When the voltage V is negative, the holes in the p-type semiconductor are attracted to and accumulate at the semiconductor surface in contact with the insulator layer. Therefore this condition is called *accumulation*. In the absence of a current flow, the carriers in the semiconductor are in a state of equilibrium and the Fermi level appears as a straight line. The Maxwell–Boltzmann statistics relates the equilibrium hole concentration to the intrinsic Fermi level:

$$p_0 = n_i \, e^{(E_i - E_F)/kT} \tag{2.2}$$

---

[2] The electron affinity of a semiconductor is the difference in potential between an electron at the vacuum level and an electron at the bottom of the conduction band.
[3] The barrier height is simply the difference between the metal work function and the electron affinity of the semiconductor.

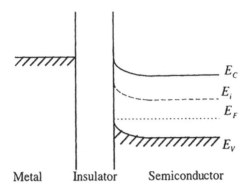

Figure 2.3   Energy bands when a negative bias is applied.

So the intrinsic Fermi level has a higher value at the surface than at a point deep in the substrate and the energy levels E_c, E_v, and E_i bend upward near the surface (Figure 2.3). The Fermi level E_F in the semiconductor is now $-qV$ below the Fermi level in the metal gate.

When the applied voltage V is positive but small, the holes in the p-type semiconductor are repelled away from the surface and leave negatively charged acceptor ions behind. A depletion region, extending from the surface into the semiconductor, is created. This is the *depletion* condition. Besides repelling the holes, the positive voltage on the gate attracts electrons in the semiconductor to the surface. The surface is said to have begun to get *inverted* from the original p-type to n-type. While V is small, the concentration of holes is still larger than the concentration of electrons. This is the *weak-inversion* condition and is important to the study of power dissipation in MOSFET circuits. The bands at this stage bend downward near the surface (Figure 2.4).

If the applied voltage is increased sufficiently, the bands bend far enough that level E_i at the surface crosses over to the other side of level E_F. This is

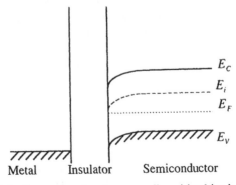

Figure 2.4   Energy bands when a small positive bias is applied.

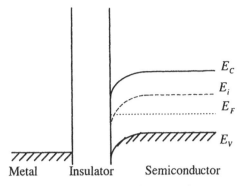

Figure 2.5   Energy bands in strong inversion.

brought about by the tendency of carriers to occupy states with the lowest total energy. The kinetic energy of electrons is zero when they occupy a state at the bottom edge of the conduction band. In the present condition of *inversion* the level E_i bends to be closer to level E_c and electrons outnumber holes at the surface. The electron density at the surface is still smaller than the hole density deep inside the semiconductor. When V is increased to the extent that the electron density at the surface n_s becomes greater than the hole density ($\approx N_A$, the concentration of acceptor impurities) in the bulk, onset of *strong inversion* is said to have occurred. This condition is depicted in Figure 2.5. As we will see in the following section, E_i at the surface now is below E_F by an amount of energy equal to $2\psi_B$, where ψ_B is the potential difference between the Fermi level E_F and the intrinsic Fermi level E_i in the bulk. The value of V necessary to reach the onset of strong inversion is called the *threshold voltage*.

2.2.1.1 Surface Space Charge Region and the Threshold Voltage

In this section we will build a mathematical model of the MIS diode. This model is known as the charge-sheet model [1, 2]. Unlike the simpler model [3] based on the depletion approximation that is accurate only in the strong-inversion and beyond regions of operation, the charge-sheet model remains valid in the weak-inversion and preceding regions of operation. The latter regions are important when power dissipated in submicrometer MOSFETs is considered.

We begin with the Poisson equation

$$\mathbf{\nabla} \cdot \mathbf{D} = \rho(x, y, z) \tag{2.3}$$

where $\mathbf{D}$, the electric displacement vector, is equal to $\varepsilon_s \mathbf{E}$ under low-frequency or static conditions; ε_s is the permittivity of Si; $\mathbf{E}$ the electric field vector; and $\rho(x, y, z)$ the total electric charge density. In a MIS diode the electric

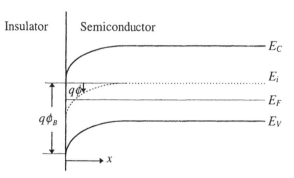

Figure 2.6 Energy bands at the insulator–semiconductor surface.

field due to the applied voltage is normal to the insulator (SiO$_2$). If the fringing fields at the edges are neglected, the variation of the electrostatic potential ϕ can be considered to be only along the x axis (Figure 2.6), that is,

$$\mathbf{E}_y = \frac{\partial \phi}{\partial y} = \mathbf{E}_z = \frac{\partial \phi}{\partial z} = 0$$

Noting that $\rho(x) = q \times [p(x) - n(x) + N_D(x) - N_A(x)]$, (2.3) is transformed to

$$\frac{d^2\phi}{dx^2} = \frac{q}{\varepsilon_S}(p_p - n_p + N_D - N_A) \tag{2.4}$$

where N_D is the concentration of donor impurities, N_A the concentration of acceptor impurities, n_p the mobile electron density, and p_p the mobile hole density. The subscript p is to emphasize that p-type semiconductor is being considered. The two carrier densities at a point x are related to the intrinsic carrier density n_i, Fermi potential ϕ_F, and the electrostatic potential $\phi(x)$ according to Boltzmann statistics:

$$p_p(x) = n_i \, e^{q(\sigma(x) - \phi_p)/kT} \tag{2.5}$$

$$n_p(x) = n_i \, e^{q(\phi(x) - \phi_p)/kT} \tag{2.6}$$

The Fermi potential ϕ_F corresponds to the Fermi energy level $E_F (= -q\phi_F)$. Electrostatic potential is a relative quantity. In the following discussion we measure electrostatic potential relative to the potential which corresponds to the intrinsic Fermi energy level in the bulk $E_i(x = \infty)$, that is, the absolute value of $\phi(x) = \phi(x) + \phi(\infty)$. We will denote the equilibrium hole and electron densities in the bulk by $p_{p^0}(= p_p(\infty)) = n_i \exp q[\phi_F - \phi(\infty)]/kT)$ and n_{p^0}, respectively. Simplifying the right-hand sides of (2.5) and (2.6), substituting in (2.4), multiplying both sides of the resulting equation by $2d\phi/dx$, rearranging, and integrating from a point deep in the bulk to an arbitrary point x [4], we get

$$\int_{\infty}^{x} 2 \frac{d\phi}{dx} \frac{d^2\phi}{dx^2} dx = \int_{\infty}^{x} \frac{d}{dx} \left(\frac{d\phi}{dx}\right)^2 dx$$

$$= \int_{0}^{\phi} 2 \frac{q}{\varepsilon_S} \left(p_{p^0} e^{q\phi(x)/kT} - n_{p^0} e^{q\phi(x)/kT} + N_D - N_A\right) d\phi$$

$$(2.7)$$

At relatively elevated temperatures, most of the donors and acceptors are ionized. So $p_{p^0} \approx N_A$ and $n_{p^0} \approx N_D = n_i^2/N_A$. Assuming Boltzmann statistics, $n_{p^0} = n_i e^{-\beta\phi_B} = p_{p^0} e^{-2\beta\phi_B}$, where $\beta = kT/q$. Then

$$\frac{d\phi}{dx} = -\mathbf{E}(x) = -\sqrt{\frac{2qN_A}{\varepsilon_S}} \sqrt{\frac{e^{-\beta\phi}}{\beta} + \phi - \frac{1}{\beta} + e^{-2\beta\phi_B} \left(\frac{e^{\beta\phi}}{\beta} - \phi - \frac{1}{\beta}\right)}$$

$$(2.8)$$

The electric field at the surface $\mathbf{E}_s$ can be obtained by substituting for ϕ the value of the potential at the surface ϕ_s. To determine the total charges in the semiconductor Q_s, we make use of Gauss's law and obtain

$$Q_s = \varepsilon_s \mathbf{E}_s = -\sqrt{\frac{2q\varepsilon_s N_A}{\beta}} \sqrt{e^{-\beta\phi_s} + \beta\phi_s - 1 + e^{-2\beta\phi_B}\left(e^{\beta\phi_s} - \beta\phi_s - 1\right)}$$

$$(2.9)$$

Since $\phi(x = \infty) = 0$, part of the applied voltage V appears across the insulator and the remaining across the semiconductor, or

$$V = \phi_i + \phi_s = \frac{Q_s}{C_i} + \phi_s = \frac{Q_s d}{\varepsilon_i} + \phi_s$$

where C_i, ε_i, and d are the capacitance, permittivity, and thickness of the insulator. At the onset of strong inversion $\phi_s = 2\phi_B$, and so

$$V_T = \frac{Q_s(\text{strong inversion}) \times d}{\varepsilon_i} + 2\phi_B \qquad (2.10)$$

Or,

$$V_T = \frac{2d}{\varepsilon_i} \sqrt{q\varepsilon_s N_A \phi_B (1 - e^{-2\beta\phi_B})} + 2\phi_B \qquad (2.11)$$

The assumptions made to facilitate the derivation of the expression for the threshold voltage are never strictly true. In particular, the work function difference ϕ_{ms} is never zero and there may be charges present in the insulator or at the insulator–semiconductor boundary. The latter include mobile ionic charges, fixed oxide charges, interface trap charges, and oxide trap charges. Let Q_T be the effective net charge per unit area. The total voltage needed to offset the effect of nonzero work function difference and the presence of the charges is referred to as the *flat-band voltage* V_{FB}. Hence,

$$V_{FB} = \phi_{ms} - \frac{Q_T d}{\varepsilon_i} \qquad (2.12)$$

The voltage V that must be applied to reach the onset of strong inversion must include the flat-band voltage as well. Therefore

$$V_T = V_{FB} + \frac{2d}{\varepsilon_i}\sqrt{q\varepsilon_s N_A \phi_B(1 - e^{-2\beta\phi_B})} + 2\phi_B \qquad (2.13)$$

2.2.1.2 Depth of Depletion Region

The MIS structure is in the *depletion* condition when a small positive bias V is applied between the metal plate and the semiconductor bulk. The *inversion* condition exists when V is large enough as to attract enough minority carriers (electrons) to the surface that their density exceeds the free-hole density in the bulk. It is once again assumed that the semiconductor region is uniformly doped. Two additional simplifying assumptions are invoked. The *depletion assumption* allows us to regard the depletion region as being completely devoid of mobile charges. In the inversion condition the attracted minority carriers are all assumed to be in a very thin inversion layer near the semiconductor surface. The *one-sided abrupt-junction assumption* allows us to regard that the carrier concentrations abruptly change to their intrinsic values at a distance W beneath the surface, where W is the depth of the depletion region. The exponential relation (as we will see shortly) between the total charge in the semiconductor Q_c and d requires negligible increase in d in order to balance the increased charge on the metal when V is increased beyond the onset of strong inversion. Hence it is assumed that d reaches a maximum value of W_m and does not increase further. Similarly the potential at the surface ϕ_s does not increase beyond $2\phi_B$.

Once again we start with the Poisson equation

$$\frac{d^2\phi}{dx^2} = \frac{q}{\varepsilon_S}(p_p - n_p + N_D - N_A)$$

The assumptions above and the fact that $N_D = 0$ in p-type semiconductor allow us to simplify the above to

$$\frac{d^2\phi}{dx^2} = \begin{cases} \dfrac{qN_A}{\varepsilon_S} & 0 \le x < d \\ 0 & x \ge d \end{cases} \qquad (2.14)$$

Integrating twice and applying boundary conditions $\phi(x = 0) = \phi_s$ and $\phi(x = W) = 0$ yields

$$\phi(x) = \phi_s\left(1 - \frac{x}{W}\right)^2 \qquad (2.15)$$

Therefore,

$$2\phi_s\frac{1}{W^2} = \frac{qN_A}{\varepsilon_S}$$

Solving for W,

$$\sqrt{\frac{2\phi_s\varepsilon_S}{qN_A}} = W \qquad (2.16)$$

When $W = W_m$, $\phi_s = 2\phi_B$. Therefore,

$$W_m = \sqrt{\frac{4\phi_B\varepsilon_S}{qN_A}} \qquad (2.17)$$

2.2.1.3 Charge in the Inversion Layer

In a previous section Q_s, the total charge in the semiconductor, was seen to depend on the parameters of the MIS structure according to the following relation:

$$Q_s = \varepsilon_s E_s = -\sqrt{\frac{2q\varepsilon_s N_A}{\beta}} \sqrt{e^{-\beta\phi_s} + \beta\phi_s - 1 + e^{-2\beta\phi_B}\left(e^{\beta\phi_s} - \beta\phi_s - 1\right)}$$

$$(2.18)$$

In this section we will determine the charge in the depletion region due to the ionized atoms left behind when the holes are repelled away by the positive potential on the metal and, in turn, the charge in the inversion layer. The inversion of the semiconductor surface does not begin until $\phi_s \ge \phi_B$. For the range of doping concentrations used in MOSFETs and for the range of temperatures of interest, $9 \le \beta\phi_B \le 16$, the other terms in (2.20) are negligible in comparison with the second and the fourth terms and can be dropped. Thus,

$$Q_s = \sqrt{\frac{2q\varepsilon_s N_A}{\beta}} \sqrt{\beta\phi_s + e^{\beta(\phi_s - 2\phi_B)}} \qquad (2.19)$$

Now the charge per unit area in the semiconductor Q_s is the sum of the charge per unit area in the inversion layer Q_i and the charge per unit area in the depletion region Q_d. The charge in the depletion region is due to the acceptor atoms using an extra electron to complete their covalent bonds. Therefore,

$$Q_d = qN_A W = \sqrt{2q\varepsilon_s N_A \phi_s} \qquad (2.20)$$

From (2.21) and (2.22),

$$Q_i = Q_s - Q_d = \sqrt{2q\varepsilon_s N_A} \left(\sqrt{\frac{\beta\phi_s + e^{\beta(\phi_s - 2\phi_B)}}{\beta}} - \sqrt{\phi_s} \right) \qquad (2.21)$$

We noted above that, in the desired range of values of parameters, $\exp(-\beta\phi_s) \ll 1 < \beta\phi_s$. Then in weak inversion, where $\phi_s < 2\phi_B$, $\beta\phi_s > e^{\beta(\phi_s - 2\phi_B)}$ and using the first two terms in the Taylor series expansion about $e^{\beta(\phi_s - 2\phi_B)} = 0$,

$$\sqrt{\beta\phi_s + e^{\beta(\phi_s - 2\phi_B)}} = \sqrt{\beta\phi_s} + \frac{1}{2\sqrt{\beta\phi_s}} e^{\beta(\phi_s - 2\phi_B)} \qquad (2.22)$$

Substituting in (2.23),

$$Q_i = \frac{\sqrt{2q\varepsilon_s N_A}}{2\beta\sqrt{\phi_s}} e^{\beta(\phi_s - 2\phi_B)} \qquad (2.23)$$

2.2.1.4 Inversion Layer Thickness

As the next step, we will derive an approximation for the thickness of the inversion layer. This is done by assuming that the charge density in the inversion layer is much higher than the density of ionic charge in the bulk and that the inversion layer is very thin. Thus $\partial E_x / \partial x$ in the inversion layer is much greater than in the bulk. Here, $E_x / \partial x$ can be approximated by considering the value of the electric field at the bottom edge of the inversion layer to be zero (Figure 2.7).

The concentration of electrons at some point in the semiconductor is exponentially dependent upon the potential, with the exponential constant being $\beta = kT/q$. This implies that the majority of the charge is contained within a distance from the surface over which ϕ drops by kT/q. To illustrate,

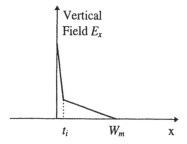

Figure 2.7   Variation of the vertical electrical field.

at the point where the potential drops kT/q below ϕ_s, the electron density will have dropped to $1/e = 0.37$ of its value at $x = 0$. Then we can approximate t_i by this distance. Furthermore, if the electric field in the inversion layer is approximated by the ratio of the potential difference across this layer ($\approx \beta$) divided by its thickness t_i,

$$\mathbf{E}_s \approx \frac{\beta}{t_i} \tag{2.24}$$

or

$$t_i \approx \frac{\beta}{\mathbf{E}_s} \tag{2.25}$$

Since in weak inversion $\phi_s < 2\phi_B$, the expression (2.21) for Q_s can be further simplified by regarding $e^{\beta(\phi_s - 2\phi_B)}$ negligible in comparison to $\beta\phi_s$. Then

$$t_i \approx \frac{\beta}{\mathbf{E}_s} \approx \frac{\beta \varepsilon_s}{Q_s} \approx \frac{\beta\sqrt{\varepsilon_s}}{\sqrt{2qN_A\phi_S}} \tag{2.26}$$

2.2.2 Long-Channel MOSFET

2.2.2.1 Body Effect

The analysis in the previous section assumes the substrate or the bulk electrode to be at zero potential and the voltages at the other terminals are measured with respect to it. When MOSFETs are used in real circuits, the terminal voltages are expressed with respect to the source terminal and the bulk, relative to the source, may be at a nonzero voltage. Since $V_{GS} = V_{GB} - V_{BS}$, if the potential at the surface when the bulk is at zero potential, is ϕ_s, it becomes $\phi_s + V_{BS}$ relative to the source terminal. If the analysis of the previous section is carried out with potentials measured relative to the source terminal, the right-hand side of Eq. (2.13) is found to be

$$V_{\text{FB}} + \frac{d}{\varepsilon_i}\sqrt{2q\varepsilon_s N_A(2\phi_B + V_{BS})(1 - e^{-2\beta\phi_B - V_{BS}})} + 2\phi_B + V_{BS} \tag{2.27}$$

Hence, relative to the source terminal

$$V_T = V_{FB} + \frac{d}{\varepsilon_i}\sqrt{2q\varepsilon_S N_A(2\phi_B + V_{BS})(1 - e^{-2\beta\phi_B - V_{BS}})} + 2\phi_B \quad (2.28)$$

The V_T value obtained from the above equation is greater than the one obtained from Eq. (2.13). This increase in the V_T when the bulk bias voltage V_{BS} is nonzero is termed the body effect.

2.2.2.2 Subthreshold Current

In n-channel MOSFETs, when the gate-to-source voltage V_{GS} is less than the threshold voltage V_T, the condition identified as *weak inversion* in the discussion of the MIS structure exists. The minority-carrier concentration in the channel is small but not zero. Figure 2.8 shows the variation of minority-carrier concentration along the length of the channel.

Let us consider that the source of an n-channel MOSFET is grounded, $V_{GS} < V_T$, and the drain-to-source voltage $|V_{DS}| \geq 0.1$ V. In this condition of weak inversion, V_{DS} drops almost entirely across the reverse-biased substrate–drain p–n junction. As a result, the variation along the channel (the y axis) in the electrostatic potential ϕ_s at the semiconductor surface is small. The y component $\mathbf{E}_y$ of the electric field vector $\mathbf{E}$, being equal to $\partial\phi/\partial y$, is also small. With both the number of mobile carriers and the longitudinal electric field small, the drift component of the subthreshold drain-to-source current $I_{D,\,st}$ is negligible. In addition, the *long* channel allows the gradient of the electric field along the channel to also be considered small and the gradual-channel approximation to be used.

Earlier we observed that if we measure potential relative to the potential deep in the bulk,

$$n_p = \frac{n_i^2}{N_A}e^{\beta\phi_s} \quad (2.29)$$

Because of the exponential dependence of the minority-carrier concentration n_p on the surface potential ϕ_s, $\partial n_p(y)/\partial y$ can be relatively large. Since

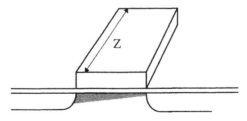

Figure 2.8 Variation of minority concentration in the channel of a MOSEFET biased in weak inversion.

diffusion current is proportional to the carrier concentration gradient, carrier diffusion can and does produce a significant $I_{D,\,st}$. The diffusion current is given by

$$I_{\text{diffusion}} = AJ_{\text{diffusion}} = AqD_n \frac{\partial n(y)}{\partial y} = Zt_i qD_n \frac{\partial n(y)}{\partial y} = ZD_n \frac{\partial Q_i(y)}{\partial y} \quad (2.30)$$

where A is the cross-sectional area of the channel, D_n the electron diffusion coefficient, Z the width of the channel, t_i the thickness of the inversion layer, and Q_i, the per-unit area charge in the inversion layer, is equal to $t_i qn(y)$.

The equilibrium electron concentration is given by

$$n_p = \left(n_i^2/N_A\right)exp(q\beta\phi_s).$$

The charge in the inversion layer in the weak-inversion condition can then be written as

$$Q_i = qt_i n(y) \approx q\frac{\beta\sqrt{\varepsilon_s}}{\sqrt{2qN_A\phi_s}} \frac{n_i^2}{N_A}e^{q\beta\phi_s} \quad (2.31)$$

If n_i^2 in the right-hand side of the above equation is replaced by its approximate value expressed as $N_A^2 \exp(-2q\beta\phi_B)$, we see that the above expression is the same as the one derived earlier for Q_i:

$$Q_i = \frac{\sqrt{2q\varepsilon_s N_A}}{2\beta\sqrt{\phi_s}}e^{\beta(\phi_s - 2\phi_B)}$$

One respect in which a MOSFET differs from a MIS diode is the presence of a potential gradient along the y axis. With the source grounded (i.e., $V_{SB} = 0$), the electron density at the source end of the channel is as given by the expression above with $\phi_s(y)$ replaced by $\phi_s(y = 0)$. At the drain end of the channel, however, V_{DS} must be considered. Then,

$$Q_i(y = 0) = q\frac{\beta\sqrt{\varepsilon_s}}{\sqrt{2qN_A\phi_s(y = 0)}} \frac{n_i^2}{N_A}e^{q\beta\phi_s(y=0)}$$

$$Q_i(y = L) = q\frac{\beta\sqrt{\varepsilon_s}}{\sqrt{2qN_A\phi_s(y = 0)}} \frac{n_i^2}{N_A}e^{q[\beta\phi_s(y=0)-V_{DS}]}$$

At temperatures higher than room temperature, the $\exp(-V_{DS}/\beta)$ term is smaller than $\exp(-4)$. Neglecting this term, the electron concentration gradient along the channel can be approximated as

$$\frac{\partial Q_i(y)}{\partial y} \approx \frac{Q_i(y = L) - Q_i(y = 0)}{L} = \frac{Q_i(y = 0)}{L}$$

Therefore,

$$I_{D,\,st} = D_n Z \frac{\partial Q_i(y)}{\partial y} = \frac{q D_n Z}{L} \frac{\beta \sqrt{\varepsilon_s}}{\sqrt{2 q N_A \phi_s(y=0)}} \frac{n_i^2}{N_A} e^{q\beta\phi_s(y=0)}$$

It is seen that in long-channel MOSFETs, the subthreshold drain–source current remains independent of the drain–source voltage. As $\phi_s(y=0)$ varies exponentially with the applied gate voltage [3], so does the drain–source current. The independence of $I_{D,\,st}$ from V_{DS} ceases even in MOSFET with L as large as 2 μm when V_{DS} is large enough that the source and drain depletion regions merge. This short-channel effect is called *punchthrough*. Punchthrough must be prevented as it causes $I_{D,\,st}$ to become independent of F. This normally means that the punchthrough current must be kept smaller than the long-channel $I_{D,\,st}$ value. In Section 2.2.3.2, methods of using *implants* to control the punchthrough current will be studied.

2.2.2.3 Subthreshold Swing

The inverse of the slope of the log $I_{D,\,st}$ versus V_{GS} characteristic is called the *subthreshold swing*. For uniformly doped MOSFETs,

$$S_{st} = \log\left(\frac{d \ln I_D}{d V_{GS}}\right)^{-1} = 2.3\beta\left(1 + \frac{C_d}{C_{ax}}\right) = 2.3\beta\left(1 + \frac{\varepsilon_s d}{\varepsilon_i W}\right) \quad (2.32)$$

where C_d is the capacitance of the gate depletion layer, C_i the capacitance of the insulator layer, ε_s the permittivity of the semiconductor, ε_i the permittivity of the insulator, d the thickness of the insulator, and W the thickness of the depletion layer. The term S_{st} indicates how effectively the flow of the drain current of a device can be stopped when V_{GS} is decreased below V_T. As device dimensions and the supply voltage are being scaled down to enhance performance, power efficiency, and reliability, this characteristic becomes a limitation on how small a power supply voltage can be used.

The parameter S_{st} is measured in millivolts per decade. For the limiting case of $d \to 0$ and at room temperature, $S_{st} \approx 60$ mV/decade. In practice, the S_{st} of a typical submicrometer MOSFET is ≈ 100 mV/decade. This is due to the nonzero oxide thickness and other deviations from the ideal condition. A S_{st} of 100 mV/decade reduces the $I_{D,\,st}$ from a value of 1 μA/μm at $V_{GS} = V_T = 0.6$ V to 1 pA/μm at $V_{GS} = 0$ V.

It can be noted from the above expression for S_{st} that it can be made smaller by using a thinner oxide (insulator) layer to reduce d or a lower substrate doping concentration (resulting in larger W). Changes in operating conditions, namely lower temperature or a substrate bias, also causes S_{st} to decrease.

2.2.3 Submicron MOSFET

Since the invention of integrated circuits (ICs), both the amount of circuitry on a chip and the speed the circuitry operates at have continued to grow exponentially. In order to accommodate larger amounts of circuitry, the dimensions of the devices—L and Z—have been made progressively smaller. Continuing to increase the speed of operation of the devices has also required the dimensions L and d of the devices to be shrunk every generation. The latter is due to the need to increase $I_{D,\mathrm{st}}$, the drain current in the saturation state of a device, so that the parasitic capacitances can be charged and discharged faster.

When ICs incorporating devices with gate length $L \leq 2$ μm were fabricated, effects in the behaviors of the devices were observed that could not be explained using the prevalent theories of long-channel devices. Of greater immediate interest to us is the fact that the threshold voltage V_T and the subthreshold current $I_{D,\mathrm{st}}$ predicted by analyses in previous sections are not in agreement with observed values for the cases of $L \leq 2$ μm. Here, V_T, expected to be independent of L, Z, and V_{DS}, decreases when L is decreased, varies with Z, and decreases when the drain–source voltage V_{DS} is increased. Also, V_T is seen to increase less rapidly with V_{BS} than in the case of longer channel lengths. In case of devices with $L > 2$ μm, $I_{D,\mathrm{st}}$ is independent of V_{DS} and increases linearly as L decreases. Also, $I_{D,\mathrm{st}}$ increases with increasing V_{DS} and increases more rapidly than linearly as L decreases for the cases of $L \leq 2$ μm.

In this section we will study the effects that are thought to cause these differences in the behavior of MOSFETs with smaller dimensions. In most cases it is not possible to establish an analytical relation between the device characteristics and device parameters. Prevalent theories attempt to provide a qualitative explanation or rely on numerical analysis.

2.2.3.1 Effects Influencing Threshold Voltage

The term V_T, expected to be independent of L, Z, and V_{DS}, decreases when L is decreased, varies with Z, and decreases when the drain–source voltage V_{DS} is increased. Also, V_T is seen to increase less rapidly with V_{BS} than in the case of longer channel lengths. In this section the short-channel-length effect, the narrow-gate-width effect, and the reverse short-channel-length effects and their impacts on the threshold voltage will be examined.

Short-Channel-Length Effect   The undesirability of the V_T decrease with decreasing L and increasing V_{DS} cannot be emphasized enough. The enhancement mode FETs in CMOS are designed to operate at 0.6 V $\leq V_T \leq 0.8$ V. Even a small decrease in V_T causes the leakage currents to become excessive. Also V_T values in the range from 0.6 to 0.8 V in MOSFETs with lightly doped substrates can only be achieved by using V_T-adjust implants to increase the doping concentration at the surface. In the presence of short-channel-length effects, an even higher doping concentration may be required to compensate for the additional V_T decrease. Higher doping concentration, however, has an adverse effect on carrier mobility, subthreshold current, and other device characteristics.

The V_T values obtained from the analyses in the previous section and observed values do not agree when $L \le 2 \ \mu$m. The simplifying assumptions made to facilitate the analyses included that the space charge under the gate is not influenced by V_{DS}. When the channel is relatively long, the drain–substrate and substrate–source depletion regions account for only a small section of the total distance between the drain and the source regions. When L is of the same order of magnitude as the width of the drain–substrate or the substrate–source depletion region, the ionic charge present in these depletion regions represents a reduction in the amount of charge the gate bias has to contribute to the total space charge necessary to bring about inversion. As a result, a smaller V_{GS} appears to suffice to turn on the device. The drain depletion region expands further into the substrate, making the turn-on voltage even smaller when the reverse bias across the drain–substrate junction is increased.

To consider the effect of V_{DS} on the space charge under the gate, the two-dimensional form of the Poisson equation needs to be solved. Exact solution of the two-dimensional Poisson equation can only be obtained numerically. To analytically solve the Poisson equation, various simplifications have been proposed. One of the first simplifications, the *charge-sharing model* [5], considers the charge in the channel to be shared among source, gate, and drain. Assuming the charge controlled by the gate lies within a trapezoidal region, the Poisson equation is reduced to a one-dimensional form and solved to obtain the shift in the threshold voltage. This simple model fails to give good quantitative agreement with observed values.

Drain–induced barrier lowering (DIBL) is the basis for a number of more complex models of the threshold voltage shift. It refers to the decrease in threshold voltage due to the depletion region charges in the potential energy barrier between the source and the channel at the semiconductor surface. In one DIBL-based model, according to Hsu et al. [6], the two-dimensional Poisson equation is reduced to a one-dimensional form by essentially approximating the $\partial^2 \phi / \partial x^2$ term as a constant. This and other DIBL-based models are capable of achieving good agreement with measured data for L as small as 0.8 μm and V_{DS} as large as 3 V.

A recent model, according to Liu et al. predicts the short-channel threshold voltage shift $\Delta V_{T, \text{sc}}$ accurately even for devices with channel length below 0.5 μm [7]. Liu et al. adopt a quasi two-dimensional approach to solving the two-dimensional Poisson equation. The electric field vector **E** is regarded as having a horizontal component E_y and a vertical component E_x. The term E_y is the drain field. The drain field has only a horizontal component. Similarly, E_x is due to the charge on the gate and is the only component of the field due to the charge on the gate. Here, E_y varies with y but not with x; E_x assumes its maximum value at the source end of the channel and then varies (decreases) with y to a minimum value at the drain end. Also, $E_x(x, y)$ has a value at the insulator–semiconductor surface given by $E_x(0, y)$ and goes to zero at the bottom edge of the depletion region, that is, $E_x(W, y) = 0$. Most importantly, it is assumed that $\partial E_x / \partial x$ at each point (x, y) can be replaced with the average of its value at $(0, y)$ and at (W, y) given by

$$\frac{\partial E_x}{\partial x} \approx \frac{E_x(0, y) - E_x(W, y)}{W} = \frac{E_x(0, y)}{W} \tag{2.33}$$

From the condition of continuity of the electric displacement vector,

$$E_x(0, y) = \frac{\varepsilon_{ox}}{\varepsilon_s} E_{ox}(y) \tag{2.34}$$

Furthermore,

$$E_{ox}(y) = \frac{V_T - V_{FB} - \phi_s(y)}{d} \tag{2.35}$$

Invoking the depletion approximation again, the charge in the depletion region is simply the ionic charge, that is, $\rho(x, y) = qN_A$. Substituting these in the Poisson equation,

$$\frac{\partial \mathbf{E}}{\partial x} + \frac{\partial \mathbf{E}}{\partial y} = -\frac{\rho(x, y)}{\varepsilon_s} \tag{2.36}$$

we get

$$\varepsilon_i \frac{V_T - V_{FB} - \phi_s(y)}{d} + \frac{\varepsilon_s W_m}{\eta} \frac{\partial E_y(y)}{\partial y} = qN_A W_m \tag{2.37}$$

where η is an empirical fitting parameter and $W = W_m$ at the onset of strong inversion. Or,

$$\frac{\varepsilon_s W_m}{\eta} \frac{\partial^2 \phi_s(y)}{\partial y^2} + \varepsilon_i \frac{V_T - V_{FB} - \phi_s(y)}{d} = qN_A W_m \tag{2.38}$$

Under boundary conditions $\phi_s(0) = V_{bi}$ and $\phi_s(L) = V_{bi} + V_{DS}$, the solution $\phi_s(y)$ to the above equation is

$$\phi_s(y) = V_{sL} + (V_{bi} + V_{DS} - V_{sL}) \frac{\sinh(y/l)}{\sinh(L/l)}$$

$$+ (V_{bi} - V_{sL}) \frac{\sinh([L - y]/l)}{\sinh(L/l)} \tag{2.39}$$

where $V_{sL} = V_{GS} - V_T$, V_{bi} is the built-in potential at the drain–substrate and substrate–source p–n junctions, and l is the characteristic length defined as

$$l = \sqrt{\frac{\varepsilon_s W_m d}{\varepsilon_i \eta}} \qquad (2.40)$$

$\Delta V_{T,sc}$ is now found by subtracting the long-channel value of ϕ_s at V_T from the minimum of $\phi_s(y)$ given by Eq. (2.39). The minimum of $\phi_s(y)$ is found by evaluating the right-hand side of Eq. (2.39) for a handful of values of y, $0 < y < L$, plotting and fitting a curve to them. Figure 2.9 shows the variation of the surface potential along the channel for channel lengths of 0.35 and 0.8 μm. For each channel length, surface potential has been plotted for $V_{DS} = 0.05$ V and $V_{DS} = 1.5$ V.

The surface potential of the device with $L = 0.8$ μm is seen to remain constant over a significant portion of the channel. This characteristic becomes more pronounced in cases of longer channel lengths. The surface potential of the device with $L = 0.35$ μm, however, does not exhibit a region where its value is unvarying. The minimum surface potential value for the device with $L = 0.35$ μm is greater than that for the device with $L = 0.8$ μm. In fact, the minimum value of the surface potential increases with decreasing channel length and increasing V_{DS}.

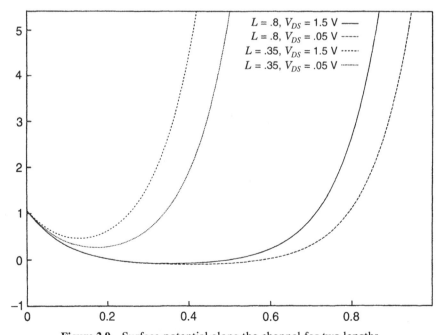

Figure 2.9   Surface potential along the channel for two lengths.

If the expression for the minimum value of $\phi_s(y)$ is subtracted from the right-hand side of Eq. (2.39), an expression for $\Delta V_{T,\text{sc}}$ is obtained. The general form of this expression is complex. When $L > 5l$, the expression for $\Delta V_{T,\text{sc}}$ can be simplified to

$$\Delta V_{T,\text{sc}} \approx \left[3(V_{\text{bi}} - 2\phi_B) + V_{DS}\right]e^{-L/l}$$
$$+ 2\sqrt{(V_{\text{bi}} - 2\phi_B)(V_{\text{bi}} - 2\phi_B + V_{DS})}\, e^{-L/2l} \qquad (2.41)$$

Equation (2.41) can be further simplified for small values of V_{DS} to obtain

$$\Delta V_{T,\text{sc}} \approx \left[2(V_{\text{bi}} - 2\phi_B) + V_{DS}\right](e^{-L/2l} + 2\,e^{L/l}) \qquad (2.42)$$

The values of V_T computed from the above two equations have been compared with measured values [7] and have been found to be in good agreement.

The fitting parameter η in the expression [Eq. (2.37)] for the characteristic length l makes the expression unsuitable for determining the exact value of l. The exact value of l needs to be obtained from measurements of V_T carried out on fabricated devices.

To facilitate empirical determination of l, it can be expressed in an alternative form by relating it to the minimum channel length L_{min} a MOSFET must have so that it exhibits long-channel characteristics. Brew et al. give an empirical expression for L_{min} [8]:

$$L_{\text{min}} = 0.41\left(W_j dW_m^2\right)^{1/3} \qquad (2.43)$$

If it is assumed that L_{min} is $4l$, then

$$l = 0.1\left(W_j dW_m^2\right)^{1/3} \qquad (2.44)$$

It can be shown that for n-channel MOSFETs with an n^+ polysilicon gate, to maintain V_T, given by

$$V_T = V_{\text{FB}} + 2\phi_B + \frac{d}{\varepsilon_i}\frac{4\varepsilon_s\phi_B}{W_m} \qquad (2.45)$$

at a certain value (e.g., 0.7 V), it is necessary that

$$W_m = \frac{d}{\varepsilon_i}\frac{4\varepsilon_s\phi_B}{V_T - V_{\text{FB}} - 2\phi_B} \approx \frac{2\varepsilon_s}{\varepsilon_i}d \qquad (2.46)$$

From (2.44) and (2.46),

$$l = 0.0007 W_j^{1/3} d \qquad (2.47)$$

and for n-channel MOSFETs with a p^+ polysilicon gate [where, assuming $V_T = 1.2$ V, $W_m \approx 4(\varepsilon_s/\varepsilon_i)d$],

$$l = 0.0011 W_j^{1/3} d \qquad (2.48)$$

Dependence on V_{BS}   Equation (2.15), which gives the threshold voltage of a long channel MOSFET, can be rewritten as

$$V_T = V_{FB} + \gamma\sqrt{(2\phi_B + V_{BS})} + 2\phi_B \qquad (2.49)$$

where the $e^{-2\beta\phi_s - V_{BS}}$ term has been dropped as being negligibly small and $\gamma = (d/\varepsilon_i)\sqrt{2q\varepsilon_s N_A}$. For shorter channel lengths and higher drain biases, V_T is less sensitive to V_{BS} than specified by the above equation. Here, V_T becomes altogether independent of V_{BS} for all values of V_{BS} when $L = 0.7$ μm [9] and for large values of V_{BS} in all cases.

Narrow-Gate-Width Effects   In general, the three narrow-gate-width effects discussed in this section have a smaller impact on V_T than the short-channel effects discussed earlier. The first two effects cause V_T to increase and are exhibited in MOSFETs fabricated with either raised field–oxide isolation structures or semirecessed local oxidation (LOCOS) isolation structures. The third effect causes V_T to decrease and is exhibited in MOSFETs fabricated with fully recessed LOCOS or trench isolation structures.

To understand the reason behind the first effect, the channel can be viewed as a rectangle in a horizontal cross section. Two parallel edges border the drain and the source and therefore, fall on depletion regions. The other two edges do not have depletion region under them. The presence of charges under the first two edges brings about a decrease in the amount of charge to be contributed by the voltage on the gate, so the absence of a depletion region under the other two edges implies a larger V_{GS} is required to invert the channel. The effect is an increase in V_T [10].

The second effect arises from the higher channel doping along the edges along the width dimension [11]. The higher doping is due to the channel stop dopants [boron and intentional in case of n-type MOS transistors (NMOST) and a result of oxidation of the piled-up phosphorus in case of p-type MOS transistors (PMOSTs)] encroaching under the gate. Due to the higher doping, a higher voltage must be applied to the gate to completely invert the channel.

In MOSFETs with trench or fully recessed isolation, when the gate is biased, the field lines from the region of the gate overlapping the channel are focused by the edge geometry [12]. Thus an inversion layer is formed at the edges at a lower voltage than that required for the center and gives rise to the third effect.

Reverse Short-Channel Effect   Experimental measurements of V_T with decreasing channel lengths do not bear out the steady decrease expected from the theories outlined in the previous two sections. *Reverse short-channel effect* is the name given to the phenomenon whereby, as the channel length is reduced from $L \sim 3~\mu$m, V_T initially increases until $L \sim 0.7~\mu$m [13]. As L is decreased below 0.7 μm, V_T decreases at a faster rate than predicted by the theories. Researchers have sought and proposed new explanations [13–19], and the effects continue to attract further research.

2.2.3.2   *Subsurface Drain-Induced Barrier Lowering (Punchthrough)*

The depletion regions at the drain–substrate and the substrate–source junctions extend some distance into the channel. Were the doping to be kept constant as L is decreased, the separation between the depletion region boundaries decreases. Increase in the reverse bias across the junctions also leads to the boundaries being pushed farther away from the junction and nearer to each other. When the combination of the channel length and junction reverse biases is such that the depletion regions merge, punchthrough is said to have occurred. In submicrometer MOSFETs a V_T-adjust implant is used to raise the doping at the surface of the semiconductor to a level above that in the bulk of the semiconductor. This causes greater expansion in the portion of the depletion regions below the surface (due to the smaller doping there) than at the surface. Thus punchthrough is first established below the surface.

Any increase in the drain voltage beyond that required to establish punchthrough lowers the potential energy barrier for the majority carriers in the source. A larger number of these carriers thus come to have enough energy to cross over and enter the substrate. Some of these carriers are collected by the drain. The net effect is an increase in the subthreshold current $I_{D,\mathrm{st}}$. Furthermore if $\log(I_{D,\mathrm{st}})$ is plotted versus V_{GS}, the slope of the curve (S_{st}) becomes smaller (i.e., the curve becomes flatter) if subsurface punchthrough has occurred [20].

While S_{st}, or rather an increase in its measured value, serves as an indication of subsurface punchthrough, the device parameter most commonly used to characterize punchthrough behavior is the punchthrough voltage V_{PT} defined as the value of V_{DS} at which $I_{D,\mathrm{st}}$ reaches some specific magnitude with $V_{GS} = 0$. The parameter V_{PT} can be roughly approximated as the value of V_{DS} for which the sum of the widths of the source and the drain depletion regions becomes equal to L [21]:

$$V_{\mathrm{PT}} \propto N_B(L - W_j)^3 \qquad (2.50)$$

where the bulk doping concentration is represented by N_B to distinguish it from the surface doping concentration N_A.

The undesirability of subsurface punchthrough currents for low-power devices cannot be emphasized enough. As these currents flow when the device is off, even a tiny current represents an unproductive leakage. Several techniques have been developed for eliminating subsurface punchthrough.

The obvious one is to suitably select N_B and N_A, on the one hand, to achieve the V_T adjustment, and, on the other hand, to increase the doping in the substrate to reduce depletion region widths. The rule of thumb proposed by Klassen [22] is $N_B > N_A/10$. While this approach has the advantage of requiring only one implant, it fails to satisfy the requirements when $L < 1$ μm. Therefore, other approaches use additional implants, either to form a layer of higher doping at a depth equal to that of the bottom of the junction depletion regions [23] or to form a tip or halo at the leading edge of (toward the drain for the source and vice-versa) the drain and the source regions [24, 25].

2.2.4 Gate-Induced Drain Leakage

A large field exists in the oxide in the region where the n^+ drain of a MOSFET is underneath its gate and the drain and the gate are at V_{DD} and ground potential, respectively (Figure 2.10).

In accordance with Gauss's law, a charge $Q_s = \varepsilon_{ox} E_{ox}$ is induced in the drain electrode. The charge Q_s is provided by a depletion layer in the drain. Because the substrate is at a lower potential for the minority carriers, any minority carriers that may have accumulated and formed an inversion layer at the surface of the drain underneath the gate are swept laterally to the substrate. For this reason the nonequilibrium surface region is called "incipient inversion layer" and the nonequilibrium depletion layer is called "deep depletion layer."

If the field in the oxide, E_{ox}, is large enough, the voltage drop across the depletion layer suffices to enable tunneling in the drain via a near-surface trap. Several trap-assisted tunnelings are thought to be possible [26]. Whatever the mechanism, the minority carriers emitted to the incipient inversion layer are laterally removed to the substrate, completing a path for a gate-induced drain leakage (GIDL) current. In CMOS circuits this leakage current contributes to standby power. The GIDL can be controlled by increasing

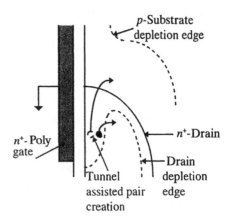

Figure 2.10 Gate-induced drain leakage in MOSFET.

the oxide thickness (reducing the field for a given voltage), increasing the doping in the drain (to limit the depletion width and the tunneling volume), or eliminating traps (assuming voltages and fields low enough that trap-free band-to-band tunneling is not possible).

2.3 POWER DISSIPATION IN CMOS

The first ICs ever fabricated used a PMOS process. This was due to the simplicity of fabrication of a p-channel enhancement mode MOS field-effect transistor (PMOST) with threshold voltage $V_{T_p} < 0$. But soon thereafter the PMOS was replaced by the NMOS as the process of choice for fabricating ICs. The drain–source current in a PMOST results from flow of positively charged holes and in an NMOST from the flow of negatively charged electrons. Since the electron mobility μ_n is always larger than hole mobility μ_p and the switching time for an inverter is inversely proportional to the carrier (hole or electron) mobility, an NMOS inverter has a $\mu_n/\mu_p = 2.5$ times faster transient response than a PMOS inverter. This fact caused the move to the NMOS process. Today NMOS is no longer the dominant process, CMOS is. This change has been brought about by the very consideration central to us, that of power dissipation.

The NMOS inverter dissipates significant amount of power even when its input is not changing (for this reason this component of power dissipation is called *static* power dissipation). Static power dissipation in CMOS is due to leakage currents and is small in comparison to other components. This advantage of CMOS over NMOS has proven to be important enough that the shortcomings of CMOS are overlooked. The CMOS process is more complex than the NMOS, the CMOS requires use of guard-rings to get around the latch-up problem, and CMOS circuits require more transistors than the equivalent NMOS circuits.

Figure 2.11 shows a CMOS inverter. When $V_{\text{in}} = 0$, the NMOST is in cutoff and when $V_{\text{in}} = V_{dd}$, the PMOS is in cutoff. Hence with the input stable at either 0 or V_{dd}, no current flows from V_{dd} to ground. A very small

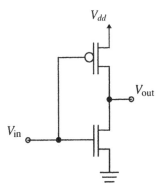

Figure 2.11 A CMOS inverter.

amount of power dissipation, though, does take place. This is due to the leakage currents. With $V_{in} = 0$, the PMOST is conducting and the NMOST is in cutoff. The output capacitor is charged up to V_{dd}. As V_{in} is increased beyond V_{T_n}, the NMOST starts to come out of cutoff and enters the saturation region. The output capacitor starts to discharge through the NMOST and V_{out} decreases. At this time, a conducting path exists for current to flow directly from V_{dd} to ground. When the gate-to-source voltage V_{gs} and the drain-to-source voltage V_{ds} of the PMOST satisfy $|V_{gs}| < |V_{ds}| - |V_{T_p}|$, the PMOST also enters saturation. Here, V_{out} begins to drop steeply until the increasing V_{in} makes the gate-to-source voltage of the NMOST V_{T_n} greater than its V_{ds}. Then the NMOST becomes nonsaturated. Finally when V_{in} has increased to the point that $V_{dd} - V_{in} < |V_{T_p}|$, the PMOST goes into cutoff and the flow of current from V_{dd} to ground ceases. Hence we saw that the current flow consisted of two components, one due to the output capacitor discharging through the NMOST and the second straight from V_{dd} to ground.

Changing V_{in} from 0 to V_{dd} elicits a similar behavior. The current flow consists of a V_{dd}-to-ground component and an output capacitor charging component. In either cases the two components of the current flow give rise to power dissipation, referred to as *dynamic* power dissipation, as it occurs when the gate output is changing. The component of power dissipation due to the flow of current from V_{dd} direct to ground is called *short-circuit* power dissipation. The power dissipated in charging and discharging the load capacitances is called *switching* power.

At this point, two facts about the dynamic component of power dissipation are of interest: It is by far the dominant component and it is proportional to the product of load capacitance C_L and the square of the supply voltage V_{dd}. Hence reduction of dynamic power voltages V_{T_n} and V_{T_p} become critical parameters. The threshold voltages place a limit on the minimum supply voltage that can be used without incurring unreasonable delay penalties. As low as possible values of threshold voltages are desirable. Unfortunately, if the threshold voltage is too low, the static component of the power due to subthreshold currents becomes significant. Detailed analysis of subthreshold leakage is given in Chapter 5. Thus design of devices and selection of supply and threshold voltage requires trade-offs.

2.3.1 Short-Circuit Dissipation

In static CMOS circuits, while $V_{T_n} < V_{in} < V_{DD} - |V_{T_p}|$, where V_{in} is the voltage at a changing, say the ith, input and the voltage at remaining inputs are steady, both the NMOST subnetwork and the PMOST subnetwork conduct and a short-circuit path exists for direct current flow from V_{DD} to the ground terminal. The mean short-circuit power dissipation is then given by $I_{mean}V_{DD}$. Analysis of any static CMOS gate using a circuit analysis program shows that the short-circuit dissipation of the gate varies with the output load and the input signal slope [27]. The short-circuit dissipation decreases in both absolute terms and as a fraction of total dissipation as the

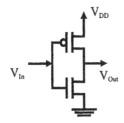

Figure 2.12 An inverter.

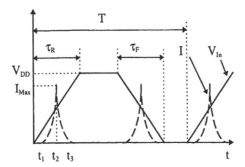

Figure 2.13 Short-circuit current.

output load is increased. Exact analysis of the short-circuit dissipation for a loaded inverter [28] is complex. The following analysis, while simplified, offers insights into the short-circuit dissipation and gives an upper bound.

For simplicity a symmetrical inverter (i.e., $\beta_N = \beta_P$ and $V_{T_n} = -V_{T_p}$; Figure 2.12) and a symmetrical input signal (i.e., rise time $\tau_R =$ fall time $\tau_F = \tau$; Figure 2.13) are considered. The input signal is also periodic with period equal to T. During the interval from t_1 to t_2, the short-circuit current increases from 0 to I_{max}. As, for the NMOST, $V_{DS} > V_{GS} - V_{T_n}$, it will be in saturation. The simple square-law formula then gives the drain current:

$$I = \beta/2(V_{in} - V_T)^2 \quad \text{for } 0 \le I \le I_{max} \tag{2.51}$$

Due to the assumption of the symmetry of the inverter, this current will reach its peak when $V_{in} = V_{DD}/2$ and its waveform will be symmetric about the vertical axis $t = t_2$.

The mean current is given by dividing by T the result of integrating the instantaneous current from $t = 0$ to $t = T$:

$$I_{mean} = \frac{1}{T}\int_0^T I(t)\,dt = 2\frac{2}{T}\int_{t_1}^{t_2}\frac{\beta}{2}(V_{in}(t) - V_T)^2\,dt \tag{2.52}$$

Assuming the rising and falling portions of the input voltage waveform to be linear ramps,

$$V_{in}(t) = \frac{V_{DD}}{\tau}t \tag{2.53}$$

Solving $V_{in}(0) = 0$, $V_{in}(t_1) = V_{DD}$, and $V_{in}(t_2) = 0$, we obtain

$$t_1 = \frac{V_T}{V_{DD}}\tau \quad \text{and} \quad t_2 = \frac{\tau}{2} \tag{2.54}$$

Substituting (2.51), (2.53), and (2.54) into (2.52), we get

$$I_{mean} = 2\frac{2}{T}\int_{(V_T/V_{DD})\tau}^{\tau/2} \frac{\beta}{2}\left(\frac{V_T}{\tau}t - V_T\right)^2 dt \tag{2.55}$$

Let $\theta = (V_T/\tau)t - V_T$; then

$$I_{mean} = \frac{2\beta}{T}\int_{\tau/2}^{(V_T/V_{DD})\tau} \theta \, d\theta \tag{2.56}$$

Solving,

$$I_{mean} = \frac{1}{12}\frac{\beta}{V_{DD}}(V_{DD} - V_T)^3 \frac{\tau}{T} \tag{2.57}$$

Therefore, the short-circuit power dissipation of an unloaded inverter is

$$P_{SC} = \frac{\beta}{12}(V_{DD} - V_T)^3 \frac{\tau}{T} \tag{2.58}$$

We see from Eq. (2.58) that P_{SC} depends upon frequency ($= 1/T$), the supply voltage, and the rise and fall times of the input signal.

The short-circuit power dissipation characteristic of an inverter with a load capacitance C_L can be studied using simulation [27]. It is seen that if the input and the output signal have equal rise and fall times, the short-circuit dissipation is small. However, if the inverter is lightly loaded, causing output rise and fall times that are relatively shorter than the input rise and fall times, the short-circuit dissipation increases to become comparable to dynamic dissipation. Therefore, to minimize dissipation, an inverter should be designed in such a way so that the input rise and fall times are about equal to the output rise and fall times.

2.3.2 Dynamic Dissipation

For an inverter, the average dynamic dissipation can be obtained by summing the average dynamic dissipation in the NMOST and the PMOST (Figure 2.14). Assuming that the input V_{in} is a square wave having a period T and that the rise and fall times of the input are much less than the repetition period, the dynamic dissipation is given by

$$P_D = \frac{1}{T}\int_0^{T/2} i_N(t)V_{out}\, dt + \frac{1}{T}\int_{T/2}^T i_P(t)(V_{DD} - V_{out})\, dt \tag{2.59}$$

Since $i_N(t) = C_L \, dV_{\text{out}}/dt$ and analogously, for $i_P(t)$,

$$P_D = \frac{C_L}{T} \int_0^{V_{DD}} V_{\text{out}} \, dV_{\text{out}} + \frac{C_L}{T} \int_{V_{DD}}^0 (V_{DD} - V_{\text{out}}) \, d(V_{DD} - V_{\text{out}}) = \frac{C_L V_{DD}^2}{T}$$

$$(2.60)$$

The important thing to note in the above expression for the dynamic power is that it is proportional to switching frequency and the square of the supply voltage but is independent of the device parameters. Since $2/T$ is the average number of transitions per second, $C_L V_{DD}^2/2$ is the energy transferred per transition. In the following we will consider another method of computing the energy per transition.

2.3.2.1 *Energy per Transition—Method 2*

Since,

$$P(t) = \frac{dE}{dt} = V_{DD} \times i_{DD}(t) \tag{2.61}$$

assuming that the input voltage is a step applied at $t = 0$ and that leakage current is negligible,

$$i_{DD}(t) = C_L \frac{dV_o}{dt} \tag{2.62}$$

Hence the energy transferred out of the power-supply during a low-to-high transition at the gate output is given by

$$E_{0 \to 1} = \int_0^{t_d} P(t) \, dt = V_{DD} C_L \int_0^V dV_o = C_L V_{DD} V \tag{2.63}$$

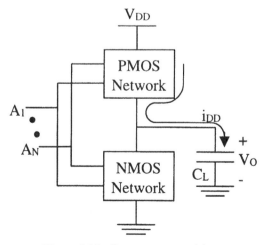

Figure 2.14 Energy per transition.

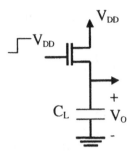

Figure 2.15   Switching energy per transition: partial swing.

where V is the maximum voltage that the load capacitor charges to. Equation (2.61) follows from the assumption that C_L is independent of V_o and does not vary with time. When $V = V_{DD}$, $E_{0 \to 1} = C_L V_{DD}^2$. When energy stored in a capacitor with capacitance C_L and voltage V_{DD} across its plates is $C_L V_{DD}^2/2$, the rest of the energy, another $C_L V_{DD}^2/2$, is converted into heat. This is because of the finite source-drain channel resistance of the PMOSTs that provide the path for the charging current. The remaining $C_L V_{DD}^2/2$ energy stored in the load capacitor is converted into heat when the gate output makes a high-to-low transition and the capacitor discharges through the on-resistance of the NMOSTs. During a high-to-low transition, no additional energy is transferred out of the power-supply. Hence the $E_{0 \to 1}$ in (2.61) is the power dissipated during a pair of transitions, one low-to-high and one high-to-low.

In many instances, a circuit node may not charge to full V_{DD}. Nodes in networks of pass transistors are such cases (Figure 2.15). In Figure 2.15 $V_o \leq V_{DD} - V_t$. Hence, from (2.63), energy transferred during a low-to-high transition of V_o, $E_{0 \to 1} = C_L V_{DD}(V_{DD} - V_t)$.

In dynamic circuits, during the evaluate phase, redistribution of the net charge stored on all the node capacitances occurs. Transient currents flow from one floating circuit node to another through conducting MOSTs and power is consumed. Consider the example circuit in Figure 2.16. During the evaluate phase the node V_o discharges from V_{DD} to $V_{DD} - \Delta V$. During the precharge phase, it is charged back from $V_{DD} - \Delta V$ to V_{DD}. The discharge of node V_o occurs as a result of charge sharing with the node V_{int} via the conducting MOST with input A. Both the nodes V_o and V_{int} are floating due to the absence of conducting paths to either the V_{DD} or the ground terminal.

Assuming $V_{int} \leq V_{DD} - V_t$ (so that it is assured that the MOST with input A is conducting), $V_{DD} - \Delta V = V_{int}$. Since at the start of the evaluate phase, capacitor C_{int} is completely discharged, at the end of the evaluate phase the charge stored in C_{int} is equal to the charge given up by C_L

$$C_{int}(V_{DD} - \Delta V) = C_L \Delta V \tag{2.64}$$

Therefore,

$$\Delta V = \frac{C_{int}}{C_L + C_{int}} V_{DD} \tag{2.65}$$

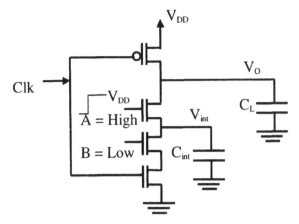

Figure 2.16 Charge sharing in dynamic circuits.

Energy transferred during a complete cycle, E_{tot}, comprising of a precharge and an evaluate phase, is given by

$$E_{tot} = V_{DD}Q = V_{DD}C_L\Delta V = \frac{C_L C_{int}}{C_L + C_{int}}V_{DD}^2 = (C_L\|C_{int})V_{DD}^2 \quad (2.66)$$

2.3.3 The Load Capacitance

It can be observed from (2.60) and (2.66) that the power dissipation of a CMOS inverter is directly proportional to the load capacitance. Besides gates that are in-chip output buffers and that drive chip output and input pins through printed wiring board (PWB) interconnects, all gates in the interior of a chip only drive other gates through on-chip interconnects. The load capacitance of such a gate comprises of a number of parasitic components some of which are shown in Figure 2.17.

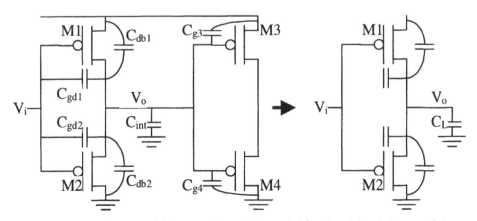

Figure 2.17 An inverter driving another and its model (on the right of the arrow) for computing the parasitic load capacitance.

TABLE 2.1   **Average Gate Capacitance of a MOS Transistor in the Three Regions of Operation**

Operation Region	C_{gb}	C_{gs}	C_{gd}
Cutoff	$C_{ox}WL_{exx}$	0	0
Triode	0	$C_{ox}WL_{exx}/2$	$C_{ox}WL_{exx}/2$
Saturation	0	$(2/3)C_{ox}WL_{exx}$	0

The overall load capacitance is modeled as the parallel combination of four capacitances—the gate capacitance C_g, the overlap capacitance C_{ov}, the diffusion capacitance C_{diff}, and the interconnect capacitance C_{int}.

2.3.3.1 The Gate Capacitance

The gate capacitance is the largest of the four components. It is the equivalent capacitance of three capacitors in parallel and is given by

$$C_g = C_{gs} + C_{gd} + C_{gb} \tag{2.67}$$

where C_{gb} is the sum of the gate-to-bulk capacitances of the two MOSTs in the load inverter and the other two capacitances are the sum of the gate-to-drain/source capacitances of the MOSTs in the load inverter. The value of an individual component capacitance depends on the region of operation of the respective MOST. The values are given in Table 2.1.

The W in Table 2.1 is the sum of the channel widths of the PMOST and the NMOST of the load inverter, $C_{ox} = \varepsilon_{ox}/t_{ox}$ and L_{eff} is the identical channel width of the two MOSTs.

2.3.3.2 The Overlap Capacitance

The overlap capacitances arise from the unwanted lateral diffusion of the drain and source impurities into the channel region under the gate (Figure 2.18). The gate-drain overlap capacitances of the driver inverter need to be considered in addition to the load inverter. Because of the Miller effect, the gate-drain overlap capacitances of the driver inverter appear to be larger than a similar sized load inverter. The gate-drain overlap capacitances of the MOSTs in the driver inverter are given by

$$C_{gd1} = C_{gd2} = 2C_{ox}x_dW \tag{2.68}$$

whereas the gate-to-source/drain capacitances or the MOSTs in the load inverter are given by

$$C_{gd3} = C_{gd4} = C_{gs3} = C_{gs4} = C_{ox}x_dW \tag{2.69}$$

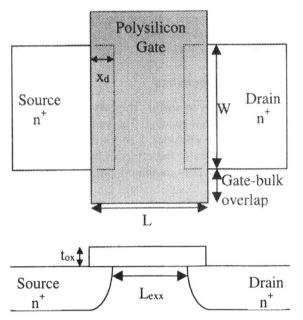

Figure 2.18   Overlap capacitances of a MOS field-effect transistor.

The total overlap capacitance is simply the sum of all the above:

$$C_{ov} = C_{gd1} + C_{gd2} + C_{gd3} + C_{gd4} + C_{gs3} + C_{gs4} \qquad (2.70)$$

2.3.3.3 Diffusion Capacitance

The overall diffusion capacitance is the sum of the diffusion capacitances of the two MOSTs in the driver inverter. Diffusion capacitance for a MOST itself has two components, the bottomwall area capacitance and the sidewall capacitance. The bottomwall capacitance is given by

$$C_{bw} = K_{eq}C_{jo} = \frac{-\phi_0^m}{(V_H - V_L)(1 - m)} \left[(\phi_0 - V_H)^{1-m} - (\phi_0 - V_L)^{1-m} \right] C_{jo}$$

$$(2.71)$$

where C_{jo} is the junction capacitance per unit area under zero bias, $\phi_0 \approx 0.6$ V is the built-in potential for silicon p–n junction, $m \approx \frac{1}{2}$ is the grading coefficient, $V_H \approx V_{DD}$, and $V_L \approx 0$. Observe that C_{jo} and hence C_{bw} increase as V_{DD} decreases.

The sidewall capacitance has contributions from each of the four walls and is given by

$$C_{sw} = WC_{jsw_g} + (W + 2L)C_{jsw_i} \qquad (2.72)$$

where C_{jsw_g} and C_{jsw_i} are the gate and the insulation side per unit gate-width capacitances. C_{jsw_i} depends on the type of insulation and is high for LOCOS and very low for shallow trench isolation (STI).

2.3.3.4 Interconnect Capacitance

As the feature size continues to shrink, there is a corresponding pressure to reduce metal-widths and metal spacing. The small oxide thickness and metal spacing make it necessary to consider, besides the parallel-plate capacitance between the metal and substrate, the effect of fringing fields and of coupling between neighboring wire. The combined parallel-plate and fringing-field capacitance is given by

$$
\begin{aligned}
C_{int_p} &= \varepsilon_{ox} \left[\frac{W}{t_{ox}} - \frac{H}{2t_{ox}} + \frac{2\pi}{\ln\left(1 + \frac{2t_{ox}}{H}\left\{1 + \sqrt{1 + \frac{H}{t_{ox}}}\right\}\right)} \right] \times L \\[2mm]
&\approx C_{p-p} = \frac{\varepsilon_{ins}}{t_{ins}} WL \quad \text{when } \frac{H}{t_{ox}} \ll 1 \text{ and } \frac{W}{t_{ox}} \gg 1 \\[2mm]
&\approx C_x \quad\quad\quad\quad\quad \text{when } \frac{W}{H} \sim 0 \text{ and } \frac{W}{t_{ox}} \to 0 \quad\quad (2.73)
\end{aligned}
$$

where W, H, and L are the width, length and height of the metal wire.

The mutual coupling between two wires and that between a wire and the ground plane contribute to the interconnect capacitance. While the mutual-coupling capacitance decreases with increasing design rule, both the parallel-plate capacitance and the ground-coupling capacitance increase linearly. As a result the total interconnect capacitance first decreases with increasing design rule and then increases [47].

2.4 LOW-POWER VLSI DESIGN: LIMITS

Since the fabrication of the first transistor in the late 1940s, microelectronics has advanced at a breathtaking rate from one transistor per chip to nearly 50 million transistors per chip today. This growth has come about as a net effect of a decline by a factor of $1/50$ in the minimum feature size, a growth by a factor of 170 in the die area, and an improvement in the number of devices per minimum feature area by a factor of 100. In 1975 Gordon Moore of Intel made the astute observation that, for the preceding two decades, the number of transistors per chip had been doubling every year. This observation has become known as Moore's law, and the exponential nature of growth implied by the law remains valid two decades later today, albeit the rate has slowed to 1.5 times per year. The growth in number of transistors per chip has been accompanied by an increase in reliability while the price has remained virtually unchanged. What is of more immediate interest to us is that in the same time period the power–delay product Pt_d or the energy E consumed (transformed from electrical energy to heat and dissipated away) by a binary transition has declined by a factor of $1/10^5$. The decline in the power–delay product per binary transition has been, largely but not entirely, a by-product of scaling. It has nevertheless been an indispensable by-product for without it the problem of heat removal would have become serious enough to come in the way of continued scaling. Microelectronics being undoubtedly the most important technology in the present information era, much attention [29–32] is being paid to the limits upon this scaling that may arise from power dissipation or from any of the other considerations. This section considers the different limits that apply to continued scaling, the state of the VLSI circuit technology vis-à-vis the limits, and ways of realizing and possibly going beyond these limits.

Researchers began to ponder over limits as far back as 1983 [33]. More recently Nagata [29] reviewed physical limitations on MOS devices and the ways in which the limitations constrain the way the devices scale. Hu also considers scaling of MOS devices [30] but with focus on reliability constraints. Meindl described a hierarchy of limits [31] that would govern the realization of billion-transistor chips. Most recently Taur et al. [32] addressed the challenges in further scaling the MOS devices into the sub-100-nm region in light of fundamental physical effects and practical considerations.

2.4.1 Principles of Low-Power Design

To deduce the limits on low-power design, it is important to understand the different aspects—from fundamental physical laws to practical considerations—that govern it. There are three key principles [31, 34] of low-power design: (1) using the lowest possible supply voltage, (2) using the smallest geometry, highest frequency devices but operating them at the lowest possible frequency, (3) using parallelism and pipelining to lower required frequency of operation, (4) power management by disconnecting the power source when the system is idle, and (5) designing systems to have lowest requirements on subsystem performance for the given user level functionality.

2.4.2 Hierarchy of Limits

Meindl [31] defines a hierarchy of limits that has five levels: (1) fundamental, (2) material, (3) device, (4) circuit, and (5) systems. At each level two types of limits exist: (1) from theoretical considerations and (2) from practical considerations.

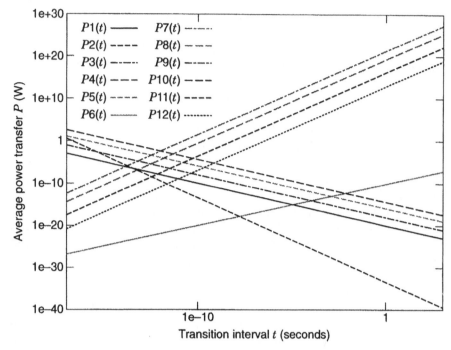

Figure 2.19   Average power transfer during a switching transistor versus the transition interval.

In order to graphically represent the various limits we are about to study, the average power transfer P during a binary switching transition have been plotted versus the transition time t_i in Figure 2.19. Limits imposed by interconnections have been graphically represented by plotting the square of the reciprocal of interconnect length $(1/L^2)$ versus the response time τ in Figure 2.20. For logarithmic scales, diagonal lines in the P-versus-t_d plane represent loci of constant switching energy. Similarly, for logarithmic scales, diagonal lines in the length- $(1/L^2)$ versus-τ plane represent loci of constant distributed resistance–capacitance product for an interconnect.

2.4.3   Fundamental Limits

The fundamental limits are independent of devices, materials, and circuits. They are derived from the basic principles of thermodynamics, quantum mechanics, and electromagnetics.

The limit from thermodynamic principles results from the need to have, at any node with an equivalent resistor R to the ground, the signal power P_s exceed the available noise power P_{avail}:

$$P_s = \gamma P_{avail} = \gamma \frac{\bar{e}_n^2}{4} \frac{1}{R} = \gamma \frac{4kTRB}{4} \frac{1}{R} = \gamma kTB$$

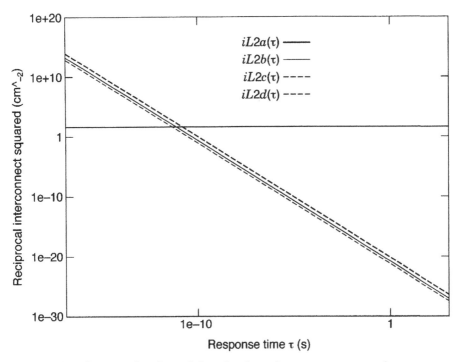

Figure 2.20 Square of reciprocal length of an interconnect versus interconnect circuit response time.

where $\gamma \geq 1$ is some constant factor, $\bar{e}_n^2$ is open-circuit mean-square voltage across the equivalent resistor, k is Boltzmann's constant, T is the absolute temperature, and B is the bandwidth of the node. The corresponding loci is labeled $P1(t)$ in Figure 2.19. Because of consideration that we will discuss in a later section, $\gamma = 4$ is recommended. Then at $T = 300$ K, P_s must be larger than 0.104 eV. In practice the signal power today is larger by a factor of nearly 10^7.

The quantum theoretic limit on low power comes from the Heisenberg uncertainty principle. In order to be able to measure the effect of a switching transition of duration Δt, it must involve an energy greater than $h/\Delta t$:

$$P \geq \frac{h}{(\Delta t)^2}$$

where h is the Planck's constant. The corresponding loci is labeled $P2(t)$ in Figure 2.19. Finally the fundamental limit based on electromagnetic theory results in the velocity of propagation of a high-speed pulse on an interconnect to be always less than the speed of light in free space, c_0:

$$\frac{L}{\tau} \le c_0$$

where L is the length of the interconnect and τ is the interconnect transit time. The corresponding loci is labeled $iL2a(\tau)$ in Figure 2.20.

2.4.4 Material Limits

Material limits are independent of the particular devices built with the materials and, in turn, the particular circuits composed from those devices. The attributes of a semiconductor material that determine the properties of a device built with the material are (1) carrier mobility μ, (2) carrier saturation velocity, σ_s, (3) self-ionizing electric field strength E_c, and (4) thermal conductivity K.

Semiconductor material limits that are independent of the structures and the geometry of devices can be calculated by considering a cube of the undoped material of dimension Δx that is imbedded in a three-dimensional matrix of similar cubes. The voltage difference V_0 across a pair of its opposite faces is just as large as necessary to produce an electric field equal to the self-ionizing electric field strength ε_c, that is, $V_0/\Delta x = E_c$. The limit on switching energy and the switching time can then be calculated as the amount of electrostatic energy stored in the cube and the transit time of a carrier through the cube:

$$Pt_d = E \frac{\varepsilon_m V_0^3}{2E_c} \quad \text{and} \quad t_d > \frac{V_0}{\sigma_s E_c}$$

Solving,

$$P = \tfrac{1}{2} \varepsilon_m E_c^2 \sigma_s^3 t_d^2$$

The corresponding loci is labeled $P3(t)$ in Figure 2.19.

The second material level limit arises from heat removal considerations. To derive this limit, an isolated generic device that resides in an ideal heat sink maintained at temperature T_0 is considered. The device is hemispherical in shape with a radius of $r_s = \sigma_s t_d$. The power or the rate of transfer of the heat energy from the device to the heat sink is then given by

$$P = \frac{Q}{t_d} = -KA \frac{dT}{dx} = -K\pi (\sigma_s t_d)^2 \left(-\frac{\Delta T}{\sigma_s t_d} \right) = \pi K \sigma, \Delta T t_d$$

where we have used Fourier's law of heat conduction, K is the thermal conductivity of the semiconductor material, A is the surface area through which the heat flow is occurring, and dT/dx is the temperature gradient. The corresponding locus is labeled $P4(t)$ in Figure 2.19.

An interesting use of the above limit is to compare suitability of GaAs and Si for low-power applications. Using representative values, P/t_d comes out to be 0.21 nS/W for Si and 0.69 ns/W for GaAs. This shows that while GaAs has the speed advantage demonstrated earlier, it needs to conduct away three times as much heat for the same transition time.

If we now consider an SOI structure by surrounding the above generic device in a hemispherical shell of SiO_2 of radius r_i, the thermal conductivity of the structure as a whole is given by [31]

$$K_{eq} = \left(\frac{K_{ox} K_{Si} r_i}{r_s} \right) \left[K_{Si} \left(\frac{r_i}{r_s} - 1 \right) + K_{ox} \right]$$

By substituting $K_{ox} \approx 0.1 K_{Si}$ and $r_i = 1.5 r_s, 2 r_s, 4 r_s$, we obtain $K_{eq} \approx 0.029 K_{Si}, 0.02 K_{Si}, 0.013 K_{Si}$, indicating a two-order-of-magnitude reduction in thermal conductivity.

The interconnect material limit again arises from speed-of-light considerations. The propagation time through an interconnect of length L of a material with a relative dielectric constant ε_r must satisfy

$$t_d \geq \frac{L}{c_0/\sqrt{\varepsilon_r}}$$

The corresponding locus is labeled $iL2b(\tau)$ in Figure 2.20.

2.4.5 Device Limits

The device limits are independent of the circuits that may have been composed with the devices. As the MOSFET device is used significantly more than any other, it will be considered in this section. The most important attribute of a MOSFET is its allowable minimum effective channel length L_{min} [35]. As we saw earlier in the chapter, MOSFET devices with small effective channel length exhibit the undesirable short-channel effects. Hence means of achieving L_{min} while minimizing the short-channel effects are required. In order to achieve L_{min}, both the gate oxide thickness T_{ox} and source–drain junction depth X_j should be as small as possible [36, 37]. Decreasing T_{ox} leads to increased tunneling leakage currents [29] and decreasing X_j leads to increase in parasitic source–drain conductance [38].

As we saw earlier, short-channel effects in bulk MOSFETs can be controlled by using channels with lower impurity concentration and abrupt retrograde doping profiles. Studies have shown the use of dual gates on the two sides of a channel as being effective in controlling short-channel effects [39–41]. Agarwal et al. [36] analyzed six different MOSFET structures using a combination of options from deep versus shallow junction and uniform versus low-impurity channel in case of bulk MOSFETs and single versus dual gate in case of SOI MOSFETs. The results of analysis indicate the possibility of shallow-junction retrograde channel profile bulk MOSFETs with channel length as short as 60 nm and dual-gate or delta MOSFETs with channel length as short as 30 nm.

Besides the channel length, the thickness of the oxide layer and its permittivity are important parameters for controlling the short-channel-length effects as indicated by the expression for the threshold voltage shift given below (derived earlier in the chapter):

$$\Delta V_T \approx e^{L \varepsilon_{ox} / \pi \varepsilon_{Si} T_{ox}}$$

MOSFET device leakage current and its overall reliability are affected by factors other than the threshold voltage shift, for example, bulk punchthrough, gate-induced drain barrier lowering, and impact ionization. In determining the MOSFET device limits, the latter need to be considered too.

The energy transferred during a switching transition is stored on the gate of a MOSFET prior to the transition. Therefore, the minimum effective channel length translates into a limit on the switching energy given by

$$E = \tfrac{1}{2} C_0 L_{min}^2 V_0^2$$

Since

$$E = P t_d = P \frac{L_{min}}{v_{sat}}$$

$$P = \tfrac{1}{2} (C_0 L_{min}) \frac{V_0^2}{L_{min}^2} v_{sat}^3 t_d^2$$

If the minimum transition times corresponding to material and device limits are computed for even conservative value of L_{min}, and T_{ox} equal to 100 and 3 nm, respectively, it is seen that the difference between them is small [loci labeled $P4(t)$ and $P7(t)$ in Figure 2.19]. This indicates that the MOSFET device limits are already pushing against the material limits of silicon.

At the device level, a global interconnect can be modeled as a canonical distributed resistance–capacitance network. When such a network is driven by an ideal voltage source that applies a unit step function, the 0–90% response time of the network is given by [31]

$$\tau = RC = \frac{\rho}{H_\rho} \frac{\varepsilon}{H_\varepsilon} L^2$$

where ρ/H_p is the conductor sheet resistance in ohms per square, $\varepsilon/H_\varepsilon$ is the sheet capacitance in farads, per square centimeter, and L is the length of the interconnect. The above expression specifies a limit on the minimum response time of an interconnect given its length.

2.4.6 Circuit Limits

The circuit level limits are independent of the architecture of a particular system. There are four principal circuit level limits.

To be able to distinguish between the "zero" and "one" logic levels with very nearly zero error is the most basic requirement of a digital logic gate. For a static CMOS logic gate this means that at the transition point of the static transfer characteristics of the gate (i.e., where output voltage is equal to the input voltage), the incremental voltage gain a_v must exceed unity in absolute value. A CMOS inverter can only satisfy this requirement if its supply voltage is larger than a minimum limit $V_{dd,min}$ [42]

$$V_{dd} \geq V_{dd,min} = \frac{2kT}{q}\left(1 + \frac{C_{fs}}{C_0 + C_d}\right)\ln\left(2 + \frac{C_0}{C_d}\right) \geq \frac{\beta kT}{q}$$

where C_{fs} is channel fast surface state, C_0 is gate oxide capacitance, C_d is channel depletion region capacitance, and β is typically between 2 and 4. At $T = 300$ K, $V_{dd,min} \approx 0.1$ V. This also is the reason for selecting the constant $\gamma = 4$ in the fundamental limit on switching energy $E = \gamma kT$.

In practice, a value of $V_{dd} = 0.1$ V cannot be used because the threshold voltage V_t would need to be so small that the drain leakage current in the off state of the MOSFET would be unacceptably large. In considering logic and memory circuit behavior, $V_{dd} = 1.0$ V appears to be a good compromise for small dynamic and static power dissipation.

The second generic circuit limit for CMOS technology is the often discussed switching energy per transition

$$E = Pt_d = \tfrac{1}{2}C_{ro}V_0^2$$

where C_{ro} is taken as the total load capacitance of a ring oscillator stage, including output diffusion capacitance, wiring capacitance, and input gate capacitance for an inverter that occupies a substrate area of $100F^2$ ($F =$ minimum feature size $= 0.1$ μm).

The third generic circuit limit is on the intrinsic gate delay and is given by the time taken to charge/discharge the load capacitance C_{ro}. Hence,

$$t_d = \frac{1}{2}\frac{C_c V_0}{I_{ds}}$$

Assuming carrier velocity saturation in the MOSFET, an approximate value for the drain saturation current I_{ds} is

$$I_{ds} = ZC_0 v_s (V_g - V_t)$$

where Z is the channel width, V_t is the threshold voltage, and the gate voltage $V_g = V_0$. Therefore,

$$t_d = \frac{1}{2} \frac{C_c}{Zv_sC_0} \frac{V_0}{(V_0 - V_t)}$$

Combining with the expression for energy per switching transition,

$$P = 4 \left(\frac{V_0}{C_c} \right)^2 (Zv_sC_0)^3 t_d^2$$

which, given one of P, the power consumption during a transition, or t_d, the duration of the transition, specifies a lower bound on the other. The corresponding locus is labeled $P9(t)$ in Figure 2.19.

The fourth generic circuit limit considers a global (i.e., extending from one corner of the chip to the other) interconnect represented as a distributed resistance-capacitance network. The response time of this interconnect circuit is [43]

$$\tau \approx (2.3R_{tr} + R_{int})C_{int}$$

where R_{tr} is the output resistance of the driving transistor and R_{int} and C_{int} are the total resistance and capacitance, respectively, of the global interconnect. The circuit should be designed so that $R_{int} < 2.3R_{tr}$ to ensure the delay due to wiring resistance is not excessive. Then,

$$\tau \approx 2.3R_{tr}C_{int} = 2.3R_{tr}c_{int}L$$

where c_{int} is the capacitance per unit length of the interconnect. If we model the interconnect as a nearly lossless transmission electron microscopy (TEM) mode line, then the distributed capacitance is given by $c_{int} = 1/vZ_0$, where $v = c_0/\sqrt{\varepsilon_r}$ is the wave propagation velocity of the line, ε_r is the relative permittivity of its dielectric, $Z_0 = \sqrt{\mu_0/\varepsilon_0\varepsilon_r}$ is its characteristic impedance, and $c_0 = \sqrt{1/\varepsilon_0\mu_0}$ is the velocity of light in free space.

The global interconnect response time limit is then given by

$$\tau \approx 2.3R_{tr}C_{int} = 2.3 \left(\frac{R_{tr}}{Z_0} \right) \left(\frac{L}{v} \right)$$

The corresponding locus is labeled $iL2c(\tau)$ in Figure 2.20.

2.4.7 System Limits

System limits depend on all the other limits and are the most restrictive ones in the hierarchy. There are five generic system limits that are given rise to by (1) the architecture of the chip, (2) the power–delay product of the CMOS technology used to implement the chip, (3) the heat removal capacity of the chip package, (4) the clock frequency, and (5) its physical size. To illustrate these limits, a broadly applicable architecture is selected. The system is a systolic array [44] of 1024 identical square macrocells measuring L to a side. Communication is assumed to occur only among neighboring macrocells at the shared boundary. A five-level clock distribution H-tree delivers an unskewed clock signal to each macrocell. The maximum Manhattan distance the clock signal has to travel within a macrocell is L and the same for a logic signal is $2L$.

Before we can progress further, a set of boundary conditions need to be assumed: (1) the number of gates N_g in each macrocell is 1 billion divided by 1024, (2) the feature size used is a conservative 0.1 μm, (3) the package cooling coefficient is 50 W/cm^2, (4) the clock frequency is 1 GHz, and (5) the system is implemented as a single chip.

Using Rent's rule [45], the average length of an interconnect in gate pitches is

$$\bar{R}_{rl} = \frac{2}{9}\left[\frac{7\left(N_g^{p-0.5}-1\right)}{4^{p-0.5}-1} - \frac{1N_g^{p-0.5}}{1-4^{p-0.5}}\right]\frac{1-4^{p-1}}{1-N_g^{p-1}}$$

where $p = 0.45$ (determined empirically for microprocessors [43]) is Rent's exponent. Evaluation returns $\bar{R}_{rl} \approx 6$ under the present set of assumptions. The total wire length a gate has to drive is then

$$l_{rl} = M\bar{R}_{rl}\sqrt{A_{rl}}$$

where M is the fanout of the gate and A_{rl} the gate area. Typical values for them are $M = 3$ and $A_{rl} = 200F^2$. The gate area is also assumed to be limited by the transistor packing density. The assumption causes stringent demands to be placed on the local wiring area, which requires a logic gate dimension [43]

$$[A_{rl}]^{1/2} = \bar{R}_{rl}M\frac{p_w}{e_w n_w}$$

where n_w is the number of wiring levels, p_w is the wiring pitch, and e_w is the wiring efficiency factor. The gate logic area is found to be logic limited for values $n_w = 4$, $p_w = 0.2$ μm, and $e_w = 0.75$ as well as $n_w = 6$, $p_w = 0.2$ μm, and $e_w = 0.5$.

The system switching energy limit is defined by a composite gate that characterizes the critical path within a macrocell. The critical path is assumed to pass through n_{cp} random logic gates and include a total interconnect length corresponding to the corner-to-corner Manhattan distance $2L$

[46]. Then the prorated capacitance loading per gate is $C_{rl} + C_{cc}/n_{cp}$ where C_{rl} includes the MOSFET diffusion capacitance, wiring capacitance for an interconnect of length l_{rl}, and the MOSFET gate capacitance and C_{cc} is the capacitance of the corner-to-corner interconnect. The switching energy of the composite gate therefore is

$$E = Pt_d = \frac{1}{2}\left[C_{rl} + \frac{C_{cc}}{n_{cp}}\right]V_0^2$$

The effective propagation delay time of the composite gate can be analogously formulated as

$$t_d = t_{drl} + \frac{T_{cc}}{n_{cp}}$$

where t_{drl} is the delay time of a random logic gate and T_{cc} is the response time of the corner-to-corner interconnect.

The system heat removal limit is defined by the requirement that the average power dissipation of a composite gate $\bar{P}$ must be less than the cooling capacity of the packaging or

$$\bar{P} = \frac{aE}{T_c} \leq QA$$

where E is the average switching energy of the composite gate, a the probability that a gate switches during a clock interval, $T_c = 1/f_c$ is the clock interval, Q is the package cooling coefficient, and A is the substrate area occupied by the critical path composite gate. The substrate area A has a prorated share of the area A_{cc} due to the corner-to-corner driver in addition to the random logic gate area A_{rl}. Let

$$\frac{A_{rl}}{A_{cc}} = \frac{C_{rl}}{C_{cc}}$$

Hence,

$$A = A_{rl}\left(1 + \frac{C_{cc}}{n_{cp}}C_{rl}\right)$$

Since the number of gates on the critical path is assumed to be n_{cp} if the small clock skew is taken into account using a factor $s_{cp} > 1$, the clock period can be expressed as

$$T_c = s_{cp}n_{cp}t_d$$

Combining the above four expressions,

$$P \leq \frac{s_{cp}n_{cp}}{a}QA_{rl}\left(1 + \frac{C_{cc}}{n_{cp}C_{rl}}\right)$$

gives the limit on P that is a result of the finite cooling capacity of the package. Combining with the expression for the switching energy,

$$P \le \left(\frac{1}{2}C_{rl}V_0^2\right)^{-2}\left(1 + \frac{C_{cc}}{n_{cp}C_{rl}}\right)t_d^2\left(\frac{s_{cp}n_{cp}}{a}QA_{rl}\right)^3$$

The corresponding locus is labeled $P11(t)$ in Figure 2.19. Typical values for some of the parameters used are $C_{cc} = 100$ fF, $C_{rl} = 3.28$ fF, $s_{cp} = 1.11$, $R_{int} = 2.3R_{tr}$, $H_\rho = H_\varepsilon = 0.3$ μm, and $L^2 = N_g A_{rl}$.

At the system level, the longest interconnect obviously needs to be focused on for the $(1/L)^2$-versus-t_d limit. The system cycle time limit is given by

$$T_c \ge T_{cs} + T_{cp} = T_{cs} + T_{cc} + n_{cp}t_{drl}$$

where T_{cs} is the maximum clock skew within a macrocell and T_{cc} is response time of the global interconnect of length $2L$.

2.4.8 Practical Limits

The basis for practical limits is the opinion that beyond a certain point in scaling, the cost of designing, manufacturing, testing, and packaging will cause the cost per function to level off and begin to increase. To facilitate further analysis, the number of transistors per chip N can be expressed as $N = F^{-2} \cdot D^2 \cdot PE$.

The optimistic predictions for the minimum feature size F are to reach 0.0625 μm by the second decade of the millennium, for the chip area D to reach over $(50 \text{ mm})^2$, and for the packing efficiency PE to reach one transistor per minimum feature area [31]. This would make 100 billion transistor chips economically viable in addition to being technically possible.

2.4.9 Quasi-Adiabatic Microelectronics

In any thermodynamic system that proceeds from one equilibrium process to another, the entropy of a closed system either remains unchanged or increases. In a computational process, it is only those steps that discard information or increase disorder that have a lower limit on the energy dissipation. During an adiabatic process no loss or gain of heat occurs—consequently the intriguing prospect of inventing quasi-adiabatic computational technology and reducing the power dissipation to levels beyond the limits of nonadiabatic computation. In a following chapter of this book the quasi-adiabatic circuits will be examined in detail.

The P-versus-t_d limits can also be formulated for quasi-adiabatic microelectronics [31]. At the fundamental level of the hierarchy, $E_d = \gamma kT$, when the rise/fall time of input voltage $T_d = 2t_d$, the propagation or the response time is t_d. At the material level $E_d = \varepsilon_{Si}V/2E_{c0}^3$, and at the device level $E_d = C_0 L_{min}^2 V_0^2/2$. Unlike the unchanging materials and device structures, the circuit configurations used for quasi-adiabatic operations must change significantly. Without identifying specific circuit configurations and system architectures, the respective limits cannot be analyzed.

2.5   CONCLUSIONS

In this chapter we studied the physics of power consumption in CMOS VLSI circuits. We derived the threshold voltage of an MIS diode and then the threshold voltage and the subthreshold current of a long-channel MOSFET. When a low supply voltage is used to minimize the dynamic component of overall power consumption, correspondingly low threshold voltage has to be used to ensure sufficient noise margin. However, we saw that the subthreshold current increases exponentially with reducing threshold voltage and results in a larger static component to overall power consumption. Submicrometer MOSFET was considered next. The threshold voltage shift due to the various short-channel effects was examined and an expression for it was derived using a model based on DIBL.

The different components of overall power consumption in CMOS VLSI circuits were discussed. The dynamic component of power consumption itself comprises switching and short-circuit subcomponents. The dynamic component, though still the dominant component, may not be so as supply voltage continues to decrease.

The hierarchy of theoretical limits on microelectronics summarized in the preceding discussion indicates that the trend of ever more transistors per chip should not continue unabated through the first few decades of the new millennium. The paramount issue from practical consideration is whether there will continue to be sufficient economic incentives to risk the ever-growing capital investments required for further reduction of the cost per function of chips.

REFERENCES

[1]   G. Baccarani et al., "Analytical IGFET Model Including Drift and Diffusion Currents," *IEE J. Solid State Electron, Devices*, vol. 2, p. 62, 1978.

[2]   J. R. Brews, "A Charge-Sheet Model of the MOSFET," *Solid State Electron.*, vol. 21, p. 345, 1978.

[3]   S. M. Sze, *Physics of Semiconductor Devices*, Wiley-Interscience, New York, 1969.

[4]   C. G. B. Garett and W. H. Brattain, "Physical Theory of Semiconductor Surfaces," *Phys. Rev.*, vol. 99, p. 376, 1955.

[5]   L. D. Yau, "A Simple Theory to Predict the Threshold Voltage in Short-Channel IGFETs," *Solid-State Electron.*, vol. 17, p. 1059, 1974.

[6]   F.-C. Hsu et al., "An Analytical Breakdown Model for Short-Channel MOFETs," *IEEE Trans. Electron. Dev.*, vol. 30, p. 571, 1983.

[7] Z.-H. Liu et al., "Threshold Voltage Model for Deep-Submicrometer MOSFETs," *IEEE Trans. Electron. Dev.*, vol. 40, p. 86, 1993.

[8] J. R. Brews et al., "Generalized Guide for MOSFET Miniaturization," *IEEE Electron. Dev. Lett.*, vol. 1, p. 2, 1980.

[9] G. W. Taylor, *Solid State Electron.*, vol. 22, p. 701, 1979.

[10] G. Merkel, "A Simple Model of the Threshold Voltage of Short and Narrow Channel MOSFETs," *Solid State Electron.*, vol. 23, p. 1207, 1983.

[11] C. R. Ji and C. T. Shah, "Two-Dimensional Analysis of the Narrow-Gate Effect in MOSFETs," *IEEE Trans. Electron. Dev.*, vol. 30, p. 635, 1983.

[12] S. S. Chung and T.-C. Li, "An Analytical Threshold Voltage Model of the Trench-Isolated MOS Devices with Nonuniformly Doped Substrates," *IEEE Trans. Electron. Devices*, vol. 13, p. 614, 1992.

[13] C. Y. Lu and J. M. Sung, "Reverse Short-Channel Effects on Threshold Voltage in Submicron Salicide Devices," *IEEE Electron Dev. Lett*, vol. 10, p. 446, 1989.

[14] M. Orlowski et al., "Submicron Short-Channel Effects Due to Gate Reoxidation Induced Lateral Diffusion," *IEDM Tech. Dig.*, vol. 32, p. 632, 1987.

[15] N. D. Arora and M. S. Sharma, "Modelling the Anomalous Threshold Voltage Behavior of Submicron MOSFETs," *IEEE Electron. Dev. Lett.*, vol. 13, p. 92, 1992.

[16] H. Hanafi et al., "A Model for Anomalous Short-Channel Behavior in MOSFET," *IEEE Electron. Dev. Lett.*, vol. 14, p. 575, 1993.

[17] D. Sadana et al., "Enhanced Short-Channel Effects in NMOSFETs Due to Boron Redistribution Introduced by Arsenic Source and Drain Implant," *IEDM Tech. Dig.*, vol. 37, p. 849, 1992.

[18] H. Jacobs et al., "MOSFET Short Channel Effect Due to Silicon Interstitial Capture in Gate Oxide," *IEDM Tech. Dig.*, vol. 38, p. 307, 1993.

[19] C. S. Rafferty et al., "Explanation of Reverse Short-Channel Effect by Defect Gradients," *IEDM Tech. Dig.*, vol. 38, p. 311, 1993.

[20] J. Zhu et al., "Punchthrough Current for Submicrometer MOSFETs in CMOS VLSI," *IEEE Trans. Electron. Dev.*, vol. 35, p. 145, 1988.

[21] C. Hu, "Future CMOS Scaling and Reliability," *Proc. IEEE*, vol. 81, p. 682, 1993.

[22] F. M. Klassen, "Design and Performance of Micron-Sized Devices," *Solid State Electron.*, vol. 21, p. 565, 1978.

[23] T. Shibata et al., "An Optimally Designed Process for Submicrometer MOSFETs," *IEEE Trans. Electron. Dev.*, vol. 29, p. 531, 1982.

[24] C. F. Codella and S. Ogura, "Halo Doping Effects in Submicron DI-LDD Device Design," *IEDM Tech. Dig.*, p. 230, 1985.

[25] A. Hori et al., "A Self Aligned Pocket Implantation (SPI) Technology for 0.2 μm—Dual Gate CMOS," *IEDM Tech. Dig.*, vol. 36, p. 642, 1991.

[26] J. R. Brews, "Subthreshold Behavior of Uniformly and Non-Uniformly Doped Long-Channel MOSFETs," *IEEE Trans. Electron. Devices*, vol. ED-26, no. 9, p. 1282, 1979.

[27] H. Veendrick, "Short Circuit Dissipation of Static CMOS Circuitry and its Impact on the Design of Buffer Circuits," *IEEE J. Solid-State Circuits*, vol. 19, no. 4, Aug. 1984.

[28] L. Bisdounis, O. Koufopavlou, and S. Nikolaidis, "Accurate Evaluation of CMOS Short Circuit Power Dissipation for Short Channel Devices," *International Symposium on Low Power Electronics & Design*, Monterey, CA, Aug. 1996, pp. 181–192.

[29] M. Nagata, "Limitations, Innovations and Challenges of Circuits and Devices into a Half Micrometer and Beyond," *IEEE J. Solid-State Circuits*, vol. 27, pp. 465–472, 1992.

[30] C. Hu, "MOSFET Scaling in the Next Decade and Beyond," Proc. International Semiconductor Device Research Symp., *Semicon. Int.*, pp. 105–114, June 1994.

[31] J. D. Meindl, "Low Power Microelectronics: Retrospect and Prospect," *Proc. IEEE*, vol. 83, no. 4, pp. 619–635, 1995.

[32] Y. Taur et al., "High-Performance 0.1 mm, CMOS Devices with 1.5 V Power Supply," *IEDM Tech. Dig.*, vol. 38, pp. 127–130, 1993.

[33] J. D. Meindl, "Theoretical, Practical, and Analogical Limits in ULSI," *IEDM Tech. Dig.*, vol. 28, pp. 8–13, 1983.

[34] A. P. Chandrakasan et al., "Low Power CMOS Digital Design," *IEEE J. Solid-State Circuits*, vol. 27, pp. 473–484, 1992.

[35] K. N. Ratnakumer et al., "Short-Channel MOSFET Threshold Voltage Model," *IEEE J. Solid-State Circuits*, vol. SC-17, pp. 937–947, 1982.

[36] B. Agarwla et al., "Opportunities for Scaling MOSFETs for GSI," *Proc. ESSDERC 1993*, vol. 25, pp. 919–926, September 1993.

[37] C. Fiegna et al., "A New Scaling Method for 0.1–0.25 Micron MOSFET," *Symp. VLSI Tech. Dig.*, pp. 33–34, May 1993.

[38] M. Ono et al., "Sub-59 nm Gate Length N-MOSFET's with 10 nm Phosphorous S/D Junctions," *IEDM Tech. Dig.*, vol. 38, pp. 119–121, 1993.

[39] D. Hisamoto et al., "A Fully Depleted Lean Channel Transistor (DELTA-) a Novel Vertical Ultrathin SOI MOSFET," *IEEE Electron. Devices Lett.*, vol. 11, pp. 36–38, 1990.

[40] D. J. Frank et al., "Monte-Carlo Simulation of a 30 nm Dual Gate MOSFET: How Short Can Si Go?" *IEDM Tech. Dig.*, vol. 37, pp. 553–556, 1992.

[41] T. Tanaka et al., "Ultrafast Low Power Operation of P + N + Double-Gate SOI MOSFETS," *Symp. VLSI Tech. Dig.*, pp. 11–12, 1994.

[42] R. M. Swanson et al., "Ion-Implanted Complimentary MOS Transistors in Low-Voltage Circuits," *IEEE J. Solid-State Circuits*, vol. SC-7, pp. 146–152, 1972.

[43] B. Bakoglu, *Circuits, Interconnections and Packaging for VLSI*, Addison-Wesley, Reading, MA, pp. 198–200, 1990.

[44] H. S. Stone et al., "Computer Architecture in the 1990s," *IEEE Computer*, vol. 24, pp. 30–38, Sept. 1991.

[45] B. S. Landrum and R. L. Russo, "On Pin Versus Block Relationship for Partitioning of Logic Graphs," *IEEE Trans. on Computers*, vol. C-20, pp. 1469–1479, Dec. 1971.

[46] A. Masaki, "Possibilities of Deep Submicrometer CMOS for Very High-Speed Computer Logic," *Proc. IEEE*, vol. 81, pp. 1311–1324, 1993.

[47] M. Shoji, *CMOS Digital Circuit Technology*, Prentice Hall, New Jersey, 1988.

CHAPTER 3

POWER ESTIMATION

When designing VLSI circuits, the designers accurately estimate the silicon area and the expected performance before the circuit goes into fabrication. However, with the requirements for low power, the designers are faced with the added burden of qualifying their designs with respect to power dissipation. Hence, work has started in earnest to accurately estimate power dissipation at different levels of design abstraction. In this chapter we will describe in detail algorithms for estimation of both average and maximum power in CMOS circuits. The readers should note that average power determines battery life while maximum or peak power is related to circuit reliability and the proper design of power and ground lines.

In Chapter 2, we noted that the power dissipation in CMOS circuits is mainly due to dynamic (switching and short-circuit) current and the sub-threshold leakage current. With the scaling down of device sizes and transistor threshold, the subthreshold current will increase and can become a sizable component of total power dissipation. However, in the current-day technology, the dominant component power is still the switching component (about 80% of the total power dissipation). Therefore, in this chapter we will devote a considerable effort in estimation of switching power in CMOS circuits. In Chapter 5 we will consider estimation of leakage power in CMOS circuits when the supply voltage and the transistor thresholds are low. If we ignore internal capacitances of a logic gate, the average power dissipation due to charging and discharging of load capacitance of a logic gate is given by

$$P_{avg} = \tfrac{1}{2}V_{dd}^2 Cf \tag{3.1}$$

where V_{dd} is the supply voltage, C is the load capacitance of the gate under consideration, and f is the frequency of operation. For aperiodic signals, the frequency of operation can be estimated by the average number of signal transitions per unit time and is represented by *signal activity A*. A more rigorous definition of signal activity will be given later. From Eq. (3.1), we observe that power dissipation in CMOS circuits is input data dependent. Since the same supply voltage applies to each logic gate of the design, accurate estimation of power dissipation involves determination of signal activities associated with the internal node capacitances (switched capacitance). In this chapter we will also show that signal activity at circuit nodes is associated with electromigration and hot-electron degradation and, hence, is also a measure of reliability.

Let us first consider average power estimation. Average power estimation involves accurately estimating the average switched capacitances at the internal nodes of a circuit. Accurate estimation of switched capacitance at the architectural or behavioral level is quite difficult, while they can be more accurately estimated at the transistor circuit level. This chapter will describe in detail the different methodologies to estimate power dissipation at the transistor, logic, and higher levels. Before we describe techniques to accurately estimate power dissipation, let us consider the following question.

Why estimate power dissipation? An accurate estimate of power dissipation during various phases of VLSI design can verify the power dissipation requirements and thereby avoid complicated and expensive design changes that might be required due to power constraint violations. Power estimates can also be used during the circuit, logic, and high-level synthesis to accurately trade off power versus other design parameters such as area, performance, and noise margin. Power is also dissipated due to glitching activity in a circuit. Glitches occur due to different delays through different paths of the circuit. Let us consider the example of Figure 3.1 where each gate has a unit delay. A hazardous (spurious or glitch) transition occurs at the output of the AND gate (as shown in the figure) due to different delays through two

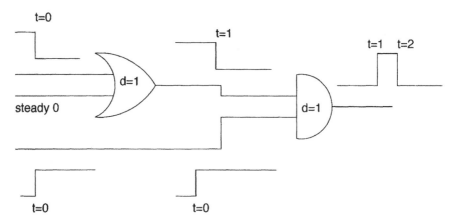

Figure 3.1 Spurious transition at a node.

different paths converging at the inputs to the AND gate. The width of the glitch depends on the delays through the logic gates and interconnects. It should also be noted that the glitches can die while propagating through a logic gate if the width of the glitch is much smaller than the inertial delay of the logic gate. Since a glitch causes extra switching activity it adds to the total power dissipation. Hence, if one can accurately predict the glitching activity at various nodes of a circuit, the power dissipation due to the glitches can be accurately estimated. Such estimates might be helpful in reducing the glitching activity during circuit or logic synthesis.

In Chapter 2 we noted that the majority of power dissipation in CMOS circuits is due to signal transitions at circuit nodes. Therefore, accurate methods are required to estimate the switching activity at the internal nodes of logic circuits to determine average power dissipation. The problem of determining when and how often transitions occur at a node in a digital circuit is difficult because they depend on the applied input vectors and the sequence in which they are applied. Hence, the easiest solution is to use circuit simulation. Based on a given set of inputs, the power supply current can be monitored to determine the average power dissipation. The major problem with this approach is that circuit simulation is too slow for large circuits. Besides being slow, the technique is strongly input pattern dependent. Hence, pattern-independent probabilistic techniques are required to quickly estimate the average number of transitions per circuit node.

The two alternative flows for the estimation of power dissipation are shown in Figure 3.2 [1]. The conventional circuit simulation technique considers a large number of input patterns. Based on each of the input patterns, circuit simulation is conducted to determine the current waveforms from the supply voltage. Average power dissipation is then calculated by determining

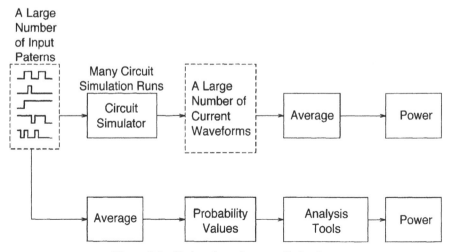

Figure 3.2   Estimation of power dissipation.

the average current from the power supply. However, a major problem is to determine a priori the number of simulation runs required. And simulation of large circuits is very expensive from a computation point of view. In the pattern-independent probabilistic technique, based on the set of real-life inputs, one determines the input signal distributions in terms of "some" probability values. Then, an analysis tool, the details of which will be given later in this chapter, determines the average power dissipation based on the input signal distributions. It should be observed that the analysis tool runs once as opposed to the simulation-based technique where the number of runs is equal to the number of different inputs. The analysis tool should provide reasonably good estimates of power dissipation with little computational effort. We will consider probabilistic and statistical analysis tools to estimate power dissipation. However, before we consider such tools, we will have to understand how digital signals are modeled.

3.1 MODELING OF SIGNALS

In this section we will briefly describe the concepts of signal probability and activity that are used to model the digital signals [3]. Let $g(t)$, $t \in (-\infty, +\infty)$, be a *stochastic process* that takes the values of logical 0 or logical 1, transitioning from one to the other at *random* times. A stochastic process is said to be *strict-sense stationary* (SSS) if its statistical properties are invariant to a shift of the time origin. More importantly, the mean of such a process does not change with time. If a constant-mean process $g(t)$ has a finite variance and is such that $g(t)$ and $g(t + \tau)$ become uncorrelated as $\tau \to \infty$, then $g(t)$ is *mean ergodic*. We use the term *mean ergodic* to refer to regular processes that are *mean ergodic* and satisfy the two conditions of finite variance and decaying autocorrelation. For such a signal we can define the following terms:

Definition 3.1 (Signal Probability) *The* signal probability *of signal $g(t)$ is given by*

$$P(g) = \lim_{T \to \infty} \frac{1}{2T} \int_{-T}^{+T} g(t) \, dt$$

Definition 3.2 (Signal Activity) *The* signal activity *of a logic signal $g(t)$ is given by*

$$A(g) = \lim_{T \to \infty} \frac{n_g(T)}{T}$$

where $n_g(t)$ is the number of transitions of $g(t)$ in the time interval between $-T/2$ and $+T/2$.

If the primary inputs to the circuit are modeled as *mutually independent SSS mean-ergodic 0-1 processes*, then the probability of signal $g(t)$ assuming the logic value 1 at any given time t becomes a constant, independent of time and is referred to as the *equilibrium signal probability* of random quantity $g(t)$ and is denoted by $P(g = 1)$, which we refer to simply as *signal probability*. Hence, $A(g)$ becomes the expected number of transitions per unit time.

Najm [3] has shown that the above convergence results are true if $g(t) \triangleq g(t + \tau)$, where τ is a random variable. He has also shown that the model is suitable for logic circuits. Let us consider a multi-input, multi-output logic module M that implements a Boolean function. Here, M can be a single logic gate or a higher level circuit block. We assume that the inputs to M, $g_1, g_2, \ldots, g_n$ are mutually independent processes each having a signal probability $P(g_i)$ and a signal activity $A(g_i)$, $i \leq n$. The calculation of signal probability at the output can be described as follows. If P_1 and P_2 are the input signal probabilities to a two-input AND gate, the output signal probability is given by $P_1 P_2$, whereas, for an OR gate the output signal probability is $1 - (1 - P_1)(1 - P_2) = P_1 + P_2 - P_1 P_2$. In this derivation we have assumed that the input signals are mutually independent. For an inverter, the output signal probability is simply $1 - P_1$, where P_1 is the input signal probability. Figure 3.3 shows the propagation of signal probabilities through the basic gates such as AND, OR, and NOT gates. It can be noted that one can only assume that the primary input signals are mutually independent. Due to reconvergent fanout, the internal signals can be correlated. Hence, accurate methods of estimating the signal probabilities at the internal nodes of a logic circuit are required for correlated signals. In the following section, we describe a general symbolic method to estimate signal activity at the internal nodes of a circuit.

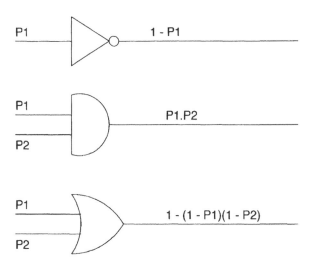

Figure 3.3 Propagation of signal probabilities through basic gates.

3.2 SIGNAL PROBABILITY CALCULATION

We will find out later in this chapter that signal probability is used for accurate estimation of signal activity. Therefore, it is essential to accurately calculate signal probability for further use in estimating *activity*. A general algorithm for estimating signal probability at the internal nodes of a logic circuit was proposed by Parker and McCluskey [2]. The algorithm is given below:

Algorithm: Compute Signal Probabilities
   Inputs: Signal probabilities of all the inputs to a circuit.
   Output: Signal probabilities for all nodes of the circuit.
   Step 1: For each input signal and gate output in the circuit, assign a unique variable.
   Step 2: Starting at the inputs and proceeding to the outputs, write the expression for the output of each gate as a function (using standard expressions for each gate type for probability of its output signal in terms of its mutually independent primary input signals) of its input expressions.
   Step 3: Suppress all exponents in a given expression to obtain the correct probability for that signal.

The primary inputs are assumed to be independent of each other, and the symbolic probability expressions for the internal nodes are expressed in terms of signal probabilities of the primary inputs. However, the internal nodes of a circuit can be correlated due to reconvergent fanout. Reconvergent fanout can produce expressions for the signal probability of internal nodes having exponents greater than 1. Intuitively, in probability expressions involving independent primary inputs, such exponents cannot be present. Hence, step 3 of the above algorithm suppresses exponents to handle signal correlations. A Boolean function f, representing an internal node or an output node of a logic circuit, can always be written in a canonical sum of products of primary inputs. It can be observed from the above algorithm that the signal probability $P(f)$ can be expressed as a sum of *primary input signal probability product terms* $\sum_{i=1}^{p} \alpha_i(\prod_{k=1}^{n} P^{m_{i,k}}(x_k))$, where n is the number of the independent inputs to the circuit and α_i is some integer. The exponent $m_{i,k}$ is either 1 or 0. The sum has p products terms. For convenience, this form will be referred to as the *sum of probability products* of f. Also for convenience, $\bar{P}(x_i)$ is defined as $P(\bar{x}_i)$ and equals $1 - P(x_i)$. Therefore, $P(f)$ can be expressed as $\sum_{i=1}^{p} \alpha_i(\prod_{k=1}^{n} p^{m_{i,k}}(x_k)\bar{P}^{l_{i,k}}(x_k))$, where $m_{i,k}$ and $l_{i,k}$ are either 1 or 0 but both cannot be 1 simultaneously. Since

$$P(x_k)\bar{P}(x_k) = P(x_k)(1 - P(x_k)) = P(x_k) - P^2(x_k)$$

and equals 0 after exponent suppression, the product term $P^{m_{i,k}}(x_k)\bar{P}^{l_{i,k}}(x_k)$ will be eliminated from $P(f)$ if $m_{i,k} = l_{i,k} = 1$. Let f be written in a canonical sum of products of primary inputs as follows: $f = \sum_{i=1}^{P}(\prod_{k=1}^{n} s_k)$, where s_k is either x_k or $\bar{x}_k$. Since the product terms inside the summation are mutually independent, we have $P(f) = \sum_{i=1}^{P}(\prod_{k=1}^{n} P(s_k))$. This expression is defined as the *canonical sum of probability products* of f.

The following example shows the use of the algorithm to calculate signal probability.

Example 3.1   Given $y = x_1 x_2 + x_1 x_3$, where x_i, $i = 1, 2, 3$ are mutually independent. If $z = x_1 \bar{x}_2 + y$, then $P(z)$ can be determined as follows: Determine $P(y)$. Without suppressing the exponents,

$$P(y) = P(x_1 x_2) + P(x_1 x_3) - P(x_1 x_2)P(x_1 x_3)$$

But we know that $P(x_1 x_2) = P(x_1)P(x_2)$ and $P(x_1 x_3) = P(x_1)P(x_3)$. Hence, after the exponent of $P(x_1)$ is suppressed,

$$P(y) = P(x_1)P(x_2) + P(x_1)P(x_3) - P(x_1)P(x_2)P(x_3)$$

Now the symbolic probabilities of all the inputs of z are known. So without suppressing the exponents (and assuming that all signals are independent),

$$P(z) = P(x_1 \bar{x}_2) + P(y) - P(x_1 \bar{x}_2)P(y)$$
$$= P(x_1)\overline{P(x_2)} + P(x_1)P(x_2) + P(x_1)P(x_3) - P(x_1)P(x_2)P(x_3)$$
$$- P(x_1)\overline{P(x_2)}(P(x_1)P(x_2) + P(x_1)P(x_3) - P(x_1)P(x_2)P(x_3))$$

where $\overline{P(x_i)} = 1 - P(x_i)$. Note that $P(x_2)\overline{P(x_2)} = P(x_2)(1 - P(x_2)) = 0$. Therefore, after suppressing the exponents, we have

$$P(z) = P(x_1)\overline{P(x_2)} + P(x_1)P(x_2) + P(x_1)P(x_3)$$
$$- P(x_1)P(x_2)P(x_3) - P(x_1)\overline{P(x_2)}P(x_3) = P(x_1)$$

In fact, $z = x_1$ by Boolean algebraic manipulation.

3.2.1 Signal Probability Using Binary Decision Diagrams

Signal probability of any arbitrary Boolean expression can be easily calculated using binary decision diagrams (BDDs) [37–39]. Binary decision diagrams graphically represent the functionality of a logic function. Let us consider a function $f(x_1, x_2, \ldots, x_n)$ of variables $x_1, x_2, \ldots, x_n$. Function f can be represented using *Shannon's expansion* [36] as follows:

$$f = x_i \cdot f(x_1, \ldots, x_{i-1}, 1, x_{i+1}, \ldots, x_n) + \bar{x}_i \cdot f(x_1, \ldots, x_{i-1}, 0, x_{i+1}, \ldots, x_n)$$

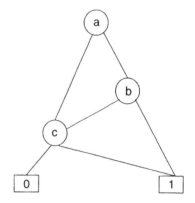

Figure 3.4 Binary decision diagram example.

The *cofactors* of f with respect to x_i and $\overline{x}_i$ respectively are defined as

$$f_{x_i} = f(x_1, \ldots, x_{i-1}, 1, x_{i+1}, \ldots, x_n)$$

and

$$f_{\overline{x}_i} = f(x_1, \ldots, x_{i-1}, 0, x_{i+1}, \ldots, x_n)$$

Functions f_{x_i} and $f_{\overline{x}_i}$ can be obtained by replacing variable x_i with logic 1 and logic 0, respectively. Each node of the BDD represents an input x_i and the edges coming out of node x_i represent the value of input x_i—either logic 0 or logic 1. By simply traversing the BDD from its root, one can determine the value of the function f by sequentially examining the values of the inputs. Let us consider an example where $f = ab + c$. The BDD of the function is shown in Figure 3.4. The leaf nodes represent the value of function f. For example, if one traverses the path of the graph by edges $a = 1$, $b = 0$, and $c = 1$, then the function equals logic 1. The tree rooted to the left of a represents function $f_{\overline{a}}$, while the tree rooted to the right of a represents function f_a. One should also notice that the ordering of the nodes of the BDD has direct implications on the complexity of the BDD.

Let us now consider the calculation of signal probability using BDDs. From Eq. 3.2.1, 3.2.1, and 3.2.1 we have

$$P(f) = \begin{cases} P\left(x_1 \cdot f_{x_1} + \overline{x}_1 \cdot f_{\overline{x}_1}\right) & (3.2) \\ P\left(x_1 \cdot f_{x_1}\right) + P\left(\overline{x}_1 \cdot f_{\overline{x}_1}\right) & (3.3) \\ P(x_1) \cdot P\left(f_{x_1}\right) + P(\overline{x}_1) \cdot P\left(f_{\overline{x}_1}\right) & (3.4) \end{cases}$$

The above equation shows that the BDD can be used to evaluate $P(f)$. The probability of the cofactors can now be represented in terms of its cofactors and so on. A depth-first traversal of the BDD, with a post order evaluation of $P(\cdot)$ at every node is required for evaluation of $P(f)$.

3.3 PROBABILISTIC TECHNIQUES FOR SIGNAL ACTIVITY ESTIMATION

The average number of transitions per unit time at a circuit node is referred to as signal activity. Now, let us consider how to estimate signal activity values at the internal nodes of a circuit when the signal probabilities and activities at the primary input signals are known. Combinational circuits will be considered first followed by a description for sequential circuits.

3.3.1 Switching Activity in Combinational Logic

Let x_i, $i = 1, \ldots, n$, represent the n inputs to a logic gate M with outputs f_j, $j = 1, \ldots, m$. Let us also assume the delay through the logic gate is decoupled from the gate function. In order to accurately evaluate signal activities, one has to consider the nature of logic signals. It was shown in Section 3.2 that spatial correlation may exist due to reconvergent fanout. Logic signals can also be temporally correlated. Temporal correlation states that the logic signals can be correlated in time. For simplicity, we will first consider temporal independence in the evaluation of signal activity. A rigorous analysis using temporal correlation will be shown later.

To understand how the propagation algorithm works, we will use the concept of *Boolean difference*. Let f_j be a function that depends on input x_i. The Boolean difference of f_j with respect to x_i is defined as follows:

$$\frac{\partial f_j}{\partial x_i} = f_j\big|_{x_i=1} \oplus f_j\big|_{x_i=0} \tag{3.5}$$

where $\oplus$ denotes the exclusive-or operation. The Boolean difference signifies the condition under which output f_j is sensitized to input x_i. Najm [3] has shown that if the primary inputs x_i, $i = 1, \ldots, n$, to logic gate M are not spatially correlated, then the signal activity at output f_j is given by

$$A(f_j) = \sum_{i=1}^{n} P\left(\frac{\partial f_j}{\partial x_i}\right) A(x_i) \tag{3.6}$$

The Boolean difference $\partial f_j / \partial x_i$ represents the condition for sensitizing input x_i to output f_j as noted in Eq. (3.5). Therefore, $P(\partial f_j / \partial x_i)$ signifies the probability of sensitizing input x_i to output f_j, while $P(\partial f_j / \partial x_i)A(x_i)$ is the contribution of switching activity at output f_j due to input x_i only. Hence, the contribution of switching activity due to all the inputs is obtained by taking the summation over all the inputs of M. Signal probability $P(\partial f_j)/\partial x_i)$ can be easily calculated using the BDD-based technique described in Section 3.2.1. Figure 3.5 shows the signal activity at the output of basic gates when the input signal probability and activity are given.

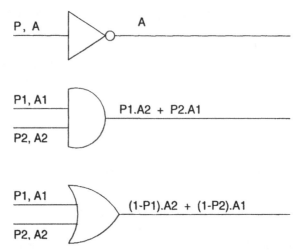

Figure 3.5   Propagation of signal activities through basic gates.

Equation (3.6) fails to consider the effect of simultaneous switching of signals at logic gate inputs and, hence, can grossly overestimate signal activity. This is due to the fact that temporal correlation among signals are not considered. Let us consider the XOR logic gate of Figure 3.6. If the signals at the input to the XOR gate are switching as shown in the figure, the output switching activity will be zero. Hence, simultaneous switching at the input to logic gates has to be considered to accurately estimate the signal activity.

Let us consider an extension of the above model to include simultaneous switching of signals [4]. When more than one primary input, say x_i and x_j, are switching simultaneously, the Boolean differences $\partial y / \partial x_i$ and $\partial y / \partial x_j$ are undefined at those time instants. Hence, Eq. (3.6) is no longer valid for this situation. Now let us try to determine an accurate expression for calculating the activity of static combinational circuits. The formulation considers both signal correlation and simultaneous switching. Let a node y of a module be observed at the leading edge of the clock. If one selects a clock cycle at random, the probability of having a transition at the leading edge of this clock cycle at node y is $A(y)/f$ and is denoted by $a(y)$. Here $A(y)$ is the *activity* at the node and f is the clock frequency. The following definition will

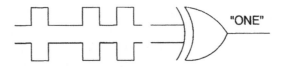

Figure 3.6   Signal activities for XOR logic.

be useful in understanding the approach to deriving the accurate expression for signal activity.

Definition 3.3 (Generalized Boolean Difference) *Let y be a Boolean expression and x_i, $i = 1, \ldots, n$, be mutually independent primary inputs of y. We define*

$$\frac{\partial y^k|_{b_{i_1}, b_{i_2}, \ldots, b_{i_k}}}{\partial x_{i_1}, \partial x_{i_2} \cdots \partial x_{i_k}} = y\Big|_{x_{i_1} = b_{i_1}, x_{i_2} = b_{i_2}, \ldots, x_{i_k} = b_{i_k}} \oplus y\Big|_{x_{i_1} = \bar{b}_{i_1}, x_{i_2} = \bar{b}_{i_2}, \ldots, x_{i_k} = \bar{b}_{i_k}}$$

where k is a positive integer, b_{i_j} is logic 1 or 0 and x_{i_j}, $j = 1, \ldots, k$, are distinct mutually independent primary inputs of y.

Let us examine the above definition closely. Because the operator $\oplus$ is commutative,

$$\frac{\partial^k y|_{b_{i_1}, b_{i_2}, \ldots, b_{i_k}}}{\partial x_{i_1} \partial x_{i_2} \cdots \partial x_{i_k}} = \frac{\partial^k y|_{\bar{b}_{i_1}, \bar{b}_{i_2}, \ldots, \bar{b}_{i_k}}}{\partial x_{i_1} \partial x_{i_2} \cdots \partial x_{i_k}} \tag{3.7}$$

It follows from the definition that if the *generalized Boolean difference* is logical 1, then the simultaneous transitions at $(x_{i_1}, x_{i_2}, \ldots, x_{i_k})$ from $(b_{i_1}, b_{i_2}, \ldots, b_{i_k})$ to $(\bar{b}_{i_1}, \bar{b}_{i_2}, \ldots, \bar{b}_{i_k})$ or from $(b_{i_1}, b_{i_2}, \ldots, b_{i_k})$ to $(b_{i_1}, b_{i_2}, \ldots, b_{i_k})$, while the rest of the primary inputs remain constant, will cause a transition at y. We define $\mathrm{Pc}(\partial^k y|_{b_{i_1}, b_{i_2}, \ldots, b_{i_k}}/\partial x_{i_1} \partial x_{i_2} \cdots \partial x_{i_k})$ as the conditional probability of having a transition at time t while $x_{i_1}, x_{i_2}, \ldots, x_{i_k}$ switch at t and the rest of the inputs do not. Here t is some leading edge of the clock. Under this condition, the probability of $x_j (\notin \{x_{i_1}, x_{i_2}, \ldots, x_{i_k}\})$ being logic 1 in the time interval $(t, t + T]$ is $[P(x_j) - a(x_j)/2]/[1 - a(x_j)]$ rather than $P(x_j)$, where T is the clock cycle. The above expression can be derived as follows: $x_j(t - T)x_j(t) = 1$ and $\bar{x}_j(t - T)\bar{x}_j(t) = 1$ signify that x_j does not switch at time t and remain logic 1 and logic 0, respectively. Similarly, $x_j(t - T)\bar{x}_j(t) = 1$ and $\bar{x}_j(t - T)x_j(t) = 1$ signify that x_j has a transition from logic 1 to logic 0 and from logic 0 to logic 1, respectively. Therefore, since x_j is SSS, the following equations hold:

$$P\big(x_j(t - T)x_j(t) + \bar{x}_j(t - T)\bar{x}_j(t)\big)$$

$$= P\big(x_j(t - T)x_j(t)\big) + P\big(\bar{x}_j(t - T)\bar{x}_j(t)\big) = 1 - a(x_j)$$

$$P\big(x_j(t - T)\bar{x}_j(t) + \bar{x}_j(t - T)x_j(t)\big)$$

$$= P\big(x_j(t - T)\bar{x}_j(t)\big) + P\big(\bar{x}_j(t - T)x_j(t)\big) = a(x_j)$$

$$P\big(x_j(t - T)\big) = P\big(x_j(t - T)x_j(t)\big) + P\big(x_j(t - T)\bar{x}_j(t)\big) \tag{3.8}$$

$$P\big(x_j(t)\big) = P\big(x_j(t - T)x_j(t)\big) + P\big(\bar{x}_j(t - T)x_j(t)\big) \tag{3.9}$$

However, since x_j is SSS,

$$P(x_j(t - T)) = P(x_j(t)) = P(x_j) \tag{3.10}$$

and

$$P(\bar{x}_j(t - T)) = P(\bar{x}_j(t)) = P(\bar{x}_j) = 1 - P(x_j) \tag{3.11}$$

Therefore, equating Eqs. (3.8) and (3.9), we obtain

$$P(x_j(t - T)\bar{x}_j(t)) = P(\bar{x}_j(t - T)x_j(t)) = \tfrac{1}{2}a(x_j) \tag{3.12}$$

In fact, one can notice that each transition from 1 to 0 will be followed by a transition from 0 to 1. Since $x_j(t - T)x_j(t) + \bar{x}_j(t - T)x_j(t) = x_j(t)$, it follows that $P(x_j) = P(x_j(t)) = P(x_j(t - T)x_j(t) + \bar{x}_j(t - T)x_j(t)) = P(x_j(t + T)x_j(t)) + P(\bar{x}_j(t - T)x_j(t))$. Hence,

$$P(x_j(t - T)x_j(t)) = P(x_j) - \tfrac{1}{2}a(x_j) \tag{3.13}$$

Similarly,

$$P(\bar{x}_j(t - T)\bar{x}_j(t)) = 1 - P(x_j) - \tfrac{1}{2}a(x_j) \tag{3.14}$$

The conditional probability of x_j being 1 in the time interval $(t, t + T]$ while it does not switch at time t, denoted as $\mathrm{Pc}(x_j)$, is given by

$$P\big(x_j(t) = 1 | \{x_j(t - T)x_j(t) + \bar{x}_j(t - T)\bar{x}_j(t) = 1\}\big)$$

$$= \frac{P\big(x_j(t)(x_j(t - T)x_j(t) + \bar{x}_j(t - T)\bar{x}_j(t)) = 1\big)}{P\big(x_j(t - T)x_j(t) + \bar{x}_j(t - T)\bar{x}_j(t) = 1\big)}$$

$$= \frac{P\big(x_j(t - T)x_j(t)\big)}{P\big(x_j(t - T)x_j(t)\big) + P\big(\bar{x}_j(t - T)\bar{x}_j(t)\big)} = \frac{P(x_j) - a(x_j)/2}{1 - a(x_j)}$$

Under the assumption that the primary inputs are *mutually independent* and the logic signals can be modeled as *SSS mean-ergodic 0–1 discrete-time stochastic processes* with logic modules having zero delays, the following theorem holds.

Theorem 3.1 (3-Inputs) *If y is a Boolean expression and x_i, $i = 1, \ldots, 3$, are* mutually independent *primary inputs of y, then*

$$
a(y) = \sum_{i=1}^{3} \text{Pc}\left(\frac{\partial y}{\partial x_i}\right)\left(a(x_i) \prod_{\substack{j \neq i \\ 1 \leq j \leq 3}} [1 - a(x_j)]\right)
$$

$$
+ \frac{1}{2}\left\{\sum_{1 \leq i < j \leq 3}\left[\text{Pc}\left(\frac{\partial^2 y|_{00}}{\partial x_i\, \partial x_j}\right) + \text{Pc}\left(\frac{\partial^2 y|_{01}}{\partial x_i\, \partial x_j}\right)\right]\right.
$$

$$
\left. \times\left(a(x_i)a(x_j) \prod_{l \in \{1,2,3\}-\{i,j\}} [1 - a(x_l)]\right)\right\}
$$

$$
+ \frac{1}{4}\left[\text{Pc}\left(\frac{\partial^3 y|_{000}}{\partial x_1\, \partial x_2\, \partial x_3}\right) + \text{Pc}\left(\frac{\partial^3 y|_{001}}{\partial x_1\, \partial x_2\, \partial x_3}\right) + \text{Pc}\left(\frac{\partial^3 y|_{010}}{\partial x_1\, \partial x_2\, \partial x_3}\right)\right.
$$

$$
\left. + \text{Pc}\left(\frac{\partial^3 y|_{011}}{\partial x_1\, \partial x_2\, \partial x_3}\right)\right]\left(\prod_{l=1}^{3} a(x_l)\right) \quad (3.15)
$$

where

$$
\text{Pc}\left(\frac{\partial y}{\partial x_i}\right), \text{Pc}\left(\frac{\partial^2 y|_{00}}{\partial x_i\, \partial x_j}\right), \ldots, \text{Pc}\left(\frac{\partial^2 y|_{011}}{\partial x_1\, \partial x_2\, \partial x_3}\right)
$$

are conditional probabilities under the condition that only some primary inputs switch and the rest do not at the leading edge of the clock cycle.

Proof: Because we assume that the module under consideration has zero delay and the primary inputs switch only at the leading edge of the clock cycle, switching times can only be discrete-time points that coincide with some leading edges of the clock signal.

At time t, which is some leading edge of the clock signal, let B_0 be the event that none of the inputs are switching. Let B_i, $i = 1, 2, 3$, be the events that only x_i is switching, $B_{i,j}$, $(i, j) = (1, 2), (2, 3), (1, 3)$, be the events that only x_i and x_j are switching and, finally, $B_{1,2,3}$ be the event that all three inputs are switching at time t.

According to the above definitions, the union of all these events is the *sample space*. All the events are also *mutually exclusive (or disjoint)*. Therefore, they form a partition of the sample space. Because x_1, x_2, x_3 are *mutually independent*,

$$
P(B_0) = [1 - a(x_1)][1 - a(x_2)][1 - a(x_3)]
$$

Similarly,

$$P(B_i) = a(x_i) \prod_{\substack{1 \le j \le 3 \\ i \ne j}} \left[1 - a(x_j) \right] \qquad i = 1, 2, 3$$

$$P(B_{i,j}) = a(x_i)a(x_j)\left[1 - a(x_l) \right] \qquad 1 \le i \le 3, i \ne j$$

and

$$P(B_{1,2,3}) = a(x_1)a(x_2)a(x_3)$$

Let A be the event that y is switching at time t. Using the *total probability theorem* of [17], we derive

$$P(A) = P(A|B_0)P(B_0) + \sum_{i=1}^{3} P(A|B_i)P(B_i)$$

$$+ \sum_{1 \le i < j \le 3} P(A|B_{i,j})P(B_{i,j}) + P(A|B_{1,2,3})P(B_{1,2,3}) \quad (3.16)$$

However, we know that if no primary inputs are switching at time t, y cannot be switching. If only x_i is switching at time t, $P(A|B_i) = P(\partial y/\partial x_i)$. Since a rising transition at any node has to be followed by a falling transition and vice versa, we have $P(x_i = \uparrow x_j = \uparrow) = P(x_i = \downarrow x_j = \downarrow) = P(x_i = \downarrow x_j = \uparrow) = P(x_i = \uparrow x_j = \downarrow) = \frac{1}{4}P(B_{i,j})$, where $\uparrow$ denotes a rising transition and $\downarrow$ denotes a falling one. Therefore, the *conditional probabilities* can be expressed as follows:

$$P(A|B_0) = 0$$

$$P(A|B_i) = Pc\left(\frac{\partial y}{\partial x_i} \right)$$

$$P(A|B_{i,j}) = \frac{1}{2}\left[Pc\left(\frac{\partial^2 y|_{00}}{\partial x_i \, \partial x_j} \right) + Pc\left(\frac{\partial^2 y|_{01}}{\partial x_i \, \partial x_j} \right) \right]$$

$$P(A|B_{1,2,3}) = \frac{1}{4}\left[Pc\left(\frac{\partial^3 y|_{000}}{\partial x_1 \, \partial x_2 \, \partial x_3} \right) + Pc\left(\frac{\partial^3 y|_{011}}{\partial x_1 \, \partial x_2 \, \partial x_3} \right) \right.$$

$$\left. + Pc\left(\frac{\partial^3 y|_{010}}{\partial x_1 \, \partial x_2 \, \partial x_3} \right) + Pc\left(\frac{\partial^3 y|_{011}}{\partial x_1 \, \partial x_2 \, \partial x_3} \right) \right]$$

After substituting the above expressions into Eq. (3.16) and using Eq. (3.7), we obtain Eq. (3.15) of Theorem 3.1.   □

The generalization of Theorem 3.1 for n-inputs is given below. The proof is very similar to the proof of Theorem 3.1 and is omitted for brevity.

Theorem 3.2 (n-Inputs)   *If y is a Boolean expression and x_i, $i = 1, \ldots, n$, are mutually independent primary inputs of y. Then*

$$a(y) = \sum_{i=1}^{n} \mathrm{Pc}\left(\frac{\partial y}{\partial x_i}\right)\left(a(x_i) \prod_{\substack{j \neq i \\ 1 \leq j \leq n}} [1 - a(x_j)]\right)$$

$$+ \frac{1}{2}\left\{ \sum_{1 \leq i < j \leq n} \left[\mathrm{Pc}\left(\frac{\partial^2 y|_{00}}{\partial x_i\, \partial x_j}\right) + \mathrm{Pc}\left(\frac{\partial^2 y|_{01}}{\partial x_i\, \partial x_j}\right)\right]\right.$$

$$\left. \left(a(x_i)a(x_j) \prod_{l \in \{1, 2, \ldots, n\} - \{i, j\}} [1 - a(x_l)]\right)\right\} + \cdots$$

$$+ \frac{1}{2^{n-1}}\left[\mathrm{Pc}\left(\frac{\partial^n y|_{00\ldots0}}{\partial x_1\, \partial x_2\, \cdots\, \partial x_n}\right) + \mathrm{Pc}\left(\frac{\partial^n y|_{00\ldots1}}{\partial x_1\, \partial x_2\, \cdots\, \partial x_n}\right) \cdots \right.$$

$$\left. + \mathrm{Pc}\left(\frac{\partial^n y|_{01\ldots1}}{\partial x_1\, \partial x_2\, \cdots\, \partial x_n}\right)\right]\left(\prod_{l=1}^{n} a(x_l)\right)$$

where

$$\mathrm{Pc}\left(\frac{\partial y}{\partial x_i}\right), \mathrm{Pc}\left(\frac{\partial^n y|_{00\ldots0}}{\partial x_1\, \partial x_2\, \cdots\, \partial x_n}\right), \ldots, \mathrm{Pc}\left(\frac{\partial^n y|_{01\ldots1}}{\partial x_1\, \partial x_2\, \cdots\, \partial x_n}\right)$$

are conditional probabilities under the condition that only some primary inputs switch and the rest do not at the leading edge of the clock cycle.

If the effect of simultaneous switching is neglected, $\mathrm{Pc}(\partial y/\partial x_i) = P(\partial y/\partial x_i)$ and $\{a(x_i)\prod_{j \neq i / 1 \leq j \leq n}[1 - a(x_j)]\}$ is equal to $a(x_i)$ and the above expression becomes identical to Eq. (3.6).   □

Example 3.2 $y = \bar{x}_1 x_2 x_3 + x_1 \bar{x}_2 x_3 + x_1 x_2 \bar{x}_3$. Determine $a(y)$?

$$\frac{\partial y}{\partial x_1} = \left(\bar{x}_2 x_3 + x_2 \bar{x}_3\right) \oplus \left(x_2 x_3\right) = x_2 + x_3$$

and by symmetry

$$\frac{\partial y}{\partial x_2} = x_1 + x_3 \qquad \frac{\partial y}{\partial x_3} = x_1 + x_2$$

$$\frac{\partial^2 y|_{00}}{\partial x_1\, \partial x_2} = 0 \oplus \bar{x}_3 = \bar{x}_3$$

Again, by symmetry,

$$\frac{\partial^2 y|_{00}}{\partial x_1 \, \partial x_3} = \tilde{x}_2, \frac{\partial^2 y|_{00}}{\partial x_2 \, \partial x_3} = \tilde{x}_1$$

It is also easy to see that

$$\frac{\partial^3 y|_{000}}{\partial x_1 \, \partial x_2 \, \partial x_3} = 0 \qquad \frac{\partial^3 y|_{001}}{\partial x_1 \, \partial x_2 \, \partial x_3} = 1$$

$$\frac{\partial^3 y|_{010}}{\partial x_1 \, \partial x_2 \, \partial x_3} = 1 \qquad \frac{\partial^3 y|_{011}}{\partial x_1 \, \partial x_2 \, \partial x_3} = 1$$

Let P_i, $a(y)$, and a_i respectively denote the signal probability of x_i, the normalized activity at y and at x_i. Ignoring simultaneous switching and using Eq. (3.6), we have

$$a(y) = (P_2 + P_3 - P_2 P_3)a_1 + (P_1 + P_3 - P_1 P_3)a_2 + (P_1 + P_2 - P_1 P_2)a_3$$

However, the accurate activity given by Theorem 3.2 can be obtained as follows:

$$\begin{aligned} a(y) = & (Pc_2 + Pc_3 - Pc_2 Pc_3)a_1(1 - a_2)(1 - a_3) \\ & + (Pc_1 + Pc_3 - Pc_1 Pc_3)a_2(1 - a_1)(1 - a_3) \\ & + (Pc_1 + Pc_2 - Pc_1 Pc_2)a_3(1 - a_1)(1 - a_2) \\ & + 1/2(1 - Pc_3)a_1 a_2(1 - a_3) \\ & + 1/2(1 - Pc_2)a_1 a_3(1 - a_2) \\ & + 1/2(1 - Pc_1)a_2 a_3(1 - a_1) + 3/4 a_1 a_2 a_3 \end{aligned}$$

Note that the probabilities Pc_i's are conditional probabilities, that is, $Pc_i = (P_i - 1/2 a_i)/1 - a_i$.

However, if Theorem 3.2 is applied to evaluate the exact activity, one must compute

$$Pc\left(\frac{\partial y}{\partial x_i}\right), Pc\left(\frac{\partial^n y|_{00 - 0}}{\partial x_1 \, \partial x_2 \, \cdots \, \partial x_n}\right), \ldots, Pc\left(\frac{\partial^n y|_{01 - 1}}{\partial x_1 \, \partial x_2 \, \cdots \, \partial x_n}\right)$$

This can be CPU time intensive. In the following section, we will show how to utilize symbolic probability to calculate accurate activity [4].

3.3.2   Derivation of Activities for Static CMOS from Signal Probabilities

We know from Eq. (3.13) that $P(y(t-T)y(t)) = P(y(t)) - \frac{1}{2}a(y) = P(y) - \frac{1}{2}a(y)$, where y is some node in a circuit. Hence, $a(y) = 2[P(y) - P(y(t-T)y(t))]$. Assume that the circuit under consideration has n independent inputs, $x_1, x_2, \ldots, x_n$. Recall from Section 3.2 that $P(y)$ can be expressed as a *sum of probability products* $\sum_{i=1}^{p} \alpha_i (\prod_{k=1}^{n} P^{m_{i,k}}(\bar{P}^{l_{i,k}}(x_k)))$, where n is the number of independent inputs to the circuit and α_i is some integer. The sum has p product terms. Let

$$U(y(t)) = \sum_{j=1}^{p} \alpha_j \left(\prod_{k=1}^{n} P^{m_{j,k}}(x_k(t)) \bar{P}^{l_{j,k}}(x_k(t)) \right) \tag{3.17}$$

Therefore,

$$U(y(t-T)) = \sum_{i=1}^{p} \alpha_i \left(\prod_{k=1}^{n} P^{m_{i,k}}(x_k(t-T)) \bar{P}^{l_{i,k}}(x_k(t-T)) \right) \tag{3.18}$$

Here, $U(y(t))$ and $U(y(t-T))$ represent the sums of probability products of $y(t)$ and $y(t-T)$. Hence, we have

$$U(y(t-T))U(y(t)) = \left[\sum_{i=1}^{p} \alpha_i \left(\prod_{k=1}^{n} P^{m_{i,k}}(x_k(t-T)) \bar{P}^{l_{i,k}}(x_k(t-T)) \right) \right]$$

$$\left[\sum_{j=1}^{p} \alpha_j \left(\prod_{k=1}^{n} P^{m_{j,k}}(x_k(t)) \bar{P}^{l_{j,k}}(x_k(t-T)) \right) \right]$$

$$= \sum_{1 \le i, j \le p} \alpha_i \alpha_j \left(\prod_{k=1}^{n} p^{m_{i,k}}(x_k(t-T)) \bar{P}^{l_{i,k}}(x_k(t-T)) \right.$$

$$\left. \times P^{m_{j,k}}(x_k(t)) \bar{P}^{l_{j,k}}(x_k(t)) \right) \tag{3.19}$$

The following theorem enables us to compute $P(y(t-T)y(t))$ and therefore $a(y)$ from the sum of probability products of $P(y)$.

Theorem 3.3   *For each product term in Eq. (3.19), if $P(x_k(t-T))P(x_k(t))$ is replaced with $P(x_k(t-T)x_k(t))$, $P(x_k(t-T))\bar{P}(x_k(t))$ with $P(x_k(t-T)\bar{x}_k(t))$, $\bar{P}(x_k(t-T))P(x_k(t))$ with $P(\bar{x}_k(t-T)x_k(t))$, and $\bar{P}(x_k(t-T))\bar{P}(x_k(t))$ with $P(\bar{x}_k(t-T)\bar{x}_k(t))$, then Eq. (3.19) becomes*

$$U(y(t-T))U(y(t))$$

$$= \sum_{1 \le i, j \le p} \alpha_i \alpha_j \left(\prod_{k=1}^{n} P(x_k^{m_{i,k}}(t-T)\bar{x}_k^{l_{i,k}}(t-T)) P(x_k^{m_{j,k}}(t)\bar{x}_k^{l_{j,k}}(t)) \right)$$

$$\tag{3.20}$$

and is equal to $P(y(t-T)y(t))$.

Proof: All sum-of-products expressions of a Boolean function y correspond to a unique canonical sum [2]. Hence, every sum of probability products of $P(y)$ can be rearranged to have the same form as the unique canonical sum. Therefore, we will assume in the following that $P(y)$ is expressed as the canonical sum of probability products for simplicity. Accordingly, Eq. (3.19) becomes

$$U(y(t-T)U(y(t))) = \sum_{1 \le i,j \le p} \left(\prod_{k=1}^{n} P(s_k(t-T))P(w_k(t)) \right) \quad (3.21)$$

where s_k and w_k are either x_k or $\bar{x}_k$.

On the other hand,

$$P(y(t-T)y(t)) = P\left(\left[\prod_{i=1}^{p} \left(\prod_{k=1}^{n} P(s_k(t-T)) \right) \right] \left[\sum_{i=1}^{p} \left(\prod_{k=1}^{n} P(w_k(t)) \right) \right] \right)$$

$$= \sum_{1 \le i,j \le p} \left(\prod_{k=1}^{n} P(s_k(t-T)w_k(t)) \right) \quad (3.22)$$

If we replace $P(s_k(t-T))P(w_k(t))$ with $P(s_k(t-T)w_k(t))$ in Eq. (3.21), the new equation is exactly the same as Eq. (3.22). $\square$

We use Theorem 3.3 and apply Eq. (3.10)–(3.14) to compute the numerical value of $P(y(t-T)y(t))$. The following example demonstrates the method.

Example 3.3 Let $y = x_1 + x_2$. Therefore, $P(y) = P(x_1) + \overline{P(x_1)}P(x_2)$, where $\overline{P(x_1)} = P(\bar{x}_1) = 1 - P(x_1)$. Calculate $P(y(t-T)y(t))$ and $a(y)$:

$$U(y(t-T))U(y(t)) = \left[P(x_1(t-T)) + \overline{P(x_1(t-T))}P(x_2(t-T)) \right]$$

$$\left[P(x_1(t)) + \overline{P(x_1(t))}P(x_2(t)) \right]$$

$$= P(x_1(t-T))P(x_1(t)$$

$$+ \left[P(x_1(t-T))\overline{P(x_1(t))}P(x_2(t)) \right]$$

$$+ \overline{P(x_1(t-T))}P(x_1(t))P(x_2(t-T))$$

$$+ \overline{P(x_1(t-T))}\overline{P(x_1(t))}P(x_2(t-T))P(x_2(t)) \right]$$

Then from Theorem 3.3, we have

$$P(y(t-T)y(t))$$

$$= P(x_1(t-T)x_1(t)) + \left[P(x_1(t-T)\bar{x}_1(t))P(x_2(t)) \right.$$

$$+ P(\bar{x}_1(t-T)x_1(t))P(x_2(t-T))$$

$$+ P(\bar{x}_1(t-T)\bar{x}_1(t))P(x_2(t-T)x_2(t)) \right]$$

$$= \left[P(x_1) - 1/2a(x_1) \right] + 1/2a(x_1)P(x_2) + 1/2a(x_1)P(x_2)$$

$$+ \left[1 - P(x_1) - 1/2a(x_1) \right] \left[P(x_2) - 1/2a(x_2) \right]$$

After rearranging and simplifying the terms, we have

$$a(y) = [1 - P(x_1)]a(x_2) + [1 - P(x_2)]a(x_1) - 1/2a(x_1)a(x_2)$$

Given the sum of probability products of y, the following algorithm computes the numerical value of $P(y(t - T)y(t))$.

Compute

$P(y(t - T)y(t))$

Input: Sum of probability products of $P(y)$.
Output: Numerical value of $P(y(t - T)y(t))$.
result = 0;
For each probability product term Q_i{
    For each probability product term Q_j{
        $P = 1$:
        For each independent input x_k{
          If (Q_i contains $P(x_k)$){
            If (Q_j contains $P(x_k)$)
              $P = P \times (P(x_k) - \frac{1}{2}a(y))$;
            Else if (Q_j contains $\bar{P}(x_k)$)
              $P = P \times \frac{1}{2}a(y)$;
            Else
              $P = P \times P(x_k)$;
          }
          Else if (Q_i contains $\bar{P}(x_k)$){
            If (Q_j contains $P(x_k)$)
              $P = P \times \frac{1}{2}a(y)$;
            Else if (Q_i contains $\bar{P}(x_k)$)
              $P = P \times (1 - P(y) - \frac{1}{2}a(y))$;
            Else
              $P = P \times (1 - P(x_k))$;
          }
          Else{
            If (Q_j contains $P(x_k)$)
              $P = P \times P(x_k)$;
            Else if (Q_j contains $\bar{P}(x_k)$)
              $P = P \times (1 - P(x_k))$;
          }
        }
    }
    result = *result* + $P \times \alpha_i \times \alpha_j$;
}
return *result*;

Here Q_i corresponds to a probability product term $U(y(t - T))$ in Eq. (3.18) and Q_j to a probability product term $U(y(t))$ in Eq. (3.17). Notice that $U(y(t - T))$ and $U(y(t))$ are expressed explicitly in terms of $P(x_j(t - T))$ and $P(x_j(t))$ in Example 3.3. Recall that $U(y(t - T))$ and $U(y(t))$ are sums of probability products of $y(t - T)$ and $y(t)$, respectively. Hence, it seems that we need memory space for storing both sums of probability products of $y(t - T)$ and $y(t)$. However, from the algorithm it is clear that only a copy of the sum of probability products of y is required, thereby making it memory efficient.

3.3.3 Switching Activity in Sequential Circuits

Accurate estimation of power dissipation or switching activity in sequential circuits is considerably more difficult than combinational circuits. This is mainly due to the fact that the signal probabilities and activities at the state input bits are not known ahead of time, unlike the combinational circuits where the switching activities at all the primary inputs are given/known. Hence, before the signal activities at the internal nodes of a sequential circuit can be estimated, one has to first determine the probabilities of being in different states of the sequential circuit in the steady state.

Let us first assume that the sequential circuits implement a finite state machine (FSM) where the states follow Markov process (future is independent of past once the present state is specified). A methodology to estimate the signal probability at the present-state input bits is proposed in [21, 25], which takes spatial correlations of signals into consideration but ignore any temporal correlations (method 1). We will first describe the estimation technique in brief. Then we will consider techniques to accurately estimate switching activity in sequential circuits using spatio temporal correlation among signals (method 2).

Let $P_{in} = [p_1, p_2, \ldots, p_n]$ be a vector representing the signal probabilities of the n independent primary inputs of a combinational logic circuit. The output signal probability vector P_{out} is given by $F(P_{in})$, where function F is determined by the combinational logic. Figure 3.7 represents an FSM, where the inputs to the combinational part of the circuit consists of a set of primary inputs I and a set of present state inputs S and the output consists of a set of primary outputs O and a set of next-state outputs N. In general, the signal probabilities of the primary inputs are known, and hence,

$$P_O = F(P_I, P_S) \tag{3.23}$$

where the signal probabilities of the S inputs be given in a vector form as P_S. Hence, the signal probability vector for the next-state inputs is

$$P_N = F'(P_I, P_S) \tag{3.24}$$

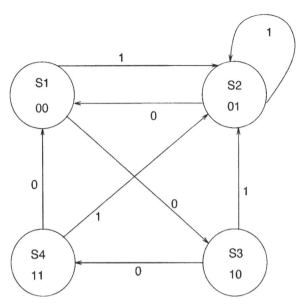

Figure 3.7   A general model for sequential logic circuit.

The above equations can be solved for P_S and P_N by noting that $P_S = P_N$ in the steady state. Iterative Newton–Raphson techniques can be used to solve the set of equations for state probabilities.

The above technique to estimate signal activities can be extended to accurately estimate signal activity under spatio-temporal correlations among signals [22–24]—method 2. In method 1, the input signals were assumed to be temporally uncorrelated. That is, the input signal at time t is independent of the same input signal at time $t + T$, where T is the clock cycle. Therefore, when input signals are temporally correlated, the activities can be very different from the ones under the assumption given in [21, 25]. Let us consider the circuit of Figure 3.8 and its state transition graph (STG) in Figure 3.7. Under the assumption of having temporally uncorrelated signals, the normalized activity a (also called transition probability) of the input signal equals $2P(1 - P)$, where P is the signal probability. However, if each input signal is temporally correlated, which means that the inputs can have normalized activities other than 0.5, the probabilities and activities at the internal nodes can be different. Table 3.1 shows the different probabilities and activities at nodes $ns1$, $ns2$, and n corresponding to different activities of the input signals. The results were obtained using logic level simulation. The signal probability of the input I is assumed to be 0.5. If the primary input is temporally uncorrelated, its normalized activity equals to $(2 \times 0.5)(1 - 0.5) = 0.5$.

Let us assume that the input signal distributions are represented by probabilities and activities of the inputs. Therefore, given an STG of an FSM,

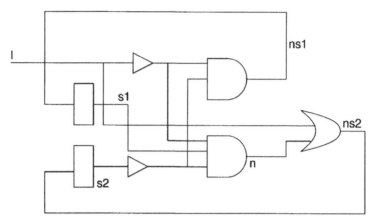

Figure 3.8   Example sequential circuit.

TABLE 3.1   Probabilities and Normalized Activities Corresponding to Different Activities of Input Signals

Normalized	$ns1$		$ns2$		n	
Activity of I	P	a	P	a	P	a
0.5	0.21	0.29	0.29	0.56	0.06	0.13
0.3	0.26	0.29	0.63	0.45	0.11	0.21
0.1	0.3	0.3	0.68	0.34	0.14	0.28

we can create an extended state transition graph (ESTG). Each state in ESTG is represented by a present state in STG and the next primary input vector (corresponding to the next state). The temporal correlation of the input signals are explicitly represented by an ESTG. In method 1, since the temporal correlation is neglected, the transition probability of the STG (the probability of transition from one state to another state) is completely determined by the present input vector. We present an exact method to estimate signal activity that similar to method 1. However, we apply the Chapman–Kolmogorov equations to the ESTG rather than to the STG. The exact method requires the solution of a linear system of equations of size at least 2^M and at most 2^{M+N}, where M is the number of latches (flip-flops) and N is the number of primary inputs. For large circuits, the method may be computation time intensive. Hence, an approximate solution method is also presented to estimate the probabilities and activities at the internal nodes of a circuit.

The approximate solution method first unrolls the circuit k times with assigned initial values at the state bits. The signal probabilities and activities of the state inputs are then calculated by applying a method similar to the one proposed for combinational logic. Since the approximate method does not assume the knowledge of an STG, it is applicable to general sequential circuits. However, this method, like the one for combinational circuits must trade off accuracy for speed.

3.3.3.1 Representation of Sequential Logic Circuits

A general model for a synchronous sequential circuit (Mealy machine) is shown in Figure 3.9. We denote the input vector to the combinational logic as U. The terms U_{-T}, U_0, and U_T represent U at time $-T$, 0, and T, respectively, where T is the system clock cycle. Input vector U consists of two parts: primary inputs I and state inputs S. Accordingly, $\langle I_{-T}, BS \rangle$, $\langle I_0, PS \rangle$, and $\langle I_T, NS \rangle$ correspond to U_{-T}, U_0, and U_T. Here I_{-T}, I_0, and I_T are primary inputs at time $-T$, 0, and T. Particularly, we denote state inputs at $-T$, 0, and T as previous state BS, present state PS, and next state NS respectively. Here PS is completely determined by $\langle I_{-T}, BS \rangle$ or U_{-T}, and NS by $\langle I_0, PS \rangle$ or U_0.

The STG is represented by a directed graph $G(V, E)$, where each vertex $S_i \in V$ represents a state of the FSM and each edge $e_{i,j} \in E$ represents a transition from S_i to S_j. Each state S_i also corresponds to an instance of the vector S (state inputs). The term $F(S_i, I) = S_j$ denotes that the machine makes a transition from state S_i to S_j due to the primary input vector I. We call $F(\cdot)$ the next state function. Therefore, we have $F(PS, I_0) = NS$ and $F(BS, I_{-T}) = PS$. We assume that the implemented sequential circuit or the STG does not contain periodic states.

3.3.3.2 Signal Activity Considering Temporal Correlation of the Input Signals

In this section, we will derive an accurate method of calculating signal activity. Some basic assumptions of the exact method will be introduced first. Based on some of the assumptions, we will extend STG to an ESTG and

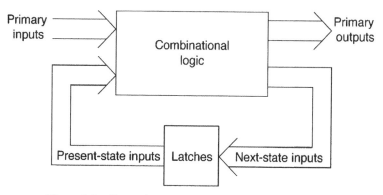

Figure 3.9 General synchronous sequential logic circuit.

derive state probabilities by applying Chapman–Kolmogorov equations to the ESTG.

Assumptions for the Accurate Method   For our analysis let us assume that the glitches can be neglected if all the paths are balanced and the logic depth is reduced. Delays are decoupled from the circuits and are assumed to be neglected. Also, we assume that primary inputs to a sequential circuit are modeled as *mutually independent (SSS) mean-ergodic 0–1 process*. Each primary input is associated with a signal probability and an activity. For combinational circuits, these assumptions will completely determine all the internal and output nodes of the circuits. However, more information is necessary for sequential circuits. The NS of the sequential circuits depends on $\langle I_0, \text{PS} \rangle$, PS on $\langle I_{-T}, \text{BS} \rangle$, and hence, NS on $\langle I_{-T}, I_0, \text{BS} \rangle$. We can extend this procedure and reach the conclusion that NS depends on $\langle I_0, I_{-T}, I_{-2T}, \ldots \rangle$, that is, the whole history of the primary inputs. This makes the calculation complicated and computationally forbidden. Therefore, we assume that for each primary input x_i, $x_i(t)$ only depends on $x_i(t - T)$, making it a discrete-time Markov chain [17] with two states, 0 and 1. Therefore, if $x_i(t - T)$ is known, from Eqs. (3.8–3.14) the probability of $x_i(t)$ is determined by its signal probability and activity as follows:

$$P\big(x_i(t)|x_i(t - T)\big) = \frac{P(x_i) - \frac{1}{2} \times a(x_i)}{P(x_i)}$$

$$P\big(\bar{x}_i(t)|x_i(t - T)\big) = \frac{\frac{1}{2} \times a(x_i)}{P(x_i)}$$

$$P\big(x_i(t)|\bar{x}_i(t - T)\big) = \frac{\frac{1}{2} \times a(x_i)}{1 - P(x_i)}$$

$$P\big(\bar{x}_i(t)|\bar{x}_i(t - T)\big) = \frac{1 - P(x_i) - \frac{1}{2} \times a(x_i)}{1 - P(x_i)} \tag{3.25}$$

As a result of the assumption of Markov chain inputs, the following equations hold:

$$P\big(x_i(t)|x_i(t - T) \cdots x_i(t - nT)\big) = P\big(x_i(t)|x_i(t - T)\big)$$

$$P\big(x_i(t)x_i(t - T) \cdots x_i(t - nT)|x_i(t - T) \cdots x_i(t - nT)\big)$$

$$= P\big(x_i(t)|x_i(t - T)\big) \tag{3.26}$$

The assumption that the primary input signal is a discrete-time Markov chain implies that there is an underlying ESTG for the FSM. We will explain it in the following section.

3.3.3.3   State Probability Calculation from ESTG

Given the STG of a sequential circuit or an FSM, we can build an ESTG and calculate the probability of a state of the ESTG. Consider the STG of Figure 3.7. Let $P(S_i)$ denote the state probability, that is, the probability of the machine making a transition from state S_i to S_j. However, $P(e_{i,j})$ depends not only on present primary inputs but also on the previous inputs, and hence it depends on S_i. As a result, $P(e_{i,j})$ is not equal to the probability of the present primary input vector I that satisfies $F(S_i, I) = S_j$, where $F(\cdot)$ denotes the next-state function mentioned. That is, $P(\mathrm{NS}|\mathrm{PS}) \neq P(I_0)$, where $P(\mathrm{NS}|\mathrm{PS})$ denotes the transition probability from PS to NS and $F(\mathrm{PS}, I_0) = \mathrm{NS}$. Since $P(e_{i,j})$ is not known, we cannot apply Chapman–Kolmogorov equations to the STG. This motivates us to create an ESTG from the STG.

The ESTG is represented by a directed graph $G(V', E')$, where each vertex $S_i' \in V'$ represents a state of the FSM and each edge $e_{i,j}' \in E'$ represents a transition from S_i' to S_j'. Each state S_i' also corresponds to an instance of the vector $\langle S, I \rangle$ (state inputs and primary inputs). Therefore, if the sequential circuit under consideration has N independent primary inputs and M state inputs, the number of the states of the ESTG can be 2^{N+M}. That is, a state S in the STG is split into 2^N states in the ESTG. Therefore, each state in the ESTG keeps track of the value of the present primary input vector.

For example, consider the STG of Figure 3.7. The corresponding ESTG is state shown in Figure 3.10, in which each state is represented by two present state bits and one present input, that is, $ps1, ps2, I_0$. The variable attached to each edge in Figure 3.10 is the next input (input at time T). The state $S1$ in the STG of Figure 3.7 is split into $S1_0$ and $S1_1$ in the ESTG.

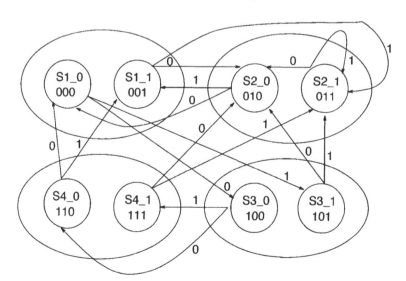

Figure 3.10   Extended state transition graph.

By definition, the machine will make a transition from state $\langle BS, I_{-T} \rangle$ to $\langle PS, I_0 \rangle$ if the present primary input vector is I_0 and from $\langle PS, I_0 \rangle$ to $\langle NS, I_T \rangle$ if the next primary input vector is I_T. Therefore, the transition probability from $\langle P, I_0 \rangle$ to $\langle NS, I_T \rangle$, denoted as $P(\langle NS, I_T \rangle | \langle PS, I_0 \rangle)$, equals $P(I_T | \langle PS, I_0 \rangle)$ if $F'(\langle PS, I_0 \rangle, I_T) = \langle NS, I_T \rangle$, where $F'(S_i', I_T) = S_j'$ denotes that the machine makes a transition from state S_i' to S_j' due to the next primary input vector. Furthermore, since the primary inputs are assumed to be Markov chain, I_T depends on I_0 only, $P(I_T | \langle PS, I_0 \rangle) = P(I_T | I_0)$. We have

$$P(\langle NS, I_T \rangle | \langle PS, I_0 \rangle) = \begin{cases} P(I_T | I_0) & \text{if } F'(\langle PS, I_0 \rangle, I_T) = \langle NS, I_T \rangle \\ 0, & \text{if } F'(\langle PS, I_0 \rangle, I_T) \neq \langle NS, I_T \rangle \end{cases}$$

(3.27)

where $P(I_T | I_0)$ is the conditional probability.

For instance, let us assume that the FSM is in STG state $S1$ (Figure 3.7) with primary input value of 1. The next STG state will be $S2$. However, in the ESTG (Figure 3.10) the state transition is different. At first, the FSM is in ESTG state $S1_1$ since the primary input value is 1 (if the primary input value is 0, it is in $S1_0$ instead). Up to this point, we know that the next ESTG state is either $S2_0$ or $S2_1$, corresponding to a STG state $S2$. But the exact next ESTG state depends on the next primary input value. Because the present input value is 1 and the next input value is 1 (0), the transition probability from $S1_1$ to $S2_1$ ($S2_0$) is $P(I_T | I_0)[P(\bar{I}_T | I_0)]$.

Since we assume that the primary inputs are mutually independent, we have

$$P(I_T | I_0) = \prod_{i=1}^{N} P(x_i(T) | x_i(0))$$

(3.28)

where x_i is one of the N primary inputs. Equation (3.28) can be calculated using Eq. (3.25). For the ESTG, we can obtain its state probabilities by solving the Chapman–Kolmogorov equation [17] as follows:

$$P(S_i') = \sum_{j \in \text{IN\_STATE}(i)} P(I_i | I_j) P(S_j') \qquad i = 1, 2, \ldots L - 1$$

$$1 = \sum_{j=1}^{L} P(S_j')$$

(3.29)

where $P(S_i') = P(\langle S_i, I_i \rangle)$ and IN_STATE(i) is the set of fanin states of S_i' in the ESTG. Let us assume that the ESTG has L states. For the STG, we can obtain its state probabilities by summing all the ESTG states that correspond to the same state in the STG. That is,

$$P(S_i) = \sum_{\forall S_j = S_i} P(\langle S_j, I_j \rangle)$$

(3.30)

3.3.3.4 *Signal Probability and Activity*

We first consider the signal probabilities and activities of the internal nodes and the primary outputs. Let y denote an internal node or a primary output. Node y can be represented by a logic function $f(\langle S, I \rangle)$ with state inputs S and primary inputs I. Let $y(0)$ and $y(T)$ denote $f(\langle PS, I_0 \rangle)$ and $f(\langle NS, I_T \rangle)$, respectively. Since y is SSS and NS $= F(\langle PS, I_0 \rangle)$, we have

$$P(y) = P(f(\langle PS, I_0 \rangle)) = P(f(\langle NS, I_T \rangle))$$

and

$$a(y) = P(y(0) \oplus y(T))$$
$$= P(f(\langle PS, I_0 \rangle) \oplus f(\langle NS, I_T \rangle))$$
$$= P(f(\langle PS, I_0 \rangle) \oplus f(\langle F(\langle PS, I_0 \rangle), I_T \rangle))$$

By summing up all the terms over all possible ESTG states, that is, by the total probability theorem [17], we have

$$P(y) = \sum_{\langle PS, I_0 \rangle \in \{\text{all ESTG states}\}} f(\langle PS, I_0 \rangle) P(\langle PS, I_0 \rangle)$$

$$a(y) = \sum_{\langle PS, I_0 \rangle \in \{\text{all ESTG states}\}} P(f(\langle PS, I_0 \rangle))$$

$$\oplus f(\langle F(\langle PS, I_0 \rangle), I_T \rangle) | \langle PS, I_0 \rangle) P(\langle PS, I_0 \rangle)$$

Notice that given a state $\langle PS, I_0 \rangle$ in the ESTG,

$$P(f(\langle PS, I_0 \rangle) \oplus f(\langle F(\langle PS, I_0 \rangle, I_T \rangle) | PS, I_0 \rangle)$$

is a function of I_T only.

Given the state encoding, each STG state is represented by state inputs $s_0, s_1, \ldots, s_{M-1}$, where M is the number of flip-flops. Similarly, each ESTG state is represented by state inputs $s_0, s_1, \ldots, s_{M-1}$ and primary input vector I_0. Therefore, for the node of the state input s_i, we can calculate its probability by summing up all the probabilities of ESTG states whose s_i value is 1 in its representation. That is,

$$P(s_i) = \sum_{PS \in S\_EN(i)} P(\langle PS, I_0 \rangle) = \sum_{Ps \in S\_EN(i)} P(PS) \qquad (3.31)$$

where S¯(i) is the set of states whose encodings have the ith bit equal to 1. In the same way, by summing all the probabilities of ESTG states whose s_i value is v at present time 0 and is $\bar{v}$ at next time T, we have

$$a(s_i) = \sum_{PS \in S\_EN(i) NS \notin S\_EN(i)} P(\langle PS, I_0 \rangle)$$

$$+ \sum_{PS \notin S\_EN(i) NS \in S\_EN(i)} P(\langle PS, I_0 \rangle) \qquad (3.32)$$

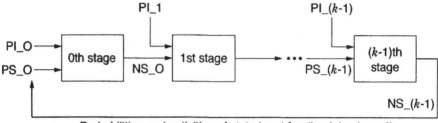

Probabilities and activities of state input feedback in phase II

Figure 3.11   Unrolling of the sequential circuits k times.

3.3.4 An Approximate Solution Method

As mentioned before, the exact method may require solving for a linear system of equations of size 2^{N+M}, where N is the number of primary inputs and M is the number of flip-flops. For large circuits with a large number of primary inputs and flip-flops, the exact method is computationally expensive. Therefore, we propose an approximate method that takes temporal correlations of primary inputs into account. We unroll the sequential circuit by k time frames as shown in Figure 3.11. As k approaches infinity, the unrolled circuit is conceptually the same as the sequential circuit. However, it is computationally expensive to unroll the circuit infinite times. As a result, we unroll it k times and approximately capture the spatial and temporal correlations. At first, we need the signal probabilities and activities of the present state bits of the zeroth stage (PS_0) to calculate the signal probabilities and activities at the internal nodes and outputs. In sequential circuits, the signal probabilities and activities at PS_0 are the same as those at the next-state bits of the $(k-1)$th stage (NS_{k-1}) if glitches are neglected. Though this is not true for the k-unrolled circuits, it implies a method to determine initial values for PS_0. We can obtain the approximate signal probabilities and activities at the PS_0 by assigning initial values to PS_0, calculating the signal probabilities and activities at NS_{k-1}, and assigning the values of NS_{k-1} to PS_0. Once we have the more accurate initial values, we can compute the signal probabilities and activities at all nodes at the $(k-1)$th stage. We will explain the details of the solution method by dividing it into two phases.

3.3.4.1 Phase I: Calculating Probabilities and Activities of State Inputs

In this phase, the circuit is unrolled k time frames, as shown in Figure 3.11. Since, the correlations among PS_0 are unknown to us, the present-state inputs of the zeroth stage are assumed to be uncorrelated and independent of the primary input signals. They have the same property described in Section 3.3.3 as that of the primary inputs. For the primary input signals (PI_0), we assume that the signal probabilities and activities are available through system level simulation of the circuit under real life inputs. For PS_0, we assign some initial values. Since we have the information of all the inputs to the unrolled circuit, we can calculate the probabilities and activities at NS_{k-1} using the method described earlier for combinational circuits. However, there are two important considerations that call for special attention. First, for combinational circuits we only need to consider primary inputs I_0 and I_T [22]. For sequential circuits, we must take the correlation among $I_0, I_T 2, I_{(k-1)T}$ into account. In [21, 25], the circuit was unrolled in a similar manner. However, the correlation of the primary inputs was not considered. The following example illustrates how correlations can be considered.

Example 3.4   Express $P(x_{2T}x_Tx_0)$, $P(x_{2T}\bar{x}_Tx_0)$, and $P(x_{2T}x_0)$ in terms of its probabilities P (P_x) and normalized activities a (a_x):

$$P(x_{2T}x_Tx_0) = P(x_{2T}x_Tx_0|x_Tx_0)P(x_Tx_0|x_0)P(x_0)$$

$$= P(x_{2T}|x_Tx_0)P(x_T|x_0)P(x_0)$$

$$= P(x_{2T}|x_T)P(x_T|x_0)P(x_0)$$

$$= \frac{P - \frac{1}{2}a}{P} \times \frac{P - \frac{1}{2}a}{P} \times P$$

$$= \frac{\left(P - \frac{1}{2}a\right)^2}{P}$$

Note that $P(x_{2T}|x_Tx_0) = P(x_{2T}|x_T)$ is based on the assumption described in Section 3.3.3. Similarly,

$$P(x_{2T}\bar{x}_Tx_0) = P(x_{2T}\bar{x}_Tx_0|\bar{x}_Tx_0)P(\bar{x}_Tx_0|x_0)P(x_0)$$

$$= P(x_{2T}|\bar{x}_Tx_0)P(\bar{x}_T|x_0)P(x_0) = P(x_{2T}|\bar{x}_T)P(\bar{x}_T|x_0)P(x_0)$$

$$= \frac{\frac{1}{2}a}{1 - P} \times \frac{\frac{1}{2}a}{P} \times P = \frac{\frac{1}{4}a^2}{1 - P}$$

Therefore, we have

$$P(x_{2T}x_0) = P(x_{2T}x_Tx_0) + P(x_{2T}\bar{x}_Tx_0) = \frac{\left(P - \frac{1}{2}a\right)^2}{P} + \frac{\frac{1}{4}a^2}{1 - P}$$

The second consideration concerns reconvergent nodes of sequential nodes that are different from those of the combinational ones. Consider the circuit of Figure 3.12, which is a two-unrolled circuit of Figure 3.8. Only from topological analysis, it seems that I^T and $ns2^0$ are independent of each other. That is, there does not exist a common ancestor of node I^T and node $ns2^0$. An ancestor y of a node z is defined as a node such that there is a

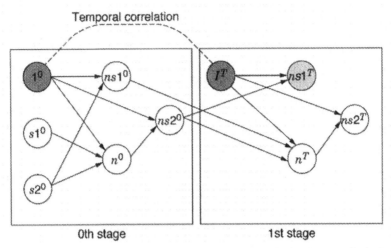

Figure 3.12   Example of a temporally reconvergent node in the unrolled circuit of Figure 3.1.

directed path from y to z. However, $ns2^0$ is topologically dependent on node I^0, and the primary inputs I^0 and I^T are temporally correlated. As a result, I^T and $ns2^0$ are not independent. Here node I is defined as a temporally reconvergent node as opposed to a topologically reconvergent node. Following the calculation of the probabilities and the activities of next-state inputs of the $(k-1)$th stage, we can calculate the total power dissipation of the circuit in phase II.

3.3.4.2 Phase II: Calculating Probabilities and Activities of Internal Nodes

In this phase, the probabilities and the activities of the state inputs of the zeroth stage are set equal to those of the next-state inputs of the $(k-1)$th stage. Under the same assumptions as those of phase I, we can (by the same method) calculate all the probabilities and the activities of the internal nodes, including the primary outputs and the next-state inputs of the $(k-1)$th stage.

Before we leave this section, it is worth mentioning that the approximate method is in fact exact for pipelined circuits. This is due to the fact that pipelined circuits do not have feedback (present-state inputs).

3.4 STATISTICAL TECHNIQUES

One of the problems with the above methods is that the estimate of circuit activity at the internal nodes of the circuit is a lower bound on the actual number of transitions. Due to the use of the zero-delay model, the glitching activity of the circuit has been ignored. The statistical techniques of power estimation that we will cover next can easily include delay models in its analysis and hence can accurately estimate both switching and glitching power.

The idea behind the statistical technique is very simple. The circuit is simulated repeatedly using a logic simulator and the switching activities at various nodes are noted. The main issue in this kind of simulation is to determine the type of input to be applied to the circuit and when to stop simulation. Burch et al. [11] used randomly generated inputs for the circuit under consideration. Statistical mean estimation techniques [16] are used in determining the stopping criteria in the Monte Carlo simulations. One of the advantages of this method is that the delay model can be used during the logic simulation and, hence, glitching power can be accurately estimated.

3.4.1 Estimating Average Power in Combinational Circuits

Average power can be estimated by simulating the circuit with a large number N of inputs. If the successive input patterns are independent, then the average power can be estimated from the node switching activity for large N. However, in order to determine the value of N, statistical mean estimation techniques can be used to determine the stopping criteria.

Burch et al. [11] experimentally determined that the power consumed by a circuit over a period t has a normal distribution. Let $\bar{p}$ and s be the measured average and the standard deviation of the random sample of the power measured over time T, respectively. Then with $(1 - \alpha) \times 100\%$ confidence we can write the following inequality:

$$|\bar{p} - P_{\text{avg}}| < \frac{t_{\alpha/2} s}{\sqrt{N}}$$

where $t_{\alpha/2}$ is obtained from a *t-distribution* with $N - 1$ degrees of freedom and P_{avg} is the true average power. Hence, one can write

$$\frac{|\bar{p} - P_{\text{avg}}|}{\bar{p}} < \frac{t_{\alpha/2} s}{\bar{p}\sqrt{N}}$$

From the above inequality one can notice that the right-hand side of the inequality should be less than ϵ, the desired percentage error, for the given confidence level $(1 - \alpha) \times 100\%$. Hence, the following condition should be satisfied:

$$\frac{t_{\alpha/2} s}{\bar{p}\sqrt{N}} < \epsilon$$

Therefore, from the above equation one can determine the required number of simulations N for obtaining a given confidence level and desired percentage error:

$$N \geq \left(\frac{t_{\alpha/2} s}{\bar{p} \epsilon} \right)^2$$

One of the main advantages of the technique is that the accuracy required by the user can be specified up-front and based on that, the number of simulation runs can be obtained. It has been found that for reasonable accuracy, the number of runs required for some of the circuits can be as low as 10. Hence, the CPU time required for the Monte Carlo simulations to estimate average power can be comparable to that of the probabilistic techniques.

The above technique provides an accurate estimate of average power dissipation; however, one of the main disadvantages is that it does not provide accurate estimates of the power consumed by individual gates of the circuit. Some of the internal gates switch very infrequently, and hence, it will take a very large number of simulation runs to estimate the power dissipation of those nodes. A simple modification to the above Monte Carlo technique can be used to accurately estimate the internal-node power dissipation by considering the absolute error bound for low-activity nodes rather than the percentage error bound [18].

Let us consider a circuit that is simulated N times for a time interval T. Let the number of transitions measured at a node every time be n_i, $1 \leq i \leq N$. Hence, the average number of transitions at the node is given by $\bar{n} = n_i/N$. Then, according to the central limit theorem [16], $\bar{n}$ has a distribution that is close to normal for large values of N. If β is the true expected value of the average number of transitions at the node under consideration and s is the measured standard deviation of N values of n_i, then for the following inequality holds with $(1 - \alpha) \times 100\%$ confidence:

$$\frac{\beta - \bar{n}}{\bar{n}} \leq \frac{z_{\alpha/2} s}{\bar{n} \sqrt{N}} \tag{3.33}$$

where $z_{\alpha/2}$ is obtained from the normal distribution. For the above inequality to be valid, N should be larger than 30. As before, if the percentage error that can be tolerated is ϵ, then the number of required simulations is

$$N \geq \left(\frac{z_{\alpha/2} s}{\epsilon \bar{n}} \right)^2$$

It is clear from the above inequality that if $\bar{n}$ is very small, then the number of simulation runs N will be very large. Xakellis et al. [18] classified the low-activity nodes as those nodes for which $\bar{n} < \beta_{\min}$, where $\beta_{\min}$ is some user-specified value. Hence, for the low-activity nodes, the following modified stopping criteria can be used:

$$N \geq \left(\frac{z_{\alpha/2} s}{\epsilon \beta_{\min}} \right)^2$$

Hence, from inequality (3.33) one can write the following condition with $(1 - \alpha) \times 100\%$ confidence:

$$|\beta - \bar{n}| \leq \frac{z_{\alpha/2} s}{\sqrt{N}} \leq \beta_{\min} \epsilon$$

Therefore $\beta_{\min} \epsilon$ is an absolute error bound for the low-activity nodes.

It should be observed that the low-activity nodes take the longest time to converge; however, they have the least effect on the average power dissipation of a circuit. Therefore, the penalty associated with a higher percentage error with the low-activity nodes is minimal in terms of average power dissipation. Electromigration and hot-electron degradation due to circuit node can be attributed to the average node activity as described in Section 3.9. Hence, the contribution of low-activity nodes toward circuit reliability is also minimal.

3.4.2 Estimating Average Power in Sequential Circuits

The above Monte-Carlo-based technique cannot be directly applied to sequential circuits because the steady-state probability for each state can be different. The problem is even more complicated due to the possible presence of "near-closed sets of states" in a state transition graph. In the presence of such a set of states, we also derive a technique to estimate power dissipation in sequential circuits.

The basic idea of Monte Carlo methods for estimating activity of individual nodes is to simulate a circuit by applying random-pattern inputs. The convergence of simulation can be obtained when the activities of individual nodes satisfy some stopping criteria. The procedure is outlined in Figure 3.13.

In sequential circuits, things are different due to the state-bit feedback. One of the approaches is to monitor the state-bit probabilities to determine the convergence [89]. In order to have $(1 - \alpha) \times 100\%$ confidence and some error ϵ (upper bound on absolute error of the state-bit probabilities), one must perform at least $K \geq \max(N_1^2, N_2^2, N_3^2)$ runs, where

$$N_1 = \frac{z_{\alpha/2}}{2\epsilon}$$

$$N_2 = \frac{z_{\alpha/2} \sqrt{2\epsilon + 0.1} + \sqrt{(\epsilon + 0.1) z_{\alpha/2}^2 + 3\epsilon}}{2\epsilon}$$

$$N_3 = \frac{\sqrt{63} + z_{\alpha/2}}{2\sqrt{\epsilon}} \tag{3.34}$$

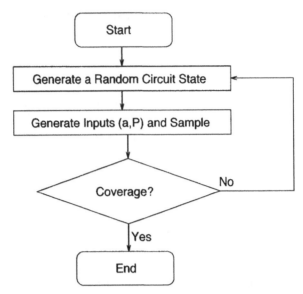

Figure 3.13   Monte-Carlo-based technique flow chart.

However, to derive each sample probability of a state bit is a problem since the probability depends on the state the sequential circuit is in. This can be resolved as follows. Assume that the state of the machine at time k (kth clock cycle) becomes independent of its initial state at time 0 as $k \to \infty$. As a result, the probability that the state bit signal $s_i(k)$ is logical 1 at time k with initial state $S(0)$ being S_0, denoted as $P(s_i(k) = 1|S(0) = S_0)$ [abbreviated as $P_k(s_i|S_0)$], has the following property:

$$\lim_{k \to \infty} P_k(s_i|S_0) = \lim_{k \to \infty} P_k(s_i = 1) = P(s_i) \tag{3.35}$$

Similarly,

$$\lim_{k \to \infty} P_k(s_i|S_i) = P(s_i) \tag{3.36}$$

That is, the probability will be independent of the initial state as $k \to \infty$. Based on this property, two runs starting from two different initial states S_0 and S_1 are performed at the same time. During the simulation, the difference and average of $P_k(s_i|S_0)$ and $P_k(s_k|S_1)$ are monitored. If both the difference and average remain within a window of width of ϵ_w for a certain user-specified number of consecutive clock cycles, it is declared that $P_k(s_i|S_0)$ and $P_k(s_i|S_1)$ have converged. This convergent value of probability is a sample value. The sample procedure is repeated K times to meet the user-specified error and confidence level, as mentioned earlier. However, there are some circuits in which the above technique may not directly apply.

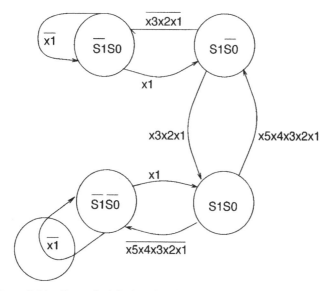

Figure 3.14   Example 3.2: Another STG of a sequential logic circuit.

Let us consider the STG of Figure 3.14. Let G_1 and G_2 denote the set of states given by $\{s1s0, \bar{s}1\bar{s}0\}$ and $\{\bar{s}1s0, \bar{s}1s0\}$, respectively. Assume that the probability of making a transition between the set of states in G_1 and the set of states in G_2 is very low. As a result, most of the samples are collected from G_1 if the initial states are $s1s0(S_0)$ or $\bar{s}1\bar{s}0(S_1)$. These sets G_1 and G_2 are called sets of near-closed states. Let y be the output node given by $y = (s1s0 + \bar{s}1\bar{s}0)x1$. Considering only the set of states in G_1, $y = (s1s0 + \bar{s}1\bar{s}0)x1 = x1$, while considering only the set of states in G_2, $y = (s1s0 + \bar{s}1\bar{s}0)x1 = 0$. Therefore, $P(y) = P(x_1)$ in G_1 and $P(y) = 0$ in G_2. That is, the probability behavior is very different in two different groups of states. Data sampled from a particular group are biased, giving errors. As a matter of fact, if we assume all the primary inputs have the same probability and normalized activity of 0.2 and 0.3, respectively, the normalized activity of y sampled from G_1 is 0.3 $[a(y, G_1)]$ and is 0.02 $[a(y, G_2)]$ from G_2.

A solution to this problem is as follows. Let us assume that we know the values of $P(G_1)$ and $P(G_2)$ of the previous example. The normalized activity $a(y)$ can be computed as follows:

$$a(y) = P(G_1) \times a(y, G_1) + P(G_2) \times a(y, G_2) \qquad (3.37)$$

If a STG is given, $P(G_1)$ and $P(G_2)$ may be computed by assuming that the primary inputs are either temporally uncorrelated (which may give errors when they are not) or Markov. A primary input is Markov if its future value depends on the present value and does not depend on its past. However, if no STG knowledge is assumed, can we find out $P(G_1)$ and $P(G_2)$? Under the assumption that the primary inputs are Markov, it turns out that we can implicitly compute $P(G_1)$ and $P(G_2)$. Based on the assumption that the primary inputs are Markov, a new STG called an extended STG (ESTG) is built by transforming the original STG as mentioned earlier. The resultant ESTG (rather than the STG) is Markov [22]. Therefore, there is a transition matrix that corresponds to the ESTG. It is found that $P^k_{\text{warmup}}(G_i)$, the probability of reaching one of the states of G_i at the end of k clock cycles ($k \geq$ a certain period of time called the warmup period), with any initial state is very close to $P(G_i)$. The error specified by the user determines the warmup period according to the following empirical inequality derived in [6]:

$$|P^k_{\text{warmup}}(G_i) - P(G_i)| \leq N_s|\lambda_2|^k \qquad (3.38)$$

where λ_2 is the second largest eigenvalue (absolute value) of the transition matrix and N_s is the number of ESTG states. The upper bound on the number of states of the ESTG is 2^{i+j}, where i and j are the number of state bits and the number of inputs, respectively. Therefore, if we specify the upper bound on the relative error ϵ_G that $P(G_i)$ can have, we have

$$k \geq \frac{\ln[N_s/\epsilon_G P(G_i)]}{\ln 1/\lambda_2} \qquad (3.39)$$

Equation (3.38) implies that, starting with a randomly generated initial state, if we simulate the circuit for a warmup period of clock cycles and then sample data, the probability of sampling data from among the states of G_i is $P(G_i)$. If we repeat the same procedure N times, we will have $N \cdot P(G_1)$ samples from G_1 and $N \cdot P(G_2)$ samples from G_2. Let $a_j(y|G_i)$ represent the jth sample taken from G_i. Hence the mean of the samples taken from G_i is $\sum_{j=1}^{N \cdot P(G_i)} a_j(y|G_i)/N \cdot P(G_i)$, denoted as $a(y|G_i)$. As a result, the mean of these samples is

$$\frac{\left(\sum_{j=1}^{N \cdot P(G_1)} a_j(y|G_1) + \sum_{j=1}^{N \cdot P(G_2)} a_j(y|G_2) \right)}{N}$$

$$= \frac{N \cdot P(G_1)a(y|G_1) + N \cdot P(G_2)a(y|G_2)}{N}$$

$$= P(G_1)a(y|G_1) + P(G_2)a(y|G_2)$$

which is not biased. However, since the STG is not given, the ESTG and its corresponding transition matrix (and hence λ_2 and N_s) cannot be derived. To be conservative, we may choose λ_2 to be 0.9 and N_s to be 2^r (r is the number of primary inputs and state bits), which is an upper bound on the number of

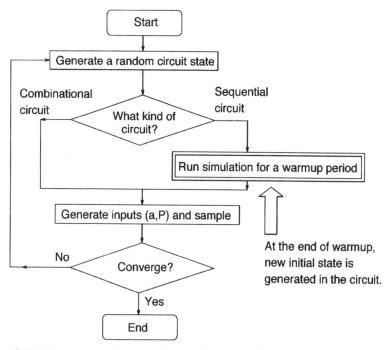

Figure 3.15 Monte-Carlo-based technique flow chart for both sequential and combinational circuits.

ESTG states. The modified version of the Monte-Carlo-based technique for sequential circuits is outlined in Figure 3.15. It is worth mentioning that the warmup period simulation does not need any delay model. What matters is the steady-state logic value of the state bits rather than the transient behavior.

Lastly, there is only one problem left. If we examine the assumption of the stopping criterion [inequality (3.35)], it is assumed that the mean a has a *normal* distribution. But in the previous example, apparently $a(y)$ has a bimodal distribution. Can we apply the same stopping criterion to the bimodal distribution in this case? In order to answer this question, we will compare the user-specified relative error [ϵ' in inequality (3.35)] with the actual relative error (ϵ_b) resulting from applying the stopping criterion [inequality (3.35)] to be bimodal distribution. Let us assume that the bimodal distribution function is a linear combination of two normal distribution functions $f_1(x)$ and $f_2(x)$. That is,

$$f(x) = P(G_1) \cdot f_1(x) + P(G_2) \cdot f_2(x) \tag{3.40}$$

where $P(G_1)$ and $P(G_2)$ are steady-state probabilities of two groups of states, as mentioned earlier (two near-closed sets), and $P(G_1) + P(G_2) = 1$.

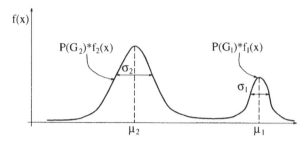

Figure 3.16 Bimodal distribution as a linear combination of two normal distributions.

The term $f_i(\cdot)$ represents the population of G_i. It is justified by assuming $T \geq 30$ that f_i is normal, where T is the number of clock cycles in each sample. It is illustrated pictorially in Figure 3.16.

Assume that in each sample we start to take data after simulating the sequential circuit for a warmup period. If a total number of N samples are taken when the stopping criterion is met, we have $N_i (\approx N \cdot P(G_i))$ samples collected from G_i. Here, N is determined by the stopping criterion [inequality (3.35)] to meet the user-specified relative error ϵ and $(1 - \alpha) \times 100\%$ confidence level. However, it is based on the assumption that the distribution is normal rather than bimodal. Without loss of generality, let us assume $P(G_1) \leq P(G_2)$. If we use N_i samples to estimate μ_i, we have s_i and m_i as the sample standard deviation and the sample mean. It can be shown that ϵ_b/ϵ', the ratio of the resultant relative error to the user-specified relative error, with the same confidence level is

$$\frac{\epsilon_b}{\epsilon'} \leq \frac{\sqrt{N_1/N}\, r_m r_r s_m t_{\alpha/2} + \sqrt{N_2/N}\, z_{\alpha/2}}{z_{\alpha/2}} \; \frac{\sqrt{N-1}}{\sqrt{r_m^2 r_{rsm}^2 (N_1 - 1) + (N_2 - 1)}}$$

$$(3.41)$$

where $r_m = m_1/m_2$, $r_{rsm} = s_1/m_1/s_2/m_2$, and $t_{\alpha/2}$ and $z_{\alpha/2}$ are obtained from the t-distribution and normal distribution [7, 16]. A few plots showing the impact of the ratio of relative error are shown in Figures 3.17–3.20. Surprisingly the ratios are less than 1.5 for most of the values of $P(G_1)$. This implies that in order to ensure the actual relative error to be less than ϵ_b, ϵ' in the stopping criterion [inequality (3.35)] has to be less than $\epsilon_b/1.5$. For example, if we assume 7.5% error to be tolerable ($\epsilon_b = 0.075$), then ϵ' must be less than $7.5/1.5 = 5\%$ ($\epsilon' = 0.05$). The worst case only occurs when $P(G_1)$ is very small (that is, when only a couple of samples are collected from G_1). Several other observations can also be made. The larger the r_m is (m_1 is greater than m_2), the higher is the ratio of the relative error when $P(G_1)$ is very small (< 0.05). This is shown in Figure 3.17 and can be explained as

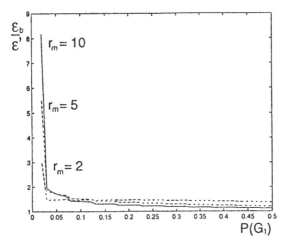

Figure 3.17 Relative error ratio with $r_{rsm} = 1$ and $N = 120$.

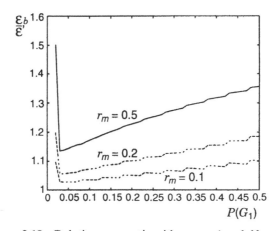

Figure 3.18 Relative error ratio with $r_{rsm} = 1$ and $N = 120$.

follows. Since only a few samples are collected from G_1 ($m_1 \gg m_2$), we may expect higher error ratio when the ratio $r_m(m_1/m_2)$ is larger. On the other hand, when m_1 is smaller than m_2 (Figure 3.18) and even when only a couple of samples are from G_1, it does not really affect the error since $P(G_1)$ is very small. It is also observed (Figure 3.19) that with more samples ($> N$) the error ratio is smaller when $P(G_1)$ is small. Another factor that affects the error ratio is r_{rsm}, which is the ratio of the relative sample standard deviation s_1/m_1 to s_2/m_2, as shown in Figure 3.20.

In the above analysis, we have assumed that there are only two near-closed sets. But there can be more than two near-closed sets in a sequential circuit. The technique can be extended to cases having multiple near-closed sets. However, the experimental results show that accurate results can be obtained for the sequential ISCAS benchmark circuits [8] under the above assumption.

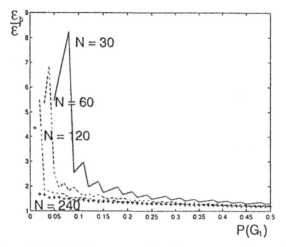

Figure 3.19 Relative error ratio with $r_{rsm} = 1$ and $r_m = 5$.

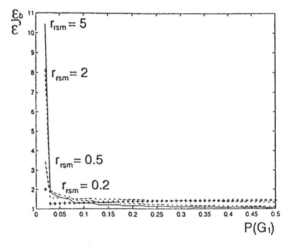

Figure 3.20 Relative error ratio with $r_m = 5$ and $N = 120$.

3.5 ESTIMATION OF GLITCHING POWER

The statistical power estimators use a logic or a timing simulator to estimate the switching activity at circuit nodes. The accuracy of the switching activity at the internal nodes is dependent on the logic or circuit simulator under consideration. Hence, if accurate delay models are used, the simulator can efficiently estimate power dissipation due to both functional and spurious transitions. In this section, we will briefly describe techniques to estimate the glitching activity in a logic circuit using logic simulation techniques. To clearly understand the technique, the following definitions for hazards will be useful.

Definition 3.4 (Static Hazard) *A static hazard is defined as the possible occurrence of a transient pulse on signal line whose static value is not supposed to change. For example, if there is a rising and a falling transition at the two inputs of an AND gate (e.g., Figure 3.1), there may be a spurious transition $0 \rightarrow 1 \rightarrow 0$ at the output. The static signal value at the output of the AND gate is 0, while depending on the input signal transition times, a static 0 hazard appears at the output.*

Definition 3.5 (Dynamic Hazard) *A dynamic hazard is the possible occurrence of a spurious transition during the occurrence of a functional $0 \rightarrow 1$ or a $1 \rightarrow 0$ transition.*

Let us first consider how static hazard is detected using logic simulation. Let $S(t)$ and $S(t + 1)$ be the values of a signal line at time instants t and $t + 1$. If $S(t)$ and $S(t + 1)$ are different, then the exact time in which the signal changed value is unknown during logic simulation. Let t' represent a time unit between time t and $t + 1$ during which the value of $S(t')$ is unknown (X) [26–29]. A sequence of signal values $S(t)S(t')S(t + 1) = 1X0$ represents one of the sequences in the set $\{100, 110\}$. The two sequences can be represented as a *fast* or a *slow* $1 \rightarrow 0$ transition. Now let us revisit the two-input AND gate with inputs A and B. If the signal values for A and B at time instants t, t', and $t + 1$ are $0X1$ and $1X0$, respectively, then bit wise ANDing of signals A and B produces signal $0X0$ at the output C (the table showing the AND operations using three-valued logic $0, 1, X$ is given in Table 3.2). The possible output sequences are 000 or 010 depending on how *fast* or *slow* the input transitions were. Hence, the static hazard can be detected by introducing the time unit t'. Therefore logic simulation can be used to detect probable static hazards by using a six-valued logic. The six-valued logic for simulating static hazard is shown in Table 3.3. The

TABLE 3.2   Three-Valued Logic Simulation for AND Gate

AND	0	1	X
0	0	0	0
1	0	1	X
X	0	X	X

TABLE 3.3   Six-Valued Logic for Static Hazard Analysis

Logic Representation	Bit Sequences at $t, t', t + 1$
0—Static 0	000
1—Static 1	111
R—Rising	0U1
F—Falling	1U0
SH0—Static 0 hazard	0U0
SH1—Static 1 hazard	1U1

TABLE 3.4 AND Operation with Six-Valued Logic

AND	0	1	R	F	SH0	SH1
0	0	0	0	0	0	0
1	0	1	R	F	SH0	SH1
R	0	R	R	SH0	SH0	R
F	0	F	SH0	F	SH0	F
SH0	0	SH0	SH0	SH0	SH0	SH0
SH1	0	SH1	R	F	SH0	SH1

TABLE 3.5 Eight-Valued Logic for Dynamic Hazard Analysis

Logic Representation	Bit Sequence at $t, t', t'', t+1$
0—Static 0	0000
1—Static 1	1111
R—Rising	{0001, 0011, 0111}
F—Falling	{1000, 1100, 1110}
SH0—Static 0 hazard	{0000, 0100, 0010, 0110}
SH1—Static 1 hazard	{1111, 1011, 1101, 1001}
DH0—Static 0 hazard	{1110, 1100, 1000, 1010}
DH1—Dynamic 1 hazard	{0001, 0011, 0111, 0101}

corresponding truth table for a two-input AND gate is shown in Table 3.4. However, it should be observed that the estimate is pessimistic because some of these hazards might not be present under certain delay conditions.

Dynamic hazards can be similarly detected by using 4-bit sequences to represent logic values. A good $1 \rightarrow 0$ transition between time t and $t+1$ would correspond to the following set consisting of 4-bit sequences—{1000, 1100, 1110} corresponding to *fast*, *medium*, and *slow* falling signals. Eight-valued logic is required for logic simulation to detect dynamic hazard. Table 3.5 shows the eight-valued logic required for dynamic hazard analysis.

It can be observed that the simulation technique described above considers all potential hazards and, hence, is an upper bound of glitches that can possibly occur. In the following section we will consider a Monte-Carlo-based technique that can efficiently estimate glitches under different delay models.

3.5.1 Monte-Carlo-Based Estimation of Glitching Power

It turns out that minor delay model inaccuracies may lead to large errors in estimated activity. Therefore, delay models are crucial to the statistical estimation of activity. Probabilistic delay models used in the estimation will be introduced to capture the uncertainty of gate delays. Based on the probabilistic delay models, we will generalize the Monte Carlo approach.

3.5.2 Delay Models

In the design phase, a designer is faced with different sources of uncertainty that affect the delays of the circuit. These sources can be grouped into two classes: systematic and random [19]. The systematic class includes approximations made to simplify the model for improving simulation time, approximations made to estimate device and interconnect parasitics prior to layout, and uncertainty in the final process center and distribution when design proceeds in parallel with process development. On the other hand, the random class includes uncontrolled variations in photolithography, die-to-die variations, wafer-to-wafer variations, lot-to-lot variations, operating temperature, power supply voltage, and so on. In [20], it has been shown that a circuit node where two reconvergent paths with different delays meet may have a large number of spurious transitions. However, even in a tree-structured circuit with balanced paths (without reconvergent fanout) there can be a large number of spurious transitions due to slight variations in delays. These variations can be caused by any of the above sources of uncertainty.

Let us consider the circuit of Figure 3.21. All gates are assumed to have the same delay. Because the tree has perfectly balanced paths, there are no glitches at all. The final output has normalized activity of 0.5 when all the

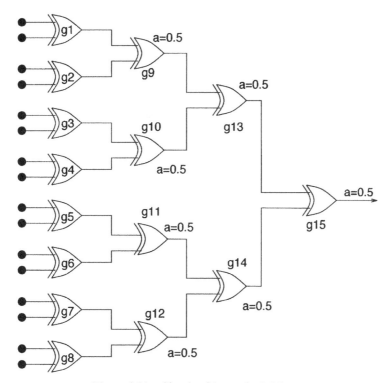

Figure 3.21 Circuit with nominal delays.

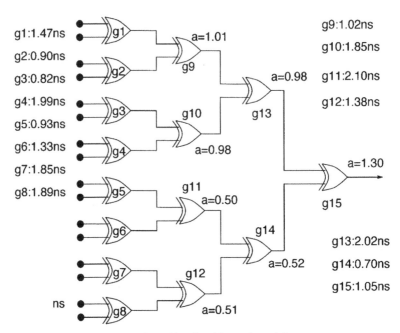

Figure 3.22   Circuit with random delays.

primary inputs are assumed to be synchronous. However, due to sources of uncertainty, the gate delays may have variations (shown in Figure 3.22). As a result, glitches do occur and the values of activities at individual nodes change. This is shown in Figure 3.22. The inertial delays are assumed to be half of the values of transport delays for the simulation. Notice that the final output-normalized activity becomes 1.30 rather than 0.5. In order to capture this random behavior in statistical design, these sources of uncertainty are represented by probability distribution, while in worst-case design, the extreme cases are taken into account.

We choose a transport delay (d) model with inertial delay (d_I). However, it should be noted that the technique is not restricted to such a delay model. The point is to model the parameters of chosen delay models as random variables in order to capture the probabilistic behavior of gate delays. The transport delay is modeled as a random variable of truncated normal distribution with mean μ_d and standard deviation σ_d, as shown in Figure 3.23. The mean is the nominal value of transport delay d and the deviation is either assigned by users or determined by feedback from the fabricated chips. Moreover, if a random delay is less than a minimum value *min*, it is discarded since in real circuits it must be larger than some positive value. Similarly, if a random delay is greater than a maximum value max, it is truncated since it can be considered as a delay fault.

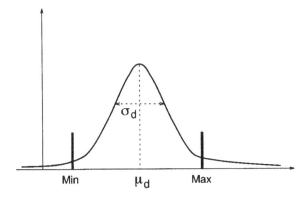

Figure 3.23   Random delay with truncated normal distribution.

3.5.2.1 *Statistical Estimation*

Recall that in a Monte-Carlo-based technique the primary input patterns are generated conforming to a given activity and probability of the input signals. In a more abstract viewpoint, we can think of activity (**a**) at a node as a function of primary input vectors **PI**. Each component of **PI** is a stochastic process. Therefore, **a** is also a stochastic process and can be expressed as follows,

$$\mathbf{a} = F(\mathbf{PI}) \tag{3.42}$$

Earlier we applied Monte-Carlo-based techniques to estimate the expected value of **a**, $E(\mathbf{a})$. However, what is missing in this approach is the information about the delay. In other words, the delays of the gates of the circuit are assumed to be some constants (deterministic). Now assume that gate delays are not deterministic and each gate delay can be represented by a random variable d_i. If D is a random vector consisting of all the random variables of gate delays, **a** can be represented as

$$\mathbf{a} = F(\mathbf{PI}, D) \tag{3.43}$$

Therefore, when applying Monte-Carlo-based techniques to estimating $F(\mathbf{PI}, D)$, delays are modeled as random variables and should be generated from time to time along the simulation. The rationale behind this is that whenever we generate a new set of delays, they correspond to another die or even the same die but with different operating conditions such as temperature and power supply voltage. In contrast to Figure 3.13, Figure 3.24 outlines the procedure of how the modified Monte-Carlo-based technique works.

The effect of delay models on power dissipation can be best understood by considering Figures 3.25 and 3.26, which plot the relative dynamic power

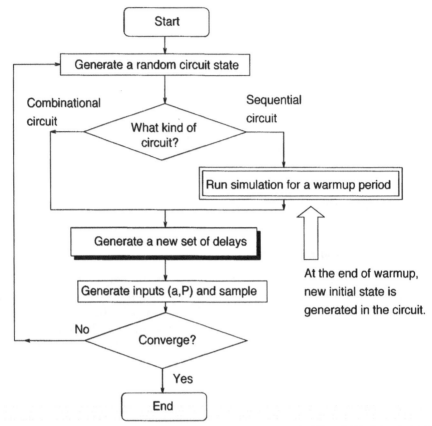

Figure 3.24   Modified Monte-Carlo-based technique flow chart.

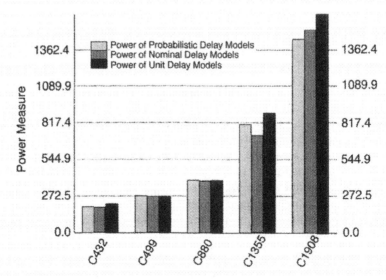

Figure 3.25   Relative dynamic power.

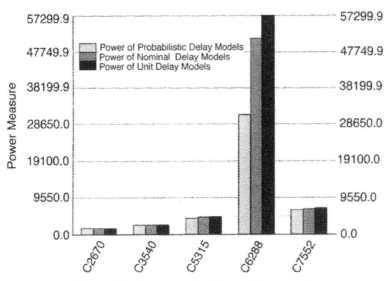

Figure 3.26 Power under different delay models.

dissipation in ISCAS benchmark circuits due to different delay models. While the different nonzero-delay models do track each other (except for one circuit, C6288, which has a depth of about 120 levels), it is clear that the nonzero-delay models can produce very different results compared to the zero-delay case. The difference is due to the glitching activity.

3.6 SENSITIVITY ANALYSIS

The techniques mentioned above can be used to accurately estimate power dissipation provided exact signal probabilities and activities of primary inputs are known. However, accurate signal probability or activity values for primary inputs may not often be available. Since power dissipation strongly depends on the input signal properties, uncertainties in specifications of input signal properties makes the estimation process difficult. In fact, average power dissipation of a circuit should be represented as a range given by $[Power_{min}, Power_{max}]$. Traditional power estimation techniques cannot deal with the complexity of the problem since it is practically impossible to try all ranges of signal properties to estimate the minimum and maximum average power dissipation when the number of primary inputs is large. Let us consider a small benchmark circuit C432 with 36 primary inputs as an example. It takes 2.3 seconds of CPU time (SPARC 5) to do one symbolic simulation [9]. If the activity of each primary input varies from a minimum to a maximum possible value, there are 2^{36} activity combinations. In order to obtain bounds for minimum and maximum average power, all 2^{36} possibilities have to be exhausted. This will take 2.3×2^{36} (more than 5000 years) of CPU time. If we use a Monte Carlo based method to simulate the circuits, it will take an even longer time. In this section we will consider a *power sensitivity method* to accurately estimate the maximum and minimum bounds for average power dissipation. The sensitivities are calculated using an efficient statistical technique and, hence, can be obtained as a by-product of the statistical power estimation process.

3.6.1 Power Sensitivity

Since power dissipation in CMOS circuits is heavily dependent on the signal properties of the primary inputs, both primary input probability and activity variations can result in uncertainties in the estimation of power dissipation. To measure this effect, we define *power sensitivity to primary input activity* S_{a_i} and *power sensitivity to primary input probability* S_{P_i} as follows:

$$S_{a_i} = \frac{\partial \text{Power}_{\text{avg}}}{\partial a_i} \tag{3.44}$$

$$S_{P_i} = \frac{\partial \text{Power}_{\text{avg}}}{\partial P_i} \tag{3.45}$$

where a_i and P_i are the activity and probability of primary input i, respectively.

$\text{Power}_{\text{avg}}$ is proportional to *normalized power dissipation measure* Φ. Therefore, we can define *normalized power sensitivity to primary input activity* ζ_{a_i} and *normalized power sensitivity to primary input probability* ζ_{P_i} in terms of Φ as follows:

$$\zeta_{a_i} = \frac{\partial \Phi}{\partial a_i} = \sum_{j \in \text{all nodes}} \text{fanout}(j) \frac{\partial a_j}{\partial a_i} \tag{3.46}$$

$$\zeta_{P_i} = \frac{\partial \Phi}{\partial P_i} = \sum_{j \in \text{all nodes}} \text{fanout}(j) \frac{\partial a_j}{\partial P_i} \tag{3.47}$$

where a_j is the activity of each internal node and primary output.

A naïve approach to estimate power sensitivity would be to simulate a circuit to obtain the average power dissipation based on nominal values of primary input signal probabilities and activities and then to assign a small property variation to only one primary input and resimulate the circuit. After all the primary inputs have been exhausted, power sensitivity can be obtained using $\Delta \text{Power}_i / \Delta \theta_i$, where θ_i can be P_i or a_i. This naïve method can easily be implemented. However, it involves $n + 1$ times of power estimation. If the number of primary inputs is large, this method can be computationally expensive. Therefore, the naïve simulation method is impractical for large circuits with a large number of primary inputs. A practical symbolic method was proposed in [9]. However, to apply this approach to large circuits, circuit partitioning is required, which can introduce error. In this section we present an efficient technique (STEPS) to estimate power sensitivities as a by-product of statistical power estimation using a Monte-Carlo-based approach.

3.6.2 Power Sensitivity Estimation

In Section 3.4 we noted that Monte-Carlo-based power estimation determines the node activities simulated by a sequence of primary input vectors. Let us consider the case for which such techniques can be used to estimate power sensitivities.

A multilevel logic circuit can be described by a set of completely specified Boolean functions. Each Boolean function f maps a primary input vector to an internal or primary output signal. The activity of an internal node or primary output j can be obtained as follows:

$$
\begin{aligned}
a_j &= P\big(f(\mathbf{V}^0) \oplus f(\mathbf{V}^\mathbf{T})\big) \\
&= \sum_{\mathbf{V}^0, \mathbf{V}^\mathbf{T}} P(\mathbf{V}^0 \mathbf{V}^\mathbf{T})\big[f(\mathbf{V}^0) \oplus \mathbf{f}(\mathbf{V}^\mathbf{T})\big]
\end{aligned}
\tag{3.48}
$$

where $\mathbf{V}^0 = (I_1^0, I_2^0, \ldots, I_n^0)$ and $\mathbf{V}^\mathbf{T} = (I_1^T, I_2^T, \ldots, I_n^T)$ are primary input vectors and I_i^0 equals i^0 or $\bar{i}^0$ and I_i^T equals i^T or $\bar{i}^T$. The superscripts denote time and T is the clock cycle. Therefore, the statistics of internal and primary output signals are completely determined by logic transitions at primary inputs. The instantaneous power dissipation of a circuit is completely determined by two consecutive input vectors. Now, the expected value of average power can be expressed as follows:

$$
\begin{aligned}
\Phi &= E[\text{Pwr}] \\
&= \sum_{\mathbf{V}^0, \mathbf{V}^\mathbf{T}} \text{Power}(\mathbf{V}^0 \mathbf{V}^\mathbf{T}) P(\mathbf{V}^0 \mathbf{V}^\mathbf{T})
\end{aligned}
\tag{3.49}
$$

The power consumption in every clock cycle is a random variable and is denoted as Pwr. Power($\mathbf{V}^0 \mathbf{V}^\mathbf{T}$) represents the power consumption due to the pair of input vectors $\mathbf{V}^0$ and $\mathbf{V}^\mathbf{T}$. For a particular pair of consecutive vectors, Power($\mathbf{V}^0 \mathbf{V}^\mathbf{T}$) is independent of the probability and activity values of primary inputs. The probability of having consecutive input vectors $\mathbf{V}^0$ followed by $\mathbf{V}^\mathbf{T}$ is represented by $P(\mathbf{V}^0 \mathbf{V}^\mathbf{T})$. Therefore, power sensitivity can be expressed as follows:

$$
\begin{aligned}
\zeta_{\theta_i} &= \frac{\partial \Phi}{\partial \theta_i} \\
&= \frac{\partial E[\text{Pwr}]}{\partial \theta_i} \\
&= \sum_{\mathbf{V}^0, \mathbf{V}^\mathbf{T}} \left(\text{Power}(\mathbf{V}^0 \mathbf{V}^\mathbf{T}) \frac{\partial P(\mathbf{V}^0 \mathbf{V}^\mathbf{T})}{\partial \theta_i} \right)
\end{aligned}
\tag{3.50}
$$

All the primary inputs are assumed to be spatially independent. Therefore, we have

$$
P(\mathbf{V}^0 \mathbf{V}^\mathbf{T}) = P\big(I_1^0 I_1^T\big) \cdots P\big(I_i^0 I_i^T\big) \cdots P\big(I_n^0 I_n^T\big)
$$

and

$$\frac{\partial P(\mathbf{V}^0 \mathbf{V}^T)}{\partial \theta_i} = P(I_1^0 I_1^T) \cdots \frac{\partial P(I_i^0 I_i^T)}{\partial \theta_i} \cdots P(I_n^0 I_n^T) \qquad (3.51)$$

From basic calculus, we have

$$\frac{\partial P(I_i^0 I_i^T)}{\partial \theta_i} = P(I_i^0 I_i^T) \frac{\partial \ln[P(I_i^0 I_i^T)]}{\partial \theta_i} \qquad (3.52)$$

Substituting Eq. (3.52) into (3.51) and then into Eq. (3.50), we obtain

$$\zeta_{\theta_i} = \sum_{\mathbf{V}^0, \mathbf{V}^T} \left(\text{Power}(\mathbf{V}^0 \mathbf{V}^T) P(\mathbf{V}^0 \mathbf{V}^T) \frac{\partial \ln(P(I_i^0 I_i^T))}{\partial \theta_i} \right)$$

$$= E\left[\text{Pwr} \frac{\partial \ln(P(I_i^0 I_i^T))}{\partial \theta_i} \right] \qquad (3.53)$$

As mentioned earlier, statistical techniques to estimate average power calculates the instantaneous power Pwr. Multiplying Pwr by a factor $\partial \ln[P(I_i^0 I_i^T)]/\partial \theta_i$, we obtain a sample of power sensitivity. The only expression left to be evaluated is $\partial \ln[P(I_i^0 I_i^T)]/\partial \theta_i$. Since I_i^0 is either i^0 or $\bar{i}^0$ and I_i^T is either i^T or $\bar{i}^T$, we have the following four combinations for $P(I_i^0 I_i^T)$: $P(i^0 i^T)$, $P(\bar{i}^0 i^T)$, $P(i^0 \bar{i}^T)$, and $P(\bar{i}^0 i^T)$. Each of the expressions can be expressed in terms of probability and activity (P_i and a_i) by Eqs. (3.12), (3.13), and (3.14) [note that $i(t - T)$ and $i(t)$ are replaced by i^0 and i^T, respectively]. Therefore, $\partial \ln[P(I_i^0 I_i^T)]/\partial a_i$ can be calculated as follows:

$$\frac{\partial \ln[P(i^0 i^T)]}{\partial a_i} = \frac{\partial \ln(P_i - \frac{1}{2}a_i)}{\partial a_i}$$

$$= \frac{-\frac{1}{2}}{P_i - \frac{1}{2}a_i} \qquad (3.54)$$

$$\frac{\partial \ln[P(\bar{i}^0 \bar{i}^T)]}{\partial a_i} = \frac{\partial \ln(1 - P_i - \frac{1}{2}a_i)}{\partial a_i}$$

$$= \frac{-\frac{1}{2}}{1 - P_i - \frac{1}{2}a_i} \qquad (3.55)$$

$$\frac{\partial \ln[P(i^0 \bar{i}^T)]}{\partial a_i} = \frac{\partial \ln(\frac{1}{2}a_i)}{\partial a_i}$$

$$= \frac{1}{a_i} \qquad (3.56)$$

$$\frac{\partial \ln\left[P(\bar{i}^0 i^T)\right]}{\partial a_i} = \frac{\partial \ln\left(\frac{1}{2}a_i\right)}{\partial a_i}$$

$$= \frac{1}{a_i} \tag{3.57}$$

and $\partial \ln[P(I_i^0 I_i^T)]/\partial P_i$ has the following four combinations:

$$\frac{\partial \ln\left[P(i^0 i^T)\right]}{\partial P_i} = \frac{\partial \ln\left(P_i - \frac{1}{2}a_i\right)}{\partial P_i}$$

$$= \frac{1}{P_i - \frac{1}{2}a_i} \tag{3.58}$$

$$\frac{\partial \ln\left[P(i^0 \bar{i}^T)\right]}{\partial P_i} = \frac{\partial \ln\left(1 - P_i - \frac{1}{2}a_i\right)}{\partial P_i}$$

$$= \frac{-1}{1 - P_i - \frac{1}{2}a_i} \tag{3.59}$$

$$\frac{\partial \ln\left[P(i^0 \bar{i}^T)\right]}{\partial P_i} = \frac{\partial \ln\left(\frac{1}{2}a_i\right)}{\partial P_i}$$

$$= 0 \tag{3.60}$$

$$\frac{\partial \ln\left[P(\bar{i}^0 i^T)\right]}{\partial P_i} = \frac{\partial \ln\left(\frac{1}{2}a_i\right)}{\partial P_i}$$

$$= 0 \tag{3.61}$$

3.6.3 Power Sensitivity Method to Estimate Minimum and Maximum Average Power

Section 3.6.2 shows that STEPS can estimate power sensitivity as a by-product of average power estimation with nominal values of signal probability and activity. Hence, the minimum and maximum average power of a circuit can easily be computed as follows:

$$\Phi_{min} = \Phi_{avg} - \sum_{i \in \text{all PI's}} \zeta_{a_i} |\Delta a_i| \tag{3.62}$$

$$\Phi_{max} = \Phi_{avg} + \sum_{i \in \text{all PI's}} \zeta_{a_i} |\Delta a_i| \tag{3.63}$$

where ϕ_{avg} is the average *normalized power dissipation measure* (proportional to the switched capacitance). It can be estimated during the average power estimation process using STEPS with nominal values of primary input signal properties. The term ζ_{a_i} is the power sensitivity to activity a_i of primary

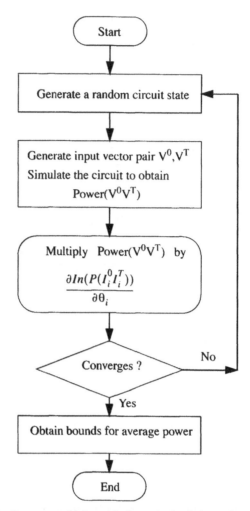

Figure 3.27 Power sensitivity technique to obtain bound average power.

input i; Δa_i is the activity variation. It should be noted that since the statistical techniques to estimate power can handle different delay models for logic gates, STEPS can also handle different delay models and includes spurious transitions in its analysis. Figure 3.27 gives the flow of the power sensitivity technique to obtain minimum and maximum average power dissipation.

Results of using the naive long run simulation method to obtain power sensitivity are shown in Figure 3.28, where the y axes represent power sensitivity to primary input activity and x axes represent primary inputs. All the primary inputs are assumed to have probability and activity values (nominal) of 0.5 and 0.26, respectively. The sample number used in this experiment was 3000, while an activity variation of 0.05 was assumed for all

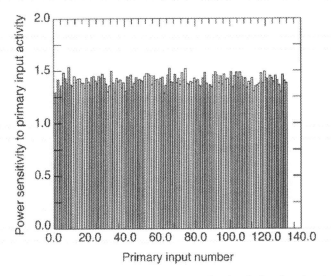

Figure 3.28   Power sensitivity ζ_{a_i} obtained by simulation for circuit i3.

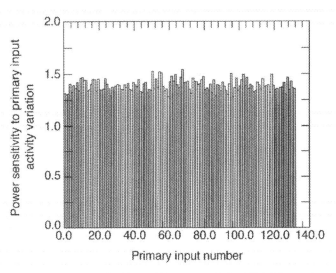

Figure 3.29   Power sensitivity ζ_{a_i} obtained by STEPS for circuit i3.

the inputs. The results obtained by STEPS are shown in Figure 3.29. Results indicate that for some circuits power dissipation is much more sensitive to some primary inputs than others. A small activity variation of such highly sensitive primary inputs will result in a dramatic change of the average power. The power sensitivity to the activity of each of the other primary inputs is less than 4. If the activities of the first and second primary inputs have a variation of ± 0.05, the power dissipation may change by 30%.

Therefore, for power conscious designs, those sensitive primary inputs have to be accurately specified for accurate estimation of average power. For some circuits, the difference among power sensitivities to different primary inputs is really small, showing that the activity variation of different primary inputs may have almost the same effect on average power dissipation.

The long run simulation method (naive technique described in Section 3.6.1) is used as a figure of merit for STEPS. To compare the results obtained by simulation with those obtained by STEPS, we compute the percentage difference using the expression

$$\frac{\sum_{\text{all PI's}} |\zeta_{a_i}(\text{SIM}) - \zeta_{a_i}(\text{STEPS})|}{\sum_{\text{all PI's}} \zeta_{a_i}(\text{SIM})}$$

where ζ_{a_i} (STEPS) is the power sensitivity obtained by STEPS and ζ_{a_i}(SIM) is the power sensitivity obtained by simulation. The comparison is shown in Table 3.6. The CPU time is also shown for a SPARC 5 workstation. Since the long run simulation method repeats the estimation procedure $n + 1$ times (n is the number of primary inputs), execution time may be unacceptably long

TABLE 3.6 Comparison of Two Methods

Circuit Chosen	PI's #	Gate #	CPU Time (s)		Diff %
			SIM	STEPS	
i1	25	33	519	19	2.1
i2	201	36	15502	78	1.4
i3	132	70	7896	57	1.2
i4	192	94	16960	95	0.2
i5	133	199	23326	223	2.0
i6	138	344	38877	328	1.8
i7	199	406	78514	422	0.3
i8	133	1183	119088	852	1.2
i9	88	353	33581	341	7.2
i10	257	2497	290744	1313	4.0
C432	36	160	8085	220	1.4
C499	41	202	11615	222	10.7
C880	60	357	5675	104	6.7
C1355	41	514	15791	426	6.8
C1908	33	880	19007	649	10.4
C2670	233	1161	160436	848	2.6
C3540	50	1667	48719	930	2.3
C5315	178	2290	—	1827	—
C6288	32	2416	46219	2110	11.9
C7552	207	3466	—	4093	—

for large n. Let us consider circuit C7552. It takes approximately 4093 of CPU time to complete one simulation run. The circuit has 207 primary inputs. It would take approximately 8.5×10^5 (9.8 days) of CPU time to obtain power sensitivities. Circuits with prohibitively long execution time are identified by dashes in the SIM and Diff% columns of Table 3.6. The column labeled Diff% represents percent difference between the results obtained by simulation and STEPS. STEPS can estimate power sensitivity simultaneously with average power, and hence, it can be several orders of magnitude faster than the naive-simulation-based approach. Consider circuit $i10$. It takes 290,744 seconds of CPU time to obtain power sensitivity to each primary input using the naive method while it takes only 1313 seconds of CPU time using STEPS, which is more than 220 times faster.

After obtaining power sensitivities, we use Eqs. (3.62) and (3.63) to compute the minimum and maximum average power for each simulated circuit. For simplicity, all primary inputs are assumed to have the same activity variation of ± 0.05. Also for simplicity, we assume that there is no probability variation for each primary input. However, the method is not limited to such assumptions. Results of the minimum and maximum average power based on the zero-delay model are shown in Figures 3.30 and 3.31. Results indicate that for some circuits minimum and maximum average power can vary widely if uncertain specifications of primary inputs exist.

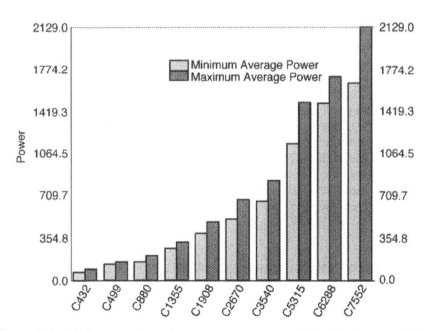

Figure 3.30 Minimum and maximum average power obtained using zero-delay model for ISCAS benchmark circuits.

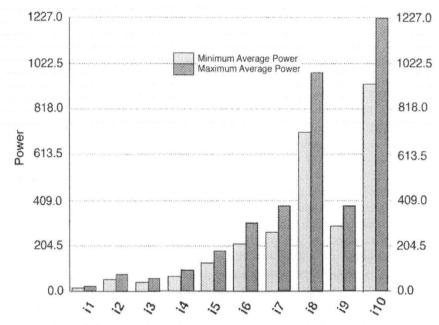

Figure 3.31 Minimum and maximum average power obtained using zero-delay model for MCNC benchmark circuits.

Consider circuit $i2$: maximum average power dissipation is 79 units, which is about 46% larger than minimum average power, which is 53 units.

All the results shown above are obtained using the zero-delay model. Note that we do not assume any delay models in deriving Eq. (3.53). Therefore, STEPS can handle different delay models for logic gates. Figures 3.32 and 3.33 shows the minimum and maximum bounds for average power dissipation obtained using the unit-delay model. Due to the delay of each logic gate, paths arriving at any internal gate may have different propagation delays. The delay mismatch of different paths causes spurious transitions. Let us consider circuit C6288. With the zero-delay model, minimum and maximum average power are 1484 and 1710 units, respectively. However, using the unit-delay model, the minimum and maximum bounds change to 38,881 and 47,574, respectively. They are 20 times greater than those obtained using the zero-delay model.

3.7 POWER ESTIMATION USING INPUT VECTOR COMPACTION

As noted earlier in this chapter, the active power dissipation in CMOS circuits is dependent on the input signal statistics. The techniques presented in the previous sections are input pattern independent or weakly pattern dependent. In this section we will consider power dissipation due to long

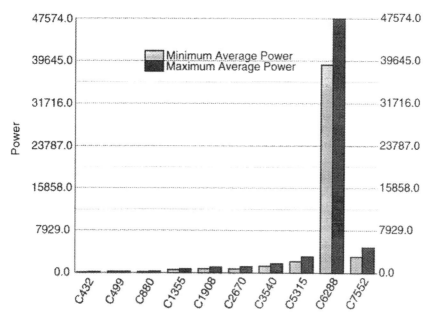

Figure 3.32   Minimum and maximum average power obtained using unit-delay model for ISCAS benchmark circuits.

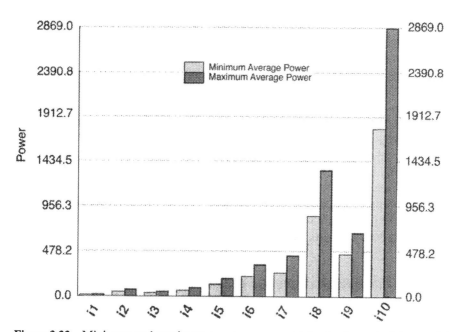

Figure 3.33   Minimum and maximum average power obtained using unit-delay model for MCNC benchmark circuits.

input sequences. Power estimation at lower levels of design abstraction for such input sequences can be very time consuming. One solution is to generate a compact vector set that is representative of the original vector set and can be simulated in a reasonable time [41]. Hence, given an input vector sequence S_1 of length L_1 with some property P_1, the problem is to generate another vector sequence S_2 of length L_2 with property P_2 such that P_1 and P_2 are nearly the same. Here, $R = L_1/L_2$ is defined as the compaction ratio. For estimation of power in digital circuits, P_1 or P_2 should represent the joint probabilities of signal lines. However, in general, using pairwise transition probabilities can be a good approximation. The pairwise transition probabilities of the input set S_1 should closely match that of vector set S_2. To measure how closely the compacted vector set resembles the original vector set, a metric C can be defined that measures the error in pairwise transition probabilities among all possible combinations of inputs:

$$C = \sum_{i=1}^{n-1} \sum_{j=i+1}^{n} \text{Diff}(x_i, x_j)$$

where

$$\text{Diff}(x_i, x_j) = \sum_{a=0}^{1} \sum_{b=0}^{1} \sum_{c=0}^{1} \sum_{d=0}^{1} \text{abs}\left(P^1_{x_{i(a \to b)} x_{j(c \to d)}} - P^2_{x_{i(a \to b)} x_{j(c \to d)}} \right)$$

where x_i is the ith input signal and $P^1_{x_{(a \to b)} y_{(c \to d)}}$ and $P^2_{x_{(a \to b)} y_{(c \to d)}}$ are the pairwise transition probabilities of signals x and y for S_1 and S_2, respectively.

Tsui et al. [41] have developed a technique to generate a sequence S_2 of length L_2 such that C is minimized given an input sequence S_1 of length L_1 and a target compaction ratio R. The problem of observing pairwise transition probabilities is reduced to that of observing pairwise signal probabilities. The technique then generates an input stream S_2 consisting of input vectors $V_1, V_2, \ldots, V_{L_2}$ using a greedy approach. The parameter V_1 is generated first followed by $V_2, \ldots, V_{L_2}$ such that the required spatio-temporal correlations are maintained. Results show that large compaction ratios (about 20 times) can be obtained while keeping the error in estimation to within 5%. The technique can be extended to sequential circuits by modeling the binary input streams as Markov sources of fixed order.

3.8 POWER DISSIPATION IN DOMINO CMOS

Unlike static CMOS logic, domino CMOS works using a precharging circuit. The output of a logic gate can be precharged high or low. Figure 3.34 shows a domino CMOS AND gate with precharge high (PZ is high when the clock is

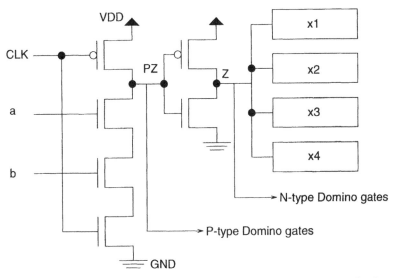

Figure 3.34   An n-type domino AND gate with n-type domino gate loads.

low). During the evaluate phase, the clock is high, and the output PZ conditionally discharges based on the signal values at inputs *a* and *b*. The three sources of power dissipation in static CMOS are switching, direct-path short-circuit current, and leakage current. Domino logic circuits do not have direct-path short-circuit currents except when static pull-up devices are used to moderate the charge redistribution problem or when clock skew is not well dealt with [34]. Ignoring power dissipation due to direct-path short-circuit current and leakage current, the average power dissipation in a static CMOS logic is given by

$$\text{Power}_{\text{avg}} = \tfrac{1}{2}V_{dd}^2 \sum_i C_i A(i)$$

where V_{dd} is the supply voltage, $A(i)$ is the *activity* at node i, and C_i is the capacitive load at that node. The summation is taken over all nodes of the logic circuit. If power dissipation due to leakage current and clock signal are ignored, the above expression also represents the power dissipation in domino logic circuits. It should be observed that $A(i)$ is proportional to $a(i)$, where $a(i)$ is the *normalized activity* obtained by dividing activity $A(i)$ by the clock frequency. Here, C_i is approximately proportional to the fanout at that node. As a result, the *normalized power dissipation measure* Φ, defined as

$$\Phi = \sum_i \text{fanout}_i \times a(i) \tag{3.64}$$

is proportional to the average power dissipation in both static and domino CMOS circuits. The parameter fanout$_i$ is the number of fanouts at node i. However, for a domino circuit, each input only drives one n-type (or p-type) rather than n- and p-type transistors. In order to compare static and domino logic circuits, we assume that the C_i of a domino circuit is half of that of a corresponding static (complementary) CMOS circuit. Furthermore, every transition at the leading edge of the clock is followed by another transition (precharge) at the falling edge of the clock. Therefore Φ of an n-type domino circuit is given by

$$\sum_i \left(\tfrac{1}{2} \times \text{fanout}_i\right)\{2 \times [1 - P(i)]\} = \sum_i \text{fanout}_i[1 - P(i)] \quad (3.65)$$

while that of a p-type is

$$\sum_i \left(\tfrac{1}{2} \times \text{fanout}_i\right)[2 \times P(i)] = \sum_i \text{fanout}_i P(i) \quad (3.66)$$

where $P(i)$ is the probability that node i is logic 1. When the power consumed by the buffer (inverter) of the n-type domino gate (Figure 3.34) is considered, the power dissipation measure Φ of an n-type domino gate is given by

$$\sum_i \left[\tfrac{1}{2} \times (\text{fanout}_i + 2)\right](2 \times [1 - P(i)]\} = \sum_i (\text{fanout}_i + 2)[1 - P(i)]$$

$$(3.67)$$

and that of the p-type is

$$\sum_i \left[\tfrac{1}{2} \times (\text{fanout}_i + 2)\right][2 \times P(i)] = \sum_i (\text{fanout}_i + 2)P(i) \quad (3.68)$$

Let us consider the domino AND gate and the static NAND gate of Figures 3.34 and 3.35 to justify these assumptions. Circuit simulator SPICE was used to simulate both circuits with different loads. The results are shown in Table 3.7. Load X_i represents an n-type domino AND gate in Figure 3.34 and a static NAND gate in Figure 3.35. In order to make the table readable, the power has been normalized so that the average power of a transition at a static CMOS node with one load is 2. Table 3.7 shows that the average power consumption of a transition at the static gate output (Z in Figure 3.35), the domino gate output (PZ in Figure 3.34), and the domino buffer output (Z in Figure 3.34) can be approximated as $2n \times p$, $2 \times p$, and $n \times p$, respectively.

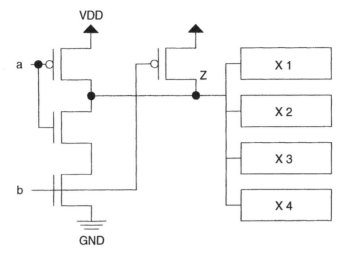

Figure 3.35   A static NAND gate with static gate loads.

TABLE 3.7   Normalized Average Power Consumption of Transition for Static and Domino Circuits

Number	Static Power (Figure 3.35)	Domino Power (Figure 3.34)	
of Loads	Z	PZ	Z
1	2	2.08	0.8
2	3.78	2.16	1.82
3	5.37	2.13	2.64
4	6.92	2.15	4.25

3.9 CIRCUIT RELIABILITY

Signal activity is also a good measure of *electromigration* and *hot-electron* degradation of a circuit. In this section let us consider how to model electromigration and hot-electron reliability in terms of signal activity estimates at the internal nodes of a CMOS circuit.

Electromigration is a major reliability problem caused by transport of atoms in a metal line due to the electron flow and it affects both MOS and bipolar components. Thermally activated ions of the conductor that normally self-diffuse in all directions are given a direction of net motion due to momentum transfer from the conducting electrons. Hence, the ions move "downstream" with the electrons [30]. A divergence of the ion flux ultimately gives rise to a circuit failure. A positive divergence leads to an accumulation of vacancies to form a void in the metal and ultimately an open circuit. A negative divergence leads to a buildup of metal, called a *hillock*, which can ultimately lead to a short circuit to adjacent or overlying metal. Electromigration is a wearout mechanism and is caused by persistent current stress. The time-to-failure is a lognormally distributed random variable. It is characterized by the *mean time to failure* (MTF), which depends on the current density in the metal line. Some recent models [31] of MTF predict that, at least under pulsed-DC conditions, the average current is sufficient to predict MTF as follows:

$$\text{MTF} = \frac{K}{J^2} \tag{3.69}$$

where K has a statistical distribution and is independent of J and J is the average current density. Hence, MTF $\alpha \, J^{-2}$. The dependence of MTF on current density at least represents a first order approximation of current stress in a wire.

Let us consider the average power dissipation given by Eq. (3.1). If the average current from the supply voltage V_{dd} is I_{dd}, then the average power dissipation can also be represented by $V_{dd}I_{dd}$. Hence, the average current is given by

$$I_{dd} = \tfrac{1}{2}V_{dd}AC$$

The above equation represents the average current through a logic gate in terms of average signal activity and average capacitance C. In turn current density is proportional to the average current. Hence, signal activity can be used to model *electromigration effect* E to the first order of approximation. For a logic gate, E is minimized if I^2 is minimized or if I is minimized. Hence, from Eq. (3.9), E can be minimized if the following expression is minimized:

$$\tfrac{1}{2}V_{dd} \sum_{i \in \text{all nodes}} C_i A_i$$

where A_i and C_i respectively represent the signal activity and capacitance associated with node i of a circuit. As V_{dd} is constant over all the gates of the circuit, the electromigration effect can be represented as:

$$E = \sum_i C_i A_i \tag{3.70}$$

where the summation is taken over all the gates.

As MOSFET devices are scaled down to small dimensions, certain physical mechanisms start to cause degradation in the device parameters, causing major reliability problems. One such mechanism is the injection of *hot carriers* into the MOS gate oxide layer [30]. Trapping of these carriers in the gate insulator layer causes degradation in the transistor transconductance and threshold voltage. Moreover, because oxide charging is cummulative over time, the phenomenon limit the useful *life* of a device. Hot carriers are electrons and holes in the channel and pinch-off regions of a transistor that have gained so much energy from lateral electric field produced by the source–drain voltage that their energy distribution is much greater than would be predicted if they were in thermal equilibrium with the lattice. The higher mobility of the electrons in silicon makes the impact worse in n-channel transistors. After the electrons have gained about 1.5 eV of energy, they can lose it via impact ionization (electron–hole pair creation), which in turn gives rise to hole substrate current. It is widely accepted that the MOSFET substrate current is a good indicator of the severity of the degradation. The average substrate current is given by [32]

$$I_{sub} = \frac{C_i}{B_i} I_{ds}(V_{ds} - V_{d,sat}) \exp\left(\frac{-B_i l_c}{V_{ds} - V_{d,sat}}\right)$$

where C_i, B_i are parameters independent of current and I_{ds} and V_{ds} respectively represent the drain–source current and voltage. The magnitude of I_{sub} is dependent on the switching activity of MOS transistors and the duration a transistor stays in the saturation region. Since hot carrier degradation occurs when a transistor is in saturation the region, which in CMOS circuits happens only during transitions, it follows that the higher the signal activity at the gate output, the more damage a transistor experiences. The duration a transistor stays in the saturation region is dependent on input slew rate and the load capacitance of a gate. The load capacitance is in turn dependent on the fanout of the gate. Hence, the first-order approximation of hot-electron degradation at the gate level (H_{gate}) can be modeled in terms of signal activity of a gate output and its fanout as follows:

$$H_{gate} = A_{gate} \cdot f_{gate}$$

where A_{gate} and f_{gate} represent the signal activity and fanout of a gate.

One can write the *age* of a transistor that has been operating for time T as follow [31]:

$$\text{Age}(T) = \int_{+T/2}^{-T/2} \frac{I_{ds}}{WH} \left(\frac{I_{sub}}{I_{ds}}\right)^m \tag{3.71}$$

where I_{ds} and I_{sub} are the MOSFET drain-to-source and substrate currents, respectively; W is the channel width, and H and m are parameters independent of transistor currents.

It can be observed from the equations for E and H_{gate} that for designing hot-carrier and electromigration-resistant circuits at the gate level, the following expression has to minimized:

$$\sum A_{gate} \cdot f_{gate} \tag{3.72}$$

summing over all the gates of the circuit. The above expression is given by R and can be thought of as the *reliability factor*. The larger is the value of R, the larger will be the unreliability of the circuit.

3.10   POWER ESTIMATION AT THE CIRCUIT LEVEL

Unlike at the logic level, more accurate information about the capacitances at the internal nodes of a complex logic gate are available at the circuit level. It is well known that a transistor is associated with a gate capacitance due to the parallel-plate capacitor being formed between the polysilicon gate and the channel and the diffusion capacitance due to the source and the drain regions of the transistor. These capacitances, though smaller than the load capacitances of a logic gate, get charged and discharged based on the switching activities of the inputs. Hence, for more accurate estimation of power dissipation, one needs to consider the switching activities due to the capacitances at the internal nodes of a logic gate. The charging/discharging mechanism for the logic gates will be apparent when we consider an example in the following section.

3.10.1   Power Consumption of CMOS Gates

Consider the CMOS gate in Figure 3.36. Input pattern $x_1 = \uparrow, x_2 = 0, x_3 = 1$ causes a falling transition at the output and so does the input pattern $x_1 = 0, x_2 = \uparrow, x_3 = 1$. But in case of the first pattern, the capacitance being discharged equals $C_y + C_{z_1}$ which is more than C_y, the capacitance being discharged in case of the second pattern. This simple example points out

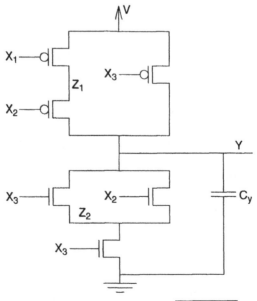

Figure 3.36   CMOS gate $y = \overline{(x_1 + x_2)x_3}$.

two things:

1. The gate presents a variable capacitance to the power/ground rails. The magnitude of this capacitance depends on the logic values at the inputs to the gate.
2. Two signals A and B are to be connected to the two equivalent inputs x_1 and x_2 of the gate in Figure 1.14 such that very often A has a transition and B stays zero then A should be connected to x_2 and B to x_1 as this results in lower power consumption than the other case.

Capacitances at the internal nodes of a CMOS gate charge and discharge resulting in additional power consumption. Some of these, but not all, happen at the same time instants as the charging and discharging of the capacitance at the gate output. For example, when the input to the gate in Figure 1 is $(x_1 = \downarrow, x_2 = 1, x_3 = 1)$ there may be a rising transition at z_1 though y remains at 0. On the other hand when the input is $(x_1 = \uparrow, x_2 = 0, x_3 = 1)$ there is a falling transition at both z_0 and y. Therefore, internal nodes of a gate need to be treated as circuit nodes distinct from (though related to) the output node of the gate. Hence, the average power dissipation is given by

$$W_{av} = V_{dd}^2 f\left(C_y d_y + \sum_i C_{z_i} d_{z_i}\right) \tag{3.73}$$

where z_i's are the internal nodes of the gate and d_{z_i} is the normalized activity of the "signal" at node z_i.

In [47] the authors show that determining $p_{z_i}^{\updownarrow}(= d_{z_i})$ is np-hard. The difficulty in determining $p_{z_i}^{\updownarrow}$ arises from the fact that for a rising transition to occur at an internal node at the beginning of a clock cycle, (i) the logic signal at the node must have had value 0 in the immediately previous clock cycle, and (ii) the conduction state of at least one path from the node to V_{dd} must have changed from off to on at the beginning of the clock cycle. The probability of a conducting path existing between any two nodes of a gate can be determined under the assumption that inputs to the gate are mutually independent. But determining the logic value at an internal node is difficult because an internal node may have been isolated from both V_{dd} and ground during the immediately previous clock cycle and possibly in all of the zero or more clock cycles immediately preceding that. In effect it is required to keep track of the past transitions at the node potentially going back to $-\infty$—an impossible task. The problem can be resolved by assuming, whenever the state of an internal node cannot be determined, that it is such that a transition occurs. This is done by using, in place of $p_{z_i}^{\uparrow}$, the probability that the number of conducting paths from z_i to V_{dd} changes from zero to a number larger than zero. The latter probability is an upper limit on $p_{z_i}^{\uparrow}$. This is well suited for the problem of determining the worst-case peak supply current. In this analysis it is required that to make a comparison between power dissipated in the same gate for two different input orderings, the errors need to be small and comparable in the two cases.

A more accurate upper limit on $p_{z_i}^\uparrow$ can be obtained from the observation that for a pair of complementary transitions to occur at an internal node, the number of conducting paths from the node to V_{dd} must change from zero to a number larger than zero followed by a similar change in that to ground. If the number of conducting paths from a node to V_{dd} changes from zero to a number larger than zero once every 40 clock cycles but that to ground changes only once every 100 clock cycles, then the node cannot have more than two transitions every 100 clock cycles. Therefore,

$$
p_{z_i}^\updownarrow = \begin{cases} p_{z_i, V_{dd}}^\uparrow & \text{if } p_{z_i, V_{dd}}^\uparrow \leq p_{z_i, V_{ss}}^\uparrow \\ p_{z_i, V_{ss}}^\uparrow & \text{otherwise} \end{cases}
$$

where $p_{z_i, z_j}^\uparrow$ is the probability that the number of conducting paths from z_i to z_j changes from zero to greater than zero.

The $p_{z_i, z_j}^\uparrow$'s can be determined by applying a reduction procedure to a graph obtained from the schematic of the gate. To each node of the gate there corresponds a vertex in the graph. Hence there is a vertex for ground, V_{dd}, y, and each of the z_i's. To each transistor with its drain and source connected to two nodes, there corresponds an edge between the corresponding vertices. Each edge e has two labels p_e and $p_e^\uparrow$. Here, p_e denotes the probability that edge e is on or conducting and $p_e^\uparrow$ denotes the probability that edge e turns from off to on. Note p_e equals p_{x_i} if e corresponds to an NMOS transistor with logic signal x_i connected to its gate and equals $1 - p_{x_i}$ otherwise. Transition probability $p_e^\uparrow$ is simply $d_{x_i}/2$ in both the cases. These graphs are a restricted class of graphs called series–parallel graphs [46].

The required path conduction probabilities can now be determined using an operation termed graph reduction in [48]. Let e_p (e_s) denote the single edge equivalent to the parallel (series) combination of two edges e_1 and e_2; then

$$p_{e_p} = p_{e_1} + p_{e_2} - p_{e_1} p_{e_2} \tag{3.74}$$

$$p_{e_s} = p_{e_1} p_{e_2} \tag{3.75}$$

$$p_{e_p}^\uparrow = p_{e_1}^\uparrow (1 - p_{e_2}) + p_{e_2}^\uparrow (1 - p_{e_1}) - p_{e_1}^\uparrow p_{e_2}^\uparrow \tag{3.76}$$

$$p_{e_s}^\uparrow = p_{e_1}^\uparrow p_{e_2} + p_{e_2}^\uparrow p_{e_1} - p_{e_1}^\uparrow p_{e_2}^\uparrow \tag{3.77}$$

Let the path from node x_i to node x_j be reduced to edge e_r; then

$$p_{x_i, x_j}^\uparrow = p_{e_r}^\uparrow$$

3.10.1.1 *Characterization of Gates in the Library*

For any given n, the total number of functions with n inputs is 2^{2^n}, a very large number even for small n. Most typical semicustom libraries will have gates corresponding to a very small subset of these and usually limited to $n \leq 6$. In that case, it is efficient to analyze each of these in advance. Subsequently the results of the analysis are used when processing a circuit containing these gates.

The characterization process can derive symbolic expressions for $p_{z_i, V_{dd}}^{\uparrow}$ and $p_{z_i, V_{ss}}^{\uparrow}$ for each internal node z_i of each gate in the library. The symbols in these expressions are the p_{x_i}'s and $p_{x_i}^{\uparrow}$'s, where x_i's denote the input signals to the gate.

In addition, the node capacitances C_{z_i} for each internal node z_i of each gate are also computed.

3.11 HIGH LEVEL POWER ESTIMATION

It is very difficult to estimate the switching activity at higher levels of design abstraction such as the system or architectural level. Landman and Rabaey [99] have proposed a power estimation technique for digital signal processing (DSP) circuits based on high-level description of the system. The DSP circuits usually use logic primitives such as adders, comparators, multipliers, and registers. Hence, estimation of switching activity at the output of such logic primitives are required, given the input signal probabilities and switching activities. Such estimation technique can be used by a high-level synthesis system for estimating power dissipation and thereby help in the synthesis of DSP systems for low power.

A stochastic modeling technique can be used for high level power estimation using high-level statistics such as mean, variance, and autocorrelation. The estimation technique follows the gate level power estimation technique. While in the gate level estimation technique, the primitives are AND, NAND, OR, NOR, and other complex gates, the high-level estimation technique involves multipliers, adder, registers, and so on. The estimation technique recognizes that there exists a direct relationship between bit level probabilities and word level statistics. It has been seen that for three common DSP input signals such as speech, music, and image, the signal and transition probabilities follow a similar pattern that can be represented by the piecewise linear curve of Figure 3.37. The figure shows that the signal probabilities of the lower order bits of a word [the least significant bits (LSBs)] are essentially uncorrelated in space and time with a signal probability of 0.5 and switching activity of 0.25 and are essentially independent of the data distributions. This is expected; for example, the LSB switches from 0 to 1 and 1 to 0 based on whether the data are even or odd. If one assumes that the odd or even data are equally likely (a valid assumption), then the result expected for

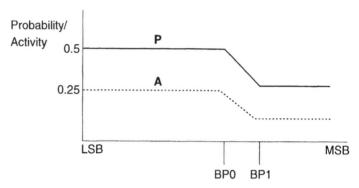

Figure 3.37   Word level data model.

the least order bits is as shown in the figure. The higher order bits [the most significant bits (MSBs)] show complete dependence because they represent the sign extensions in two's complement representation. It can be noted that there are two breakpoints (BP_0 and BP_1) in the model and the values of the breakpoints, the signal probabilities, and the transition activities can be represented in terms of the statistical parameters such as the mean (μ), the variance (σ^2), and the lag one correlation coefficient [$\rho_1 = cov(X_t, X_{t+1})/\sigma^2$]. The value of the breakpoint, BP_0, signifying the end of the low-order bits, is related to the variance of the signal distribution and is empirically given by

$$BP_0 = \log_2\left(\frac{3\sigma}{32}\right)$$

It turns out that BP_0 represents the breakpoint for the signal probability curve, while a correction term has to be added for the breakpoint of the switching activity curve. The empirical representation of BP_0 for the switching activity curve with highly correlated inputs is given by

$$BP_0 = \log_2\left(\frac{3\sigma}{32}\right) - \log_2\left(1 - \rho_1^2\right)^{0.5}$$

The probabilities and switching activities for the higher order bits (MSBs) are respectively given by

$$P_{MSBs} = F_1\left(\frac{\mu}{\sigma}\right)$$

and

$$A_{MSBs} = F_{01}\left(\frac{\mu}{\sigma}, \rho_1\right)$$

The exact probabilities and activities depend on the signal distribution. It has been noticed that DSP signals can be approximated by Gaussian processes. For such processes, the univariate and bivariate normal distribution functions are substituted for F_1 and F_{01}, respectively. The other breakpoint BP_1 for the MSBs is given by

$$\text{BP}_1 = \log_2(|\mu| + 3\sigma)$$

How about the range of bits between BP_0 and BP_1? The correlation between these midrange bits falls between the LSBs and the MSBs, and hence, a linear approximation is used in Figure 3.37.

Having considered the derivation of bit level probabilities and activities from word level statistics, one has to consider how to propagate the values of the statistical parameters through the logic primitives such as the adder and multiplier. Just like the gate level approach, Landman and Rabaey [99] considered the estimation of μ_z, σ_z^2, and ρ_1^z at the output of logic primitives, given the input statistics. If the inputs to a two-input (x and y) adder are independent, then the output statistics in terms of the input are given by

$$\mu_z = \mu_x + \mu_y \tag{3.78}$$

$$\sigma_z^2 = \sigma_x^2 + \sigma_y^2 \tag{3.79}$$

$$\rho_1^z = \frac{\rho_1^x \sigma_x^2 + \rho_1^y \sigma_y^2}{\sigma_z^2} \tag{3.80}$$

If the inputs to the adder are correlated, then the output statistics are given by

$$\mu_z = \mu_x + \mu_y \tag{3.81}$$

$$\sigma_z^2 = \sigma_x^2 + \sigma_y^2 + 2\rho_1^{xy}\sigma_x\sigma_y \tag{3.82}$$

$$\rho_1^z = \frac{\rho_1^x \sigma_x^2 + \rho_1^y \sigma_y^2 (\rho_1^{xy} + \rho_1^{yx})\sigma_x^2\sigma_y^2}{\sigma_z^2} \tag{3.83}$$

where ρ_1^{xy} and ρ_1^{yx} represent the pairwise cross-correlations between the inputs. Another common function used in DSP applications is a multiplier with an input that is a constant (C). For such a multiplier, the output statistics are given by

$$\mu_z = C\mu_x \tag{3.84}$$

$$\sigma_z^2 = C^2\sigma_x^2 \tag{3.85}$$

$$\rho_1^z = \rho_1^x \tag{3.86}$$

One may try to trade off computational complexity and accuracy by ignoring the correlations existing among signals. However, the presence of feedbacks in dataflow graphs introduces another difficulty in calculation of the statistical parameters at the internal nodes of a dataflow graph. The statistical parameters for the internal nodes may depend on each other due to feedback. An approximate solution is suggested in [99] by breaking the feedback loops.

3.12 INFORMATION-THEORY-BASED APPROACHES

Recently, information theory has also been used quite effectively to estimate power at the RT (register transfer) level of design abstraction [49]. The RT level abstraction assumes that the Boolean functionality of the circuit is known while the details of the implementation are unknown. In the following analysis we will consider an entropy-based approach to determine average power in CMOS combinational circuits [49]. It has been shown that the output entropy of Boolean functions can be used to predict the *average minimized area* of CMOS combinational circuits. Area is roughly proportional to the switched capacitance of the circuit under consideration. Hence, if entropy can be shown to track the switching activity of a circuit, then it will be easy to predict the power dissipation. Let us consider the approach in more detail.

If x is a random variable with a signal probability p, then the entropy of x is defined as

$$H(x) = p \log_2 \frac{1}{p} + (1 - p)\log_2 \frac{1}{1 - p}$$

The plot of H versus p is shown in Figure 3.38. It can be observed that entropy has a maximum when $p = 0.5$. Intuitively, the signal can carry the most information when $p = 0.5$ and can have the maximum number of transitions.

For a discrete variable x, which can take n different values, the entropy is defined as

$$H(x) = \sum_{i=1}^{n} p_i \log_2 \frac{1}{p_i}$$

where p_i is the probability that x takes the ith value x_i.

Let $Y = f(X)$ be a Boolean function where X represents the input vector with n and Y represents the output vector with m bits. Therefore, X can possibly take 2^n values while Y can take 2^m values. The input of entropy of X is defined as

$$H(X) = \sum_{i=1}^{2^n} p_i \log_2 \frac{1}{p_i}$$

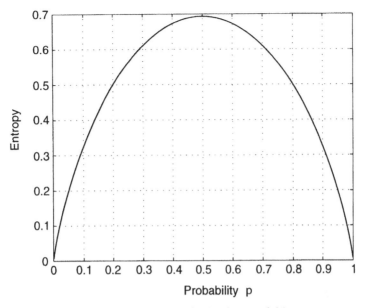

Figure 3.38   Entropy of a Boolean variable.

and the output entropy is defined as

$$H(Y) = \sum_{i=1}^{2^m} p_i \log_2 \frac{1}{p_i}$$

Given the input signal probabilities of 0.5, the output entropy of the Boolean function can be used to predict the area of its *average minimized implementation* as

$$A = K \cdot \frac{2^n}{n} H(Y)$$

where A is the area of the implementation and K is the proportionality constant [50]. For small circuits, empirical results show that $2^n H(Y)$ provides a good measure of area. Hence, it can be observed that entropy can provide a good measure of area.

Now lets consider the relationship between switching activity and entropy. If we ignore temporal correlation among signals, the probability of a signal switching from 1 to 0 or 0 to 1 between two consecutive clock cycles is given by $2p(1 - p)$, where p is the probability that the signal assumes a logic 1 within a clock cycle. The analysis assumes that logic gates have zero delays, and hence, any glitching activity is ignored. Now let us consider Figure 3.39 which shows the plot of entropy and two times activity [$4p(1 - p)$]. The plot

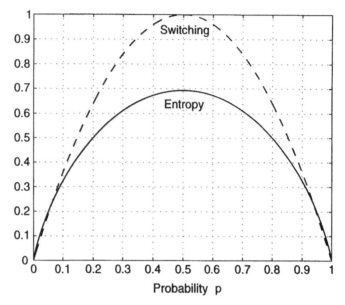

Figure 3.39   Activity vs. entropy.

clearly shows that activity tracks entropy provided temporal correlation among signals can be ignored.

Now let us consider power dissipation. Average power dissipation due to function $Y = f(X)$ can be approximated as

$$PA \sum_{i=1} NC_i \qquad (3.87)$$

where A is the average switching activity at the nodes, C_i is the capacitance at each node i, and N is the total number of nodes in the circuit. Here, A is defined as $A = 1/N \sum_{i=1}^{N} A(x_i)$. Note that $\sum_{i=1} NC_i$ represents the area (α) of the circuit, which can be estimated using the output entropy of function $Y = f(x)$ and the number of primary inputs n. Therefore, average power

$$P\alpha \cdot A \qquad (3.88)$$

can be estimated using the entropy of the inputs and outputs and the number of primary inputs and primary outputs of the circuit. The method can be extended to determine an estimate for sequential circuits. The bounds for power dissipation using the above technique can be loose, and hence, research in this area is needed to make this technique viable. The technique to estimate average power can also be used for power macromodeling.

Hence, a high-level technique to estimate power can use the following three steps:

- Determine the input/output entropies of combinational logic block by running RTL simulation of sequential circuits.
- From the input/output entropies, determine the switching activity, area, and estimate of average power.
- Combine with latch and clock power to determine the total power dissipation.

3.13   ESTIMATION OF MAXIMUM POWER

Earlier in this chapter we considered estimation of average dynamic power for static and domino CMOS circuits. Average power dissipation is important in mobile applications where battery life is important. In desktop applications, it is sometimes important to consider the maximum instantaneous power or the maximum sustained power in a circuit. Such estimation is important for the design of power and ground lines, voltage drop on lines, and package requirements. With the scaling down of device sizes, the designers are putting more transistors in a die, and hence, the instantaneous current can be large due to the simultaneous switching of a large number of logic gates. Since the power ground pins are associated with some inductance L, the voltage drop due to such a change in current is equal to $L\,di(t)/dt$, where $i(t)$ is the dynamic current from the supply line. For future designs, such voltage drop can be significant. Therefore, there is a need to accurately estimate the instantaneous current.

Estimation of maximum dynamic power (or current) requires that we are able to determine the primary input vectors of a circuit to induce maximum switching of capacitance. Such estimation is sometimes complicated by the fact that logic gates have finite delays. Hence, one has to determine the primary input vectors such that the total switched capacitance due to both functional and spurious transitions is maximized. The problem of the maximum power estimation is difficult because the problem is np-complete. However, the aim of this chapter is to determine good lower bounds of power so that the bounds can be efficiently used during the design of VLSI circuits and systems. Several research articles have appeared for estimation of maximum power [51–55, 57, 58, 60, 62]. Several approaches have been proposed to estimate the maximum power consumption for CMOS circuits. Devadas et al. [53] formulated the power dissipation of CMOS circuits as a Boolean function in term of the primary inputs. They tried to maximize the function by solving a weighted max-satisfiability problem. During the branch-and-bound process that maximizes the objective function, the lower bound can be improved successively. However, the time complexity to obtain the objective function and to optimize the objective function are both exponential functions of the number of primary inputs (PIs). Hence, this approach is only feasible for small circuits. Kriplani et al. [51, 52] addressed the problem of determining an upper bound for the maximum power dissipation. They first generated an upper bound by propagating the signal uncertainty through the circuit. The bound was then successively made tighter by considering spatial correlation of signals at the outputs of logic gates. In this approach, the time to obtain the initial upper bound is linear in the number of gates, and the branch-and-bound strategy to improve the upper bound is, of course, CPU time intensive.

To generate a lower bound, traditional simulation-based approaches search for two consecutive input binary vectors to maximize the instantaneous power dissipation. The process is CPU time intensive. Furthermore, for a circuit with large number of PIs, simulation tends to generate a loose lower bound.

In this section we will describe two approaches to determining lower bounds of maximum dynamic power in static CMOS circuits: deterministic (automatic test generation based) and simulation-based approaches.

3.13.1 Test-Generation-Based Approach

In standard CMOS circuits, the power dissipation during the active mode of operation is dominated by the dynamic current, and hence the instantaneous power dissipation due to two consecutive input binary vectors is proportional to

$$P_i = \sum_{\text{for all gates}} T(g) \cdot C(g) \tag{3.89}$$

where $C(g)$ denotes the output capacitance of gate g and $T(g)$ is a binary variable that indicates whether gate g switches or not corresponding to the two input vectors. Here, $T(g)$ equals 1 if gate g switches and is 0 if gate g does not switch. To maximize P_i efficiently, we sort the gates by the output capacitance $[C(*)]$ in nonincreasing order and then *assign* transitions to gates [i.e., let $T(*) = 1$] from the sorted list of gates with the largest output capacitance. To *justify* the transitions [i.e., to see if $T(*) = 1$ is achievable], the modified justification mechanism in a *9-V D algorithm* [65] (an ATG algorithm for stuck-at faults) is used. This algorithm was originally used to justify the fault propagation paths in combinational circuits. Experimentally, the execution time of the algorithm is approximately proportional to the number of gates in the circuit and is comparable to the time of simulating the circuit once. In addition, experiments show the quality of the estimates are superior to the results from simulation.

In CMOS circuits, the capacitive load of a logic gate can be approximated by the fanout of the gate. Hence, the dynamic power dissipation due to two consecutive input vectors is proportional to

$$P_i = \sum_{\text{for all gates}} [g(V_1) \oplus g(V_2)] \cdot F(g) \tag{3.90}$$

where V_1, V_2 denote two consecutive input binary vectors to the circuit, $g(*)$ represents the *Boolean* function of gate g in term of PI's, and $F(g)$ denotes the number of fanouts of gate g. The term P_i is a measure of the power dissipation due to two consecutive input vectors and is the objective function to be maximized.

To maximize P_i, instead of searching for appropriate V_1 and V_2, we greedily assign transitions to the gates with large fanout. The gates are first sorted by the fanout number in nonincreasing order. Then, in each iteration, we select a gate g that is untried and currently has the largest fanout number $F(g)$ to justify the assignment: $g(V_1) \oplus g(V_2) = 1$. The *justification* mechanism in the algorithm includes two processes—*backtracing* and *implication*. If the justification for assigning transitions to a gate fails, the state of the circuit from the incorrect decision will be recovered and the next gate will be tried. The gates in the circuit are assigned and justified one by one until all gates have been processed.

To recover the state of the circuit from incorrect decisions, we store all the values that have been either assigned or implied to the gates in the circuit. Each gate g is associated with a stack to store all the composite logic values a/b that have been assigned to g [a and b denotes $g(V_1)$ and $g(V_2)$, respectively]. The variables a and b can be 1, 0, or u (*unknown value*). At each gate, the top of the stack stores the most recently updated value for the gate.

For example, in Figure 3.40, the top of the stacks of gates x, y, and z are $0/u$, u/u, and $u/1$, respectively. After assigning a rising transition $(0/1)$ to x, $y(V_2)$ of y is forced to be 1, and $y(V_1)$ is still left unknown. Hence, $u/1$ is pushed into the stack of y to be gate y's current value. Then, the most

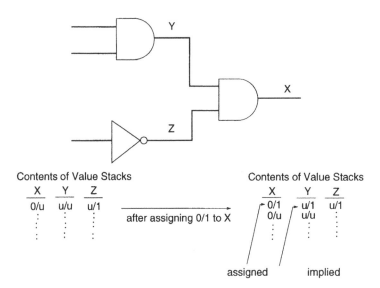

Figure 3.40   Stacks for justification and backtracking.

recently updated values for x, y, and z turn out to be $0/1$, $u/1$, and $u/1$, respectively. We know x and y are the gates whose values have changed. If it is determined later that assigning $0/1$ to x causes conflicts, the stacks of gate x and y are popped to recover the state of the circuit from the incorrect decision. To make a gate switch, either $0/1$ or $1/0$ can be tried if the current value of the gate is not in "conflict" with the new assignment. We define two composite values to be in *conflict* if they have 0 and 1 at the same position. For example, $1/0$ cannot be assigned to gate x in Figure 3.40, since $1/0$ is in conflict with $0/u$.

Experiments show the test generation approach is superior to the traditional simulation-based technique in both efficiency and the quality of the results. Within a very short CPU time, the ATG-based estimation can generate estimates superior to the results from simulating the circuit for a long period of time. Considering the speed of ATG-based estimation, it not only serves as an estimator that is superior to simulation, but also may be the only practical way to estimate the maximum power for very large circuits.

3.13.2 Approach Using the Steepest Descent

For synchronous circuits, let us assume that the switching of PIs are synchronized at the leading edge of each clock cycle. However, spurious transitions may occur at internal gates due to different propagation delays through different paths. The output of a gate may have transient pulses (spurious transitions) before it is finally stabilized. Such spurious transitions consume power and hence, such a phenomenon should be considered in the maximum power estimation.

In a CMOS circuit, the total energy dissipated due to two consecutive input vectors can be described as $E = \frac{1}{2}V_{dd}^2\sum_{\forall\text{gate}}C_{\text{load}}(g)*T(g)$, where $T(g)$ denotes the transition count at the output of gate g during the clock cycle and $C_{\text{load}}(g)$ represents the capacitive load of g. Hence, the average value of the instantaneous power over the clock cycle is $p_{\text{av}} = E/|T|$, where $|T|$ denotes the length of a clock cycle. Under the assumptions that (i) instantaneous power is a continuous function of time and (ii) $|T|$ is sufficiently small, it is reasonable to view p_{ave} as the instantaneous power during the clock cycle. $C_{\text{load}}(*)$ can be approximated by the fanout $[F(*)]$ of logic gates. Therefore, instantaneous power is proportional to

$$P(V_1,V_2) = \sum_{\forall\text{gate}} F(g) \cdot T(g) \tag{3.91}$$

where V_1 and V_2 denote the two consecutive input binary vectors applied to the circuit.

To calculate the transition count $[T(*)]$ of a gate in a clock cycle, one can associate with each gate a certain data structure to record the gate's output pattern. We call each such structure a *time sequence* (a sequence of time instants at which the gate might switch). The time sequence of a gate can be implemented as a linked list, in which each node represents a time instant at which the gate might switch. To avoid confusion, we refer to the nodes in time sequences as *elements*. Each element in a time sequence is associated with

1. a start time that denotes the corresponding switching time of the gate and
2. a logic value (0 or 1) that denotes the value at the output of the gate during a time interval beginning at the starting time of the element and ending at the starting time of the next element.

The use of these structures is described below.

Based on *static timing analysis* [64], we present a procedure with linear complexity (in terms of the number of gates in the circuit) to create the time sequences for a given circuit. By propagating the possible switching events level by level from PIs to POs, the procedure can decide all the possible switching events at each node in the circuit and equip the node with an appropriate structure of time sequence. Let us use the circuit in Figure 3.41 as an example, in which, each primary input (a, b, c, or d) is associated with two elements (with starting time 0^- and 0). Since the propagation delay of gate A is assumed to be 2, the time sequence of node e is comprised of two elements with starting time 0^- and 2. Similarly, the time sequence of node f can be built from that of primary input d. Finally, the time sequence at the fanout of gate B is built by merging the time sequences of nodes e, c, and f.

Based on the time sequences, $T(g)$ in Eq. (3.94) can be formulated as $T(g) = \sum_{i=1}^{n(g)-1}(f_g(i) \oplus f_g[i + 1])$. Here we use $n(g)$ to denote the number of elements in the time sequence of gate g and $f_g(i)$ to denote the logic value associated with the ith element of the time sequence of gate g. Hence, we

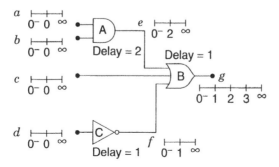

Figure 3.41   Building time sequences for a circuit.

can define our problem as follows:

$$\text{Maximize:} \quad P(V_1, V_2) = \sum_{\forall\text{gate}} F(g) \cdot \sum_{i=1}^{n(g)-1} \left(f_g(i) \oplus f_g(i+1) \right)$$

Subject to: $\mathscr{C} \equiv \{\text{spatial correlation between logic gates}\}$

Note that $f_g(j)$ (denoting the value of the jth element of gate g) is decided by the value of the corresponding elements of the fanins of gate g. If we expand $f_g(j)$ from gate g toward primary inputs gate by gate, $f_g(j)$ can be represented as a Boolean function in terms of the entries of input vectors V_1 and V_2. Here, $\mathscr{C}$ represents the structure of the logic circuit and is the constraint set in this combinatorial optimization problem.

The above model describes the problem of maximum power estimation in circuits implemented with static CMOS gates. In the following section, we discuss how to transform $P(V_1, V_2)$ into a function that is defined over the Euclidean space.

3.13.2.1 Transformation to Continuous Domain

We describe below a transformation that turns Boolean operators into arithmetic ones. After such a transformation, $P(V_1, V_2)$ is well defined even if the entries in V_1 and V_2 are real numbers.

Any Boolean function can be transformed into an arithmetic function by repeatedly applying the following fundamental rules:

1. $a \vee b = a + b - ab$,
2. $a \wedge b = a \cdot b$ (a multiplied by b), and
3. $\sim a = 1 - a$,

where $\vee$, $\wedge$, and $\sim$ on the left-hand side are Boolean operators, while $+$, $-$, and $\cdot$ on the right-hand side are arithmetic operators. Any function transformed from some Boolean function according to these three rules preserves its original values at all vertices of $[0, 1]^M$, where M is the number of dimensions of the function.

In the following discussion, we relax the entries in input vectors V_1 and V_2 from the discrete set $\{0, 1\}$ to a continuous range $[0, 1]$. To avoid ambiguity, let us use $\tilde{P}(V_1, V_2)$ to denote the function $P(V_1, V_2)$ after the relaxation. It can be noted that $\tilde{P}(*)$ is *continuous* and *differentiable* everywhere in $[0, 1]^{2n}$, where n is the number of primary inputs of the circuit. Hence, the *gradient* of $\tilde{P}(*)$ (i.e., $\nabla\tilde{P}$) is defined everywhere in $[0, 1]^{2n}$. This justifies our application of continuous optimization technique to $\tilde{P}(*)$.

Due to the smoothness and continuity of $\tilde{P}(*)$, we reach the following conclusion:

- Within the hypercube $[0, 1]^{2n}$, if point a is sufficiently close to point b, the values of $\tilde{P}(a)$ and $\tilde{P}(b)$ tend to the same. In other words, $\|\tilde{P}(a) - \tilde{P}(b)\| \to 0$ if $\|a - b\| \to 0$.

Suppose a relative maximum point v of $\tilde{P}(*)$ is sufficiently close to some vertex $\dot{v}$ of the hypercube $[0, 1]^{2n}$. The vertex $\dot{v}$ is a good choice to maximize the original combinatorial function $P(*)$ if v is sufficiently good for maximizing $\tilde{P}(*)$. This argument directly follows the above conclusion. Note that the maximum value of $\tilde{P}(*)$ over $[0, 1]^{2n}$ is an upper bound of the maximum value of $P(*)$ over $\{0, 1\}^{2n}$.

Since we are only interested in the discrete solutions, it is desired that the distance between the maxima [of $\tilde{P}(*)$] and vertices of the hypercube is as small as possible. Obviously there is no guarantee this is always the case. Hence, $\tilde{P}(*)$ is modified to favor this requirement. Some nonbiasing term $T(*)$ is subtracted from the function $\tilde{P}(*)$ to make the maximum of $\tilde{P}(*)$ move toward the vertices of the hypercube $[0, 1]^{2n}$. We call the term "nonbiasing" because the term should vanish at the vertices. In all the experiments, we determine the nonbiasing term as a summation of $2n$ concave functions: $T(X) = \sum_{i=1}^{2n} x_i(1 - x_i)$. Here n is the number of primary inputs and x_i is the ith entry of vector X. It can be noted that $T(*)$ reaches its maximum value at $X = [0.5, 0.5,, 0.5, \ldots, 0.5]$ and is also symmetric to this point. The resultant function $\hat{P}(X) = \tilde{P}(X) - wT(X) = \tilde{P}(X) - w\sum_{i=1}^{2n} x_i(1 - x_i)$ will be the objective function to be maximized. Here w is some positive real number used to adjust the shape of $\hat{P}(*)$. For convenience, the notation $\hat{P}(X) = \tilde{P}(X) - w\sum_{i=1}^{2n} x_i(1 - x_i)$ will be used in the rest of this chapter.

Let X denote the concatenation of the two input vectors V_1 and V_2. From the above discussion, the corresponding continuous problem for maximum power estimation can be defined as follows:

$$\text{Maximize:} \quad \hat{P}(X) = \sum_{\forall \text{gate}} F(g) \cdot \sum_{i=1}^{n(g)-1} \left[\tilde{f}_g(i) + \tilde{f}_g(i+1) - 2f_g(i)f_g(i+1) \right]$$

$$- w \sum_{i=1}^{2n} x_i(1 - x_i)$$

Subject to: $\mathscr{C} \equiv \{\text{spatial correlation between signals}\}$

where n is the number of primary inputs of the circuit. This constrained optimization problem can be solved based on the gradient method, which will be discussed next.

3.13.2.2 *Optimization*

We present a *steepest descent* strategy to find the maxima of the above constrained optimization problem. Starting from the neutral point (i.e., $[0.5, 0.5, 0.5, \ldots, 0.5]$) of $[0, 1]^{2n}$, the following procedure will be performed iteratively until one of the stopping criteria is met:

1. Calculate $\nabla\hat{P}$ [gradient of $\hat{P}(*)$] at the point.
2. From the point, move along the direction of $\nabla\hat{P}$ as far as possible until $\hat{P}(*)$ starts to decrease. Go to step 1.

The above iteration stops under any of the following conditions:

1. The search reaches some relative maximum point at which $\nabla \hat{P} \to 0$.
2. The search is stuck at some point on the boundary of the hypercube $[0, 1]^{2n}$, where the gradient of $\hat{P}(*)$ is normal to the boundary.

During the search, the curve gradually moves toward the boundary of $[0, 1]^{2n}$. Keep measuring the distance between the search curve and the vertices of $[0, 1]^{2n}$. The following expression is used in the experiments to represent the distance $d(P) = \sum_{i=1}^{2n} 0.5 - \|0.5 - p_i\|$. Here P is a point on the curve, p_i denotes the ith entry of P, and n is the number of primary inputs. The term $d(P) \to 0^+$ denotes that P is sufficiently close to some vertex of $[0, 1]^{2n}$. Under such a condition, the vertex will be reported as a satisfactory solution for maximizing the original combinatorial function $P(*)$ (i.e., the instantaneous power).

At each iteration of the gradient method, the gradient at a point X is calculated as $\nabla \hat{P}(X) = (\partial \hat{P}/\partial x_1, \partial \hat{P}/\partial x_2, \partial \hat{P}/\partial x_3, \ldots, \partial \hat{P}/\partial x_{2n})$, where $\partial \hat{P}/\partial x_i = (\hat{P}(x_1, x_2, \ldots, x_{i-1}, x_i + \Delta x, x_{i+1}, \ldots, x_{2n}) - \hat{P}(x_1, x_2, \ldots, x_{i-1}, x_i, x_{i+1}, \ldots, x_{2n}))/\Delta x$, $\Delta x \to 0^+$. Determining the exact expression for $\hat{P}(*)$ (in terms of primary inputs) is CPU time intensive, especially for large-scale circuits. However, it is not necessary to obtain the expression for $\hat{P}(*)$ to calculate the gradient. Instead, the value of $\hat{P}(X)$ at a point X is obtained by simulating the circuit with the input vector $V_1 = (x_1, x_2, \ldots, x_n)$ followed by $V_2 = (x_{n+1}, x_{n+2}, \ldots, x_{2n})$. Here x_i is the ith entry of X and is a real number between 0 and 1.

During the steepest descent process, it is possible that the search is stuck at some local maximum that is far away from the vertices of $[0, 1]^{2n}$. In this case, no conclusion can be drawn about which vertex is most likely to maximize the instantaneous power. Two approaches can be used to cope with this situation:

- Choose a different starting point (instead of using the neutral point) to start the steepest descent process. A different starting point may result in a local maximum that is close to some vertex of $[0, 1]^{2n}$.
- Increase w (weight of the nonbiasing term) to adjust the shape of the objective function. Using a larger w will reduce the distance between the local maxima of $\hat{P}(*)$ and the vertices of $[0, 1]^{2n}$.

The second approach is adopted to show experimental results. We need to use a different setting (larger) of w for two circuits to make the search converge out of a total of forty circuits.

Let us try to compare the results obtained using the ATPG and gradient search-based techniques. Figure 3.42 (zero-delay model) shows that the ATPG based technique and the gradient-based approach produce much

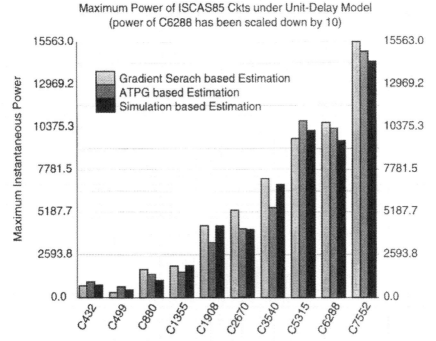

Figure 3.42 Comparison of ATPG vs. gradient vs. simulation-based technique for unit-delay model.

better estimates of maximum cycle power than the simulation-based technique. Similar results are also true for the nonzero-delay model, where the delay of each gate is modeled to be proportional to the fanout (Figure 3.43). Figures 3.44 and 3.45 show the CPU time required to estimate maximum cycle power for the ISCAS benchmark circuits. Results again show that ATPG-based technique can be orders of magnitude faster than the gradient-based or simulation-based technique. However, it should be noted that the gradient-based technique can very easily capture different kinds of delay models while the ATPG-based technique is not that versatile. One of the drawbacks of the gradient-based approach is that the optimization can get stuck in local minimas.

From the above graphs we can conclude the following significances of this continuous optimization approach:

1. Compared to the traditional simulation-based technique (SIM), this approach is superior in both speed and performance. This approach generates a tighter lower bound (1.16 times larger, on an average). For the 40 circuits, the mean value of the ratio of CPU time of the continuous optimization approach to CPU time of the simulation-based technique is 0.41.

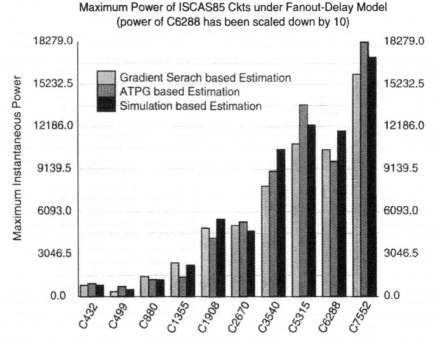

Figure 3.43 Comparison of ATPG vs. gradient vs. simulation-based technique for fanout delay model.

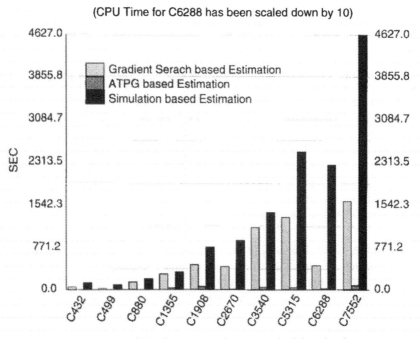

Figure 3.44 CPU time comparison for unit-delay circuits.

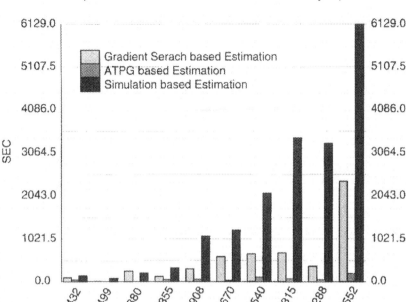

(CPU Time for C6288 has been scaled down by 10)

Figure 3.45 CPU time comparison for fanout delay circuits.

2. Although the technique does not converge as fast as the ATPG-based technique, the approach is favored under the following conditions:
   a. To estimate the power of a circuit under certain complicated delay models in which the ATPG-based technique cannot be easily applied.
   b. This approach generated better estimate of maximum power (1.19 times larger on average) than the ATPG-based technique for 24 out of 40 circuits tested. A lower bound of high quality can be decided by these two approaches together.

3.13.3 Genetic-Algorithm-Based Approach

Another approach to estimate power in CMOS circuits uses a probabilistic technique like a genetic algorithm to generate a set of input vectors that can induce maximum switching (weighted by the fanouts of logic gates) at the internal nodes of a circuit. We will not be able devote considerable effort to describe the genetic algorithm in detail. (Readers are referred to [63].) One of the salient features of the approach is that within the framework of the genetic algorithm, one can estimate good lower bounds of maximum power per cycle or the maximum sustained power under different delay models. However, it should be noted that genetic algorithms can be slow, and hence, parallel processing approaches may be required to solve the problems within reasonable time [61].

3.14 SUMMARY AND CONCLUSION

In this chapter we have presented various techniques to estimate average and maximum synamic power in CMOS digital circuits. Accurate estimation of power dissipation is critical in evaluating and synthesizing designs for low-power at various levels of design abstraction. Power dissipation is also related hot electron and electromigration reliability and hence, low-power circuits are possibly more reliable.

REFERENCES

[1]   F. Najm, "A Survey of Power Estimation Techniques in VLSI Circuits," *IEEE Trans. Very Large Scale Integration (VLSI) Syst.*, pp. 446–455, Dec. 1994.

[2]   K. P. Parker and E. J. McCluskey, "Probabilistic Treatment of General Combinatorial Networks," *IEEE Trans. Computers*, vol. C-24, pp. 668–670, 1975.

[3]   F. Najm, "Transition Density, a Stochastic Measure of Activity in Digital Circuits," *ACM/IEEE Design Automation Conference*, pp. 644–649, 1991.

[4]   T.-L. Chou, K. Roy, and S. Prasad, "Estimation of Circuit Activity Considering Signal Correlations and Simultaneous Switching," *ACM/IEEE International Conference on Computer-Aided Design*, pp. 300–303, Nov. 1994.

[5]   T.-L. Chou and K. Roy, "Estimation of Activity for Static and Domino CMOS Circuits Considering Signal Correlations and Simultaneous Switching," *IEEE Trans. on Computer-Aided Design Integrated Circuits*, pp. 1257–1265, Oct. 1996.

[6]   T.-L. Chou and K. Roy, "Accurate Estimation of Power Dissipation in CMOS Sequential Circuits," *IEEE Trans. VLSI Systems*, pp. 369–380, Sept. 1996.

[7]   T.L. Chou and K. Roy, "Statistical Estimation of Digital Circuit Activity Considering Uncertainty of Gate Delays," *IEICE (Japan) Trans. Fundamentals Electron. Commun. Computer Sci.*, special issue on VLSI Design and CAD Algorithms, accepted for publication.

[8]   F. Brglez, *IEEE Int. Symp. Circuits Syst.*, 1985.

[9]   Z. Chen, K. Roy, and T.-L. Chou, "Sensitivity of Power Dissipation to Uncertainties in Primary Input Specification," *IEEE Custom Integrated Circuits Conference*, pp. 487–490, 1997.

[10] Z. Chen, K. Roy, and T.-L. Chou, "Power Sensitivity—A New Method to Estimate Power Considering Uncertain Specifications of Primary Inputs," *ACM/IEEE International Conference on Computer-Aided Design*, pp. 40–44, 1997.

[11] R. Burch, F. Najm, P. Yang, and T. Trick, "A Monte Carlo Approach for Power Estimation," *IEEE Trans. VLSI Syst.*, pp. 63–71, Mar. 1993.

[12] S. Devadas, K. Keutzer, and J. White, "Estimation of Power Dissipation in CMOS Combinational Circuits," *IEEE Custom Integrated Circuits Conference*, pp. 19.7.1–19.7.6, 1990.

[13] R. Marculescu, D. Marculescu, and M. Pedram, "Efficient Power Estimation for Highly Correlated Input Streams," *ACM/IEEE Design Automation Conference*, pp. 628–634, 1995.

[14] R. Marculescu, D. Marculescu, and M. Pedram, "Switching Activity Analysis Considering Spatiotemporal Correlations," *IEEE International Conference on Computer-Aided Design*, pp. 294–299, 1994.

[15] A. Ghosh, S. Devadas, K. Keutzer, and J. White, "Estimation of Average Switching Activity in Combinational and Sequential Circuits," *ACM IEEE Design Automation Conference*, pp. 253–259, 1992.

[16] I. Miller and J. Freund, *Probability and Statistics for Engineers*, Prentice-Hall, Englewood Cliffs, NJ, 1985.

[17] A. Papoulis, *Probability, Random Variables, and Stochastic Processes*, 3rd ed., McGraw-Hill, New York.

[18] M. Xakellis and F. Najm, "Statistical Estimate of the Switching Activity in Digital Circuits," *ACM/IEEE Design Automation Conference*, pp. 728–733, 1994.

[19] S. G. Duvall, "A Practical Methodology for the Statistical Design of Complex Logic Products for Performance," *IEEE Trans. VLSI Syst.*, pp. 112–123, Mar. 1995.

[20] F. N. Najm and M. Y. Zhang, "Extreme Delay Sensitivity and the Worst-Case Switching Activity in VLSI Circuits," *ACM/IEEE Design Automation Conference*, pp. 623–627, 1995.

[21] J. Monteiro and S. Devadas, "A Methodology for Efficient Estimation of Switching Activity in Sequential Circuits," *ACM/IEEE Design Automation Conference*, pp. 12–17, 1994.

[22] T.-L. Chou and K. Roy, "Estimation of Sequential Circuit Activity Considering Spatial and Temporal Correlations," *IEEE International Conference on Computer Design*, Austin, TX, Oct. 1995.

[23] T.-L. Chou and K. Roy, "Accurate Estimation of Power Dissipation in CMOS Sequential Circuits," *IEEE ASIC Conference*, Austin, TX, Sept. 1995.

[24] T.-L. Chou and K. Roy, "Accurate Estimation of Signal Activity in CMOS Sequential Circuits Considering Spatio-Temporal Correlations," *IEEE/ACM International Conference on Computer-Aided Design*, Nov. 1995.

[25] C.-Y. Tsui, M. Pedram, and A. Despain, "Exact and Approximate Methods for Calculating Signal and Transition Probabilities in FSMs," *ACM/IEEE Design Automation Conference*, pp. 18–23, 1994.

[26] E. Eichelberger, "Hazard Detection in Combinational and Sequential Circuits," *IBM J. Res. Dev.*, pp. 90–99, Mar. 1965.

[27] M. Yeoli and S. Rinon, "Application of Ternary Algebra to the Study of Static Hazard," *J. of ACM*, pp. 84–97, Jan. 1964.

[28] J. Hayes, "Uncertainty, Energy, and Multiple-Valued Logic," *IEEE Trans. Computers*, 107–114, Feb. 1986.

[29] M. Abramovici, M. Breuer, and A. Friedman, *Digital Systems Testing and Testable Design*, IEEE Press, New York, 1994.

[30] M. Woods, "MOS VLSI Reliability and Yield Trends," *Proc. IEEE*, pp. 1715–1729, Dec. 1986.

[31] C. Hu, "Reliability Issues of MOS and Bipolar ICs," *International Conference on Computer Design*, pp. 438–442, 1989.

[32] P. Li and I. Hajj, "Computer-Aided Redesign of VLSI Circuits for Hot-Carrier Reliability," *International Conference on Computer Design*, 1993.

[33] K. Roy and S. Prasad, "Circuit Activity Based Logic Synthesis for Low Power Reliable Operations," *IEEE Trans. on VLSI Systems*, pp. 503–513, Dec. 1993.

[34] N. Weste and K. Eshraghian, *Principles of CMOS VLSI Design*, Addison-Wesley, Reading, MA, 1984.

[35] S. Ercolani, M. Favalli, M. Damiani, P. Olivio, and B. Ricco, "Estimation of Signal Probability in Combinational Logic Network," *1989 European Test Conference*, pp. 132–138.

[36] C. Shannon, "The Synthesis of Two-Terminal Switching Circuits," *Bell Sys. Tech. J.*, 1948.

[37] C. Lee, "Representing of Switching Circuits by Binary Decision Diagrams," *Bell Syst. Tech. J.*, pp. 985–999, July 1959.

[38] S. Akers, "Binary Decision Diagrams," *IEEE Trans. Computers*, pp. 509–516, June 1978.

[39] R. Bryant, "Graph-Based Algorithms for Boolean Function Manipulation," *IEEE Trans. Computer-Aided Design*, pp. 677–691, Aug. 1986.

[40] P. Landman and J. Rabaey, "Power Estimation for High Level Synthesis," *European Design Automation Conference*, pp. 361–366, 1993.

[41] C.-Y. Tsui, R. Marculescu, D. Marculescu, and M. Pedram, "Improving the Efficiency of Power Simulators by Input Vector Compaction," *ACM Design Automation Conference*, 1996.

[42] D. Marculescu, R. Marculescu, and K. Roy, "Sequence Compaction for Probabilistic Analysis of Finite State Machines," *ACM Design Automation Conference*, 1997.

[43] K. Chaudhary and M. Pedram, "A Near Optimal Algorithm for Technology Mapping Minimizing Area Under Delay Constraints," *Proc. 29th ACM / IEEE Design Automation Conference*, pp. 492–498, 1992.

[44] H. J. Touati, C. W. Moon, R. K. Brayton, and A. Wang, "Performance-Oriented Technology Mapping," *Proceedings of the 6th MIT Conference Advanced Research in VLSI*, E. J. Dally, Ed., pp. 79–97, 1990.

[45] T. Sakurai and A. R. Newton, "Delay Analysis of Series-Connected MOSFET Circuits," *IEEE J. Solid-State Circuits*, vol. 26, no. 2, pp. 122–131, 1991.

[46] F. Harary, "Combinatorial Problems in Graphical Enumeration," in *Applied Combinatorial Mathematics*, E. F. Beckenbach, Ed., Wiley, New York, 1984.

[47] F. N. Najm and I. Hajj, "The Complexity of Test Generation at Transistor Level," Report No. UILU-ENG-87-2280, Coordinated Science Lab, University of Illinois at Urbana Champaign, Dec. 1987.

[48] F. Najm, R. Burch, P. Yang, and I. Hajj, "Probabilistic Simulation for Reliability Analysis of CMOS VLSI Circuits," *IEEE Trans. Computer-Aided Design*, vol. 9, no. 4, pp. 439–450, 1990.

[49] M. Nemani and F. N. Najm, "Towards a High-Level Power Estimation Capability," *IEEE Trans. Computer-Aided Design*, vol. 15, no. 6, pp. 588–598, 1996.

[50] N. Pippenger, "Information Theory and Complexity of Boolean Functions," in *Mathematical Systems Theory*, vol. 10, Springer-Verlag, New York, pp. 126–167, 1977.

[51] H. Kriplani, F. N. Najm, and I. Hajj, "Pattern Independent Maximum Current Estimation in Power and Ground Buses of CMOS VLSI Circuits: Algorithms, Signal Correlations, and Their Resolution," *IEEE Trans. Computer-Aided Design*, vol. 14, no. 8, pp. 998–1012, 1995.

[52] H. Kriplani, F. Najm, and I. Hajj, "Maximum Current Estimation in CMOS Circuits," in *Proc. DAC*, 1992.

[53] S. Devadas, K. Keutzer, and J. White, "Estimation of Power Dissipation in CMOS Combinational Circuits Using Boolean Function Manipulation," *IEEE Trans. Computer-Aided Design IC's*, pp. 373–383, Mar. 1992.

[54] C.-Y. Wang and K. Roy, "Maximum Power Estimation for CMOS Circuits Using Deterministic and Statistical Techniques," *IEEE Trans. VLSI Systems*, accepted for publication.

[55] C.-Y. Wang, K. Roy, and T.-L. Chou, "Maximum Power Estimation for Sequential Circuits Using a Test Generation Based Technique," *IEEE Custom Integrated Circuits Conference*, pp. 229–232, May 1996.

[56] C.-Y. Wang, T.-L. Chou, and K. Roy, "Maximum Power Estimation for CMOS Circuits Under Arbitrary Delay Model," *IEEE International Symposium on Circuits and Systems*, May 1996.

[57] C.-Y. Wang, K. Roy, and T.-L. Chou, "Maximum Power Estimation for Sequential Circuits Using a Test Generation Based Technique," *IEEE Custom Integrated Circuits Conference*, pp. 229–232, May 1996.

[58] C.-Y. Wang and K. Roy, "COSMOS: A Continuous Optimization Approach for Maximum Power Estimation of CMOS Circuits," *ACM/IEEE International Conference on Computer-Aided Design*, 1997.

[59] S. Manne, A. Pardo, R. Bahar, G. Hachtel, F. Somenzi, E. Macii, and M. Poncino, "Computing the Maximum Power Cycles of a Sequential Circuit," in *Proc. of DAC*, 1995.

[60] M. Hsiao, E. Rudnick, and J. Patel, "Effects of Delay Models on Peak Power Estimation of VLSI Sequential Circuits," *Proc IEEE/ACM International Conference Computer-Aided Design*, 1997.

[61] M. Hsiao, E. Rudnick, and J. Patel, "K2: An Estimator for Peak Sustainable Power of VLSI Circuits," *International Symposium on Low Power Electronics and Design*, Monterrey, 1997.

[62] S. Manich and J. Figueras, "Maximizing the Weighted Switching Activity in Combinational CMOS Circuits Under Variable Delay Model," *Proc. Eur. Design Test Conf.*, pp. 597–602, 1997.

[63] D. Goldberg, *Genetic Algorithms in Search, Optimization, and Machine Learning*, Addison-Wesley, Reading, MA, 1989.

[64] R. B. Hitchcock, "Timing Verification and Timing Analysis Program," in *Proc. of DAC*, 1982.

[65] M. Abramovici, M. Breuer, and A. Friedman, *Digital Systems Testing and Testable Design*, Computer Science Press, 1990.

[66] W. Fornaciari, P. Gubian, D. Sciuto, and C. Silvano, "System-Level Power Evaluation Metrics," *IEEE International Conference on Innovative Systems in Silicon*, pp. 323–330, 1997.

[67] L. P. Yuan and S. M. Kang, "Sequential Procedure for Average Power Analysis of Sequential Circuits," *International Symposium on Low Power Electronics and Design*, pp. 231–234, 1997.

[68] J. M. Rabaey, "System-Level Power Estimation and Optimization—Challenges and Perspectives," *International Symposium on Low Power Electronics and Design*, pp. 158–160, 1997.

[69] Y. M. Jiang, K. T. Cheng, and A. Krstic, "Estimation of Maximum Power and Instantaneous Current Using a Genetic Algorithm," *IEEE Custom Integrated Circuits Conference*, pp. 135–138, 1997.

[70] W. Fornaciari, P. Gubian, D. Sciuto, and C. Silvano, "High-Level Power Estimation of VLSI Systems," *IEEE International Symposium for Circuits and Systems*, pp. 1804–1807, 1997.

[71] S. Pilli and S. S. Sapatnekar, "Power Estimation Considering Statistical IC Parametric Variations," *IEEE International Symposium on Circuits and Systems*, pp. 1524–1527, 1997.

[72] H. Choi and S. H. Hwang, "Reducing the Size of a BDD in the Combinational Circuit Power Estimation by Using the Dynamic Size Limit," *IEEE International Symposium on Circuits and Systems*, pp. 1520–1523, 1997.

[73] E. Macii, M. Pedram, and F. Somenzi, "High-Level Power Modeling, Estimation, and Optimization," *34th Design Automation Conference*, pp. 504–511, 1997.

[74] J. Zhu, P. Agrawal, and D. D. Gajski, "RT Level Power Analysis," *Asia and South Pacific Design Automation Conference*, pp. 517–522, 1997.

[75] J. E. Crenshaw and M. Sarrafzadeh, "Accurate High Level Datapath Power Estimation," *European Design and Test Conference*, pp. 590–595, 1997.

[76] H. Choi and S. H. Hwang, "Improving the Accuracy of Support-Set Finding Method for Power Estimation of Combinational Circuits," *European Design and Test Conference*, pp. 526–530, 1997.

[77] S. Gavrilov, A. Glebov, S. Rusakov, D. Blaauw, L. Jones, and G. Vijayan, "Fast Power Loss Calculation for Digital Static CMOS Circuits," *European Design and Test Conference*, 411–415, 1997.

[78] T. Uchino, F. Minami, M. Murakata, and T. Mitsuhashi, "Switching Activity Analysis for Sequential Circuits Using Boolean Approximation Method," *IEEE Symposium on Low Power Electronics*, pp. 79–84, 1996.

[79] S. Y. Huang, K. T. Cheng, K.-C. Chen, and M. Lee, "Novel Methodology for Transistor-Level Power Estimation," *IEEE Symposium on Low Power Electronics*, pp. 67–72, 1996.

[80] P. Landman, "High-Level Power Estimation," *IEEE Symposium on Low Power Electronics*, pp. 29–35, 1996.

[81] W. C. Tsai, C. B. Shung, and D. C. Wang, "Accurate Logic-Level Power Simulation Using Glitch Filtering and Estimation," *IEEE Asia-Pacific Conference on Circuits and Systems*, pp. 314–317, 1996.

[82] A. Raghunathan, S. Dey, and N. K. Jha, "Register-Transfer Level Estimation Techniques for Switching Activity and Power Consumption," *IEEE/ACM International Conference on Computer-Aided Design*, pp. 158–165, 1996.

[83] D. I. Cheng, K.-T. Cheng, D. C. Wang, and M. Marek-Sadowska, "New Hybrid Methodology for Power Estimation," *Design Automation Conference*, pp. 439–444, 1996.

[84] A. Bogliolo, L. Benini, and B. Ricco, "Power Estimation of Cell-Based CMOS Circuits," *Design Automation Conference*, pp. 433–438, 1996.

[85] G. I. Stamoulis, "Monte-Carlo Approach for the Accurate and Efficient Estimation of Average Transition Probabilities in Sequential Logic Circuits," *Custom Integrated Circuits Conference*, pp. 221–224, 1996.

[86] C. P. Ravikumar, M. R. Prasad, and L. S. Hora, "Estimation of Power from Module-Level Netlists," *IEEE International Conference on VLSI Design*, pp. 324–325, 1996.

[87] P. H. Schneider, U. Schlichtmann, and B. Wurth, "Fast Power Estimation of Large Circuits," *IEEE Design Test Computers*, vol. 13, no. 1, pp. 70–78, 1996.

[88] T. Uchino, F. Minami, T. Mitsuhashi, and N. Goto, "Switching Activity Analysis Using Boolean Approximation Method," *IEEE/ACM International Conference on Computer-Aided Design*, pp. 20–25, 1995.

[89] F. Najm, S. Goel, and I. Hajj, "Power Estimation in Sequential Circuits," *ACM/IEEE Design Automation Conference*, pp. 635–640, 1995.

[90] C. S. Ding, Q. Wu, C.-T. Hsieh, and M. Pedram, "Statistical Estimation of the Cumulative Distribution Function for Power Dissipation in VLSI Circuits," *34th ACM / IEEE Design Automation Conference*, 1997.

[91] L. P. Yuan, C. C. Teng, and S. M. Kang, "Statistical Estimation of Average Power Dissipation in Sequential Circuits," *34th ACM / IEEE Design Automation Conference*, 1997.

[92] S. Ramprasad, N. R. Shanbhag, and I. N. Hajj, "Analytical Estimation of Transition Activity from Word-Level Signal Statistics," *34th ACM / IEEE Design Automation Conference*, 1997.

[93] H. Choi and S. H. Hwang, "Time-Stamped Transition Density for the Estimation of Delay Dependent Switching Activities," *1997 International Conference Computer Design*, pp. 68–73.

[94] J. N. Kozhaya and F. N. Najm, "Accurate Power Estimation for Large Sequential Circuits," *IEEE International Conference on Computer-Aided Design*, 1997.

[95] Z. Chen and K. Roy, "A Novel Power Macromodeling Technique Based on Power Sensitivity," *35th ACM / IEEE Design Automation Conference*, 1998.

[96] J. Y. Lin, W.-Z. Shen, and J.-Y. Jou, "A Power Modeling and Characterization Method for Macrocells Using Structure Information," *IEEE International Conference on Computer-Aided Design*, 1997.

[97] L. Benini and G. D. Micheli, "Fast Power Estimation for Deterministic Input Streams," *IEEE International Conference on Computer-Aided Design*, 1997.

[98] S. Gupta and F. Najm, "Power Macromodeling for High Level Power Estimation," *34th ACM / IEEE Design Automation Conference*, 1997.

[99] P. E. Landman and J. M. Rabaey, "Architectural Power Analysis: The Dual Bit Type Method," *IEEE Trans. VLSI*, pp. 173–187, 1995.

[100] D. Marculescu, R. Marculescu, and M. Pedram, "Information Theoretic Measures of Energy Consumption at Register Transfer Level," *ACM/IEEE International Symposium on Low Power Design*, pp. 87–92, 1995.

[101] H. Mehta, R. M. Owens, and M. J. Irwin, "Energy Characterization Based on Clustering," *33rd ACM / IEEE Design Automation Conference*, pp. 702–707, 1996.

[102] A. Raghunathan, S. Dey, and N. K. Jha, "Register-Transfer Level Estimation Techniques for Switching Activity and Power Consumption," *IEEE International Conference on Computer-Aided Design*, pp. 158–165, 1996.

[103] Q. Qiu, Q. Wu, and M. Pedram, "Cycle-Accurate Macro-Models for RT-Level Power Analysis," *Proc. 1997 Int. Symp. Low Power Electron. Design*, pp. 125–130, Aug. 1997.

[104] S. R. Powell and P. M. Chau, "Estimating Power Dissipation of VLSI Signal Processing Chips: The PFA Technique," *VLSI Process.*, vol. 7, pp. 250–259, 1990.

[105] Q. Wu, C. Ding, C. Hsieh, and M. Pedram, "Statistical Design of Macromodels for RT-Level Power Estimation," *Proc. Asia South Pacific Design Automation Conf.*, pp. 523–528, Jan. 1997.

CHAPTER 4

SYNTHESIS FOR LOW POWER

In order to meet functionality, performance, cost-efficiency, and other requirements, automatic synthesis tools for IC design have become indispensable. The designs are described at the Register Transfer or higher level. A
technology-independent logic/architectural level realization is generated
next. The logic level realization is then mapped to a specific technology
library. Finally the technology-mapped circuits is optimized to ensure that all
requirements have been met. The process has multiple steps as the problem
of realizing the specified functionality with gates in a given library to meet all
the requirements is too complex to be solved in one step.

Tools have been developed to carry out each of these steps automatically.
In the beginning, the synthesis tools attempted to reduce area alone and did
not consider delay. Next came improved algorithms that, in addition to
reducing the area, ensured that the maximum delay through the circuit did
not increase or at least remained within the specified maximum bound. With
the exponential growth in the number of gates that can be accommodated on
a chip continuing unabated, the area has become less of a concern and is
increasingly being substituted for by performance and power dissipation.

As we will see later, a large fraction of the total research effort to reduce
power dissipation has been justifiably devoted to lowering the supply voltage
in conjunction with techniques to compensate for accompanying loss of
performance. However, independent of and in addition to the supply voltage
scaling, it is possible to make logic and circuit level transformations to
achieve considerable improvement in power dissipation. These transformations try to minimize dynamic power dissipation based on the switching
activity of the signals at circuit nodes. Switching activity at circuit nodes is

related to input signal probabilities and activities. The input signal probabilities and activities can be obtained by system level simulation of the function with real-life inputs, as described in Chapter 3. Hence, circuits having the same functionality but operating in different environments requiring different input signal probabilities and activities are synthesized differently for improving power dissipation. In this chapter we focus on reduction of dynamic power dissipation. Techniques to reduce leakage power during both stand-by and active modes of operation are covered later in Chapter 5.

The recent trend has been to consider power dissipation at all phases of the design cycle. The design space can be very efficiently explored starting at any behavioral and/or algorithmic level. Architectural transforms and trade-offs can be conveniently applied at that level to optimize power dissipation. However, efficient means of accurately estimating power dissipation at the behavioral level are required so that meaningful transforms can be applied. Various means of estimating the circuit activity and, hence, the average power dissipation have been considered in Chapter 3. In the following sections, synthesis and optimization procedures for low power dissipation at the algorithm, architecture, logic, and circuit level are considered.

4.1 BEHAVIORAL LEVEL TRANSFORMS

Because of the many degrees of freedom available, the behavioral level has the potential of producing a large improvement in power dissipation. At this level, since operations have not been assigned, the execution times and hardware allocation is not yet performed, a design point with given power dissipation, delay, and area can be found if one exists in the design space at all. Hence, there is a great need to systematically explore the design space. Traditionally, behavioral synthesis has targeted optimization of the amount of hardware resources required and optimization of the average number of clock cycles per task required to perform a given set of tasks. In the past few years, a handful of studies have been carried out to examine the efficacy of behavioral level techniques to reduce power dissipation. All reported techniques consider DSP circuits and fall in one of the three classes discussed in the next three sections.

4.1.1 Algorithm Level Transforms for Low Power

It has been observed that large improvements in power dissipation are possible at higher levels of design abstraction while pure combinational logic synthesis can only produce a moderate level of improvement. This is mainly due to the fact that the flexibility of making changes to random logic is rather limited. However, technology combined with innovative circuit design techniques can also produce a large improvement in power dissipation. Figure 4.1 shows a conceptual level diagram of the possible improvements in power

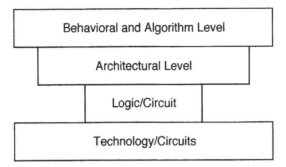

Figure 4.1 Improvements at various levels of design abstraction.

dissipation at various levels of abstraction. However, it should be clear that the improvements in power should be considered under performance constraint.

Let us consider two algorithm level techniques for improvement in power dissipation targeted for digital filters; both techniques try to reduce computation to achieve low power. The first technique uses differential coefficient representation to reduce the dynamic range of computation while the other technique optimizes the number of 1's in coefficients representation to reduce the number of additions (or switching activity). Other techniques try to use multiplier-less implementations for low-power and high-performance [7, 8]. Since digital signal processing techniques are very well represented mathematically, algorithm level techniques can be easily applied.

4.1.1.1 Differential Coefficients for Finite Impulse Response Filters

One of the most basic operations performed in DSP applications is the finite impulse response (FIR) computation. As is well known, the output of a linear time-invariant (LTI) system can be obtained by convolving in the time domain the input to the system and the transfer function of the system. For discrete-time LTI−FIR systems, this can be expressed mathematically as

$$Y_j = \sum_{n=0}^{N-1} C_n \cdot X_{j-n} \tag{4.1}$$

The notation A_r is used to denote the rth term of the sequence A. The parameters X and Y are the discrete-time input and output signals, respectively, represented as sequences. The sequence C represents the transfer function of the system. For FIR filters, C also corresponds to the set of filter coefficients and the length of the sequence N is called the number of taps or the length of the filter.

The *differential coefficients method* (DCM) [15, 16] is an algorithm level technique for realization of low-power FIR filters with a large number of taps

(N of the order of hundreds). The DCM relies on reducing computations to reduce power. The computation of the convolution, in the canonical form for the FIR filter output as given by Eq. (4.1), by using the multiply-and-accumulate sequence as defined earlier (computing the product of each coefficient with the appropriate input data and accumulating the products), will be termed *direct-form computation*. The algorithms for the DCM use various orders of differences between the coefficients (the various orders of differences will be precisely defined later on) in conjunction with stored precomputed results rather than the coefficients themselves to compute the canonical form convolution. These algorithms result in less computations per convolution as compared to direct form computation. However, they require more storage and storage accesses and hence more energy for storage operations. Net energy savings using the DCM is dependent on various parameters, like the order of differences used, energy dissipated in a storage access, and the word widths used for the digitized input data and coefficients.

The DCM can also lead to a reduction in the time needed for computing each convolution and thus one may obtain an added advantage of higher speed of computation. Analogous to the savings in energy, the speed enhancement obtained is dependent on the order of differences used and various other parameters.

4.1.1.2 Algorithm Using First-Order Differences

The *first-order differences algorithm* is based on the following observation. The notation $\{P_k\}_{t=j}$ will be used to denote the kth product term of the FIR output at the jth discrete-time instant. For convenience we will also use the shorter notation *time $t = j$* to refer to the jth discrete-time instant. Expanding Eq. (4.1) for any three consecutive samples of the output Y at times $t = j$, $j + 1$, and $j + 2$, we get

$$Y_j = C_0 X_j + C_1 X_{j-1} + C_2 X_{j-2} + \cdots + C_{N-1} X_{j-N+1} \qquad (4.2)$$

$$Y_{j+1} = C_0 X_{j+1} + C_1 X_j + C_2 X_{j-1} + \cdots + C_{N-1} X_{j-N+2} \qquad (4.3)$$

$$Y_{j+2} = C_0 X_{j+2} + C_1 X_{j+1} + C_2 X_j + \cdots + C_{N-1} X_{j-N+3} \qquad (4.4)$$

Notice that each input data multiplied by every coefficient exactly once in turn appears as a product term in the sum for successive outputs. Therefore, excepting the first, each product term in the sum for Y_{j+1} can be written as the following identity:

$$C_k X_{j-k+1} = C_{k-1} X_{j-k+1} + (C_k - C_{k-1}) X_{j-k+1} \quad \text{for } k = 1, \ldots, N-1 \qquad (4.5)$$

Since the $C_{k-1} X_{j-k+1}$ terms in identity (4.5) above have already been computed for the previous output Y_j, one needs to only compute the

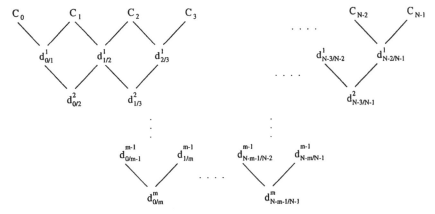

Figure 4.2 Different orders of differences.

$(C_k - C_{k-1})X_{j-k+1}$ terms and add them to the already computed $C_{k-1}X_{j-k+1}$ terms. The first product term in the sum for Y_{j+1}, which is C_0X_{j+1}, has to be computed without recourse to this scheme. Hence, all the N product terms are now available for generating Y_{j+1}.

Summarizing the above, one can say that, excepting C_0, each and every coefficient can be expressed as the sum of the preceding coefficient and the difference between it and the preceding coefficient. Therefore, each coefficient can now be expressed as the recurrence relation

$$C_k = C_{k-1} + \delta^1_{k-1/k} \quad \text{for } k = 1, \ldots, N-1 \qquad (4.6)$$

where $\delta^1_{k-1/k}$ is termed the *first-order difference* between coefficients C_k and C_{k-1}, as shown in Figure 4.2. Equation (4.6) is used to define the first-order difference $\delta^1_{k-1/k}$ between two consecutive coefficients for k in the specified range. Substituting in Eq. (4.1) the expression for C_k from Eq. (4.6) we get all the product terms $\{P_k\}_{t=j+1}$, excepting the first, for computing the FIR output Y_{j+1}. Therefore $\{P_k\}_{t=j+1}$ can be written as

$$\{P_k\}_{t=j+1} = C_k X_{j-k+1} = C_{k-1}X_{j-k+1} + \delta^1_{k-1/k}X_{j-k+1}$$

$$\text{for } k = 1, \ldots, N-1 \qquad (4.7)$$

The left-hand side of the above equation is called a *product term* in the convolution given by Eq. (4.1), whereas the terminology *partial product* will be used for the term $\delta^1_{k-1/k}X_{j-k+1}$ in the above equation. Both the partial product and the term $C_{k-1}X_{j-k+1}$ in the above equation will be called *intermediate results* used in the computation of the product term $C_k X_{j-k+1}$. The range of the subscripts of the first intermediate result in Eq. (4.7) can be changed to give an equivalent range as follows:

$$C_{k-1}X_{j-k+1} \text{ for } k = 1, \ldots, N-1 \equiv C_k X_{j-k} \text{ for } 0, \ldots, N-2 \quad (4.8)$$

They are therefore identical to the first $N - 1$ product terms of the FIR output at the immediately preceding instant of time, that is, at time $t = j$, which are

$$\{P_k\}_{t=j} + C_k X_{j-k} \quad \text{for } k = 0, \ldots, N - 2 \tag{4.9}$$

Therefore, if they are stored for reuse, one can compute all the product terms $\{P_k\}_{t=j+1}$, excepting the first, for the FIR output at time $t = j + 1$, by only computing the partial products $\delta^1_{k-1/k} X_{j-k+1}$ in Eq. (4.7) above and adding them to the appropriate stored product terms $\{P_k\}_{t=j}$ computed for the FIR output at time $t = j$. Thus we see that only one intermediate result storage variable and one extra addition per product term are needed. Furthermore, since the first coefficient C_0 along with the $N - 1$ first-order differences $\delta^1_{k-1/k}$ are used for computing the FIR output, instead of storing all the coefficients, one needs to store only C_0 and the $\delta^1_{k-1/k}$'s. The above algorithm is called the *first-order differences algorithm* for generating the FIR filter output. The additional storage accesses and additions incurred using this algorithm will be termed *overheads*.

If the differences $\delta^1_{k-1/k}$ are small compared to the coefficients C_k, then in the multiplication for computing a product term $\{P_k\}_{t=j+1}$ using this algorithm, one is trading a long multiplier (with X_{j-k} as the multiplicand) for a short one and overheads. The advantage of using this algorithm lies in the fact that if the computational cost (in terms of energy or delay) of the former is greater than the net computational cost of the latter then we make a net savings in computation as compared to using the direct form.

4.1.1.3 Algorithm Using Second-Order Differences

Higher orders of differences can also be used for expressing the coefficients as a function of differences. Let us define the second-order difference $\delta^2_{k-2/k}$ (Figure 4.2) between two consecutive first-order differences using the relation

$$\delta^2_{k-2/k} = \delta^1_{k-1/k} - \delta^1_{k-2/k-1} \quad \text{for } k = 2, \ldots, N - 1 \tag{4.10}$$

Using first- and second-order differences only, we can express the coefficients as follows:

$$C_0 = C_0$$
$$C_1 = C_0 + \delta^1_{0/1}$$
$$C_2 = C_1 + \delta^1_{0/1} + \delta^2_{0/2}$$
$$C_3 = C_2 + \delta^1_{0/1} + \delta^2_{0/2} + \delta^2_{1/3}$$
$$\vdots$$
$$C_{N-1} = C_{N-2} + \delta^1_{0/1} + \delta^2_{0/2} + \delta^2_{1/3} + \cdots + \delta^2_{N-3/N-1}$$

Hence, except C_0 and C_1, all other coefficients can be expressed compactly as follows:

$$C_k = C_{k-1} + \delta^1_{k-2/k-1} + \delta^2_{k-2/k} \quad \text{for } k = 2, \ldots, N-1 \quad (4.11)$$

which follows from substituting in Eq. (4.6) the expression for $\delta^1_{k-1/k}$ from Eq. (4.10). Multiplying both sides of Eq. (4.11) above by X_{j-k+1}, we obtain the product terms $\{P_k\}_{t=j+1}$ for computing Y_{j+1} as

$$C_k X_{j-k+1} = C_{k-1} X_{j-k+1} + \delta^1_{k-2/k-1} X_{j-k+1} + \delta^2_{k-2/k} X_{j-k+1}$$

$$\text{for } k = 2, \ldots, N-1 \quad (4.12)$$

Using terminology analogous to that used for the first-order differences algorithm, let us call the last two terms on the right-hand side of the above equation as partial products and all three terms on the right-hand side as intermediate results in the computation of the product term $\{P_k\}_{t=j+1}$. The computation of the FIR output Y_{j+1} by using the relationship in Eq. (4.12) will be termed the *second-order differences algorithm*.

One can take advantage of the above recurrence relation to compute any product term in the convolution incrementally using the minimally required storage for intermediate results as follows. We also show below that one needs just two extra storage variables and two extra additions per product term for computing the FIR output using the second-order differences algorithm.

Let $D[k]$ and $P[k]$ be the two storage variables used for intermediate results for computing the kth product term of the FIR output at time $t = j$. Since there are N product terms in the output, we will use two array variables for storage, both of array size N ($D[k]$ and $P[k]$, with $k = 0, \ldots, N-1$). Here, $D[k]$ and $P[k]$ will be used for storing the partial product $\delta^1_{k-2/k-1} X_{j-k+1}$ and the intermediate result $C_{k-1} X_{j-k+1}$, respectively, as computed using Eq. (4.12), both of which will be intermediate results for the FIR output at the next time step, that is, at time $t = j + 2$. Of course, in addition to $D[k]$ and $P[k]$, one has to store C_0, $\delta^1_{0/1}$, and the $N-2$ second-order differences $\delta^2_{k-2/k} k = 2, \ldots, N-1$, instead of the N coefficients, which would have had to be stored for direct-form computation.

Let us begin at time $t = j$, with the variables $D[0]$ and $P[0]$ both initialized to zero. The first product term $\{P_1\}_{t=j} = C_0 X_j$ of the FIR output at time $t = j$ has to be computed directly and is stored in $P[0]$ with the contents of $D[0]$ remaining unchanged at zero:

$$t = j: \quad D[0], P[0] \Leftarrow 0: \quad \text{Initialize}$$

$$P[0] \Leftarrow C_0 X_j$$

At the next time step, at time $t = j + 1$, $D[0]$ and $P[0]$ would be used for computing the second product term $\{P_2\}_{t=j+1}$ of the FIR output at time

$t = j + 1$, which is $C_1 X_j$, as follows:

$$t = j + 1: \qquad D[0] \Leftarrow D[0] + \delta^1_{0/1} X_j$$

$$P[0] \Leftarrow P[0] + D[0]$$

Thus at the end of this computation $D[0]$ contains $\delta^1_{0/1} X_j$ and $P[0]$ contains $C_1 X_j$. At the following time step, at time $t = j + 2$, $D[0]$ and $P[0]$ would be used for computing the third product term $\{P_3\}_{t=j+2}$ of the FIR output at time $t = j + 2$, which is $C_2 X_j$, as follows:

$$t = j + 2: \qquad D[0] \Leftarrow D[0] + \delta^2_{0/2} X_j$$

$$P[0] \Leftarrow P[0] + D[0]$$

Thus at the end of this computation $D[0]$ contains $\delta^1_{1/2} X_j$ and $P[0]$ contains $C_2 X_j$.

This process would be continued and $D[0]$ and $P[0]$ would accumulate results for N time steps, after which they would be reset to zero, and the process would start all over again. The remaining $N - 1$ pairs of D and P variables go through the same process, except that the initialization times of all the N pairs are unique. At any instant of time one and only one pair is initialized and at the next instant of time the next sequential (modulo N) pair is initialized.

Thus we see that using this technique only two additional variables per product term are required to store the intermediate results. Since we have N product terms, we need a total of $2N$ extra storage variables, as compared to direct-form computation. Two more additions per product term (as compared to the direct form), one each to update $D[k]$ and $P[k]$, are also needed.

4.1.1.4 Algorithm Using Generalized Mth-Order Differences

We can now generalize the algorithm to use up to mth-order differences, with the mth order difference defined as

$$\delta^m_{k-m/k} = \delta^{m-1}_{k-m+1/k} - \delta^{m-1}_{k-m/k-1} \quad \text{for } k = m, \ldots, N - 1; m = 2, \ldots, N - 1$$

$$(4.13)$$

We can obtain the recurrence relationship between the coefficients (Figure 4.2) using mth-order differences as

$$C_k = C_{k-1} + \sum_{i=1}^{m-1} \delta^i_{k-i-1/k-1} + \delta^m_{k-m/k}$$

$$\text{for } k = m, \ldots, N - 1; m = 2, \ldots, N - 1 \quad (4.14)$$

which follows by substituting in Eq. (4.6), the definition for the first-order difference in terms of the second-order difference as given by Eq. (4.10), then substituting for the second-order difference in terms of the third-order difference as defined by Eq. (4.13) above with $m = 3$, and so on, recursively, up to the mth-order difference.

We can again compute each product term incrementally as shown before for the second-order differences algorithm. However, we now need to store m intermediate results for each product term that will be stored using the same technique as for the second-order differences algorithm in the m array variables $D_1[k], D_2[k], D_3[k], \ldots, D_{m-1}[k], P[k]$. Each array will be of size N. Therefore we need a total of mN storage variables in addition to the storage requirements of the direct form. We also need m additions per product term to update the intermediate results storage variables. Therefore a total of mN more additions per convolution are needed as compared to the direct form.

4.1.1.5 Negative Differences

The differences between coefficients can be positive or negative. When the differences are negative, the differential coefficients method can still be used without any major modifications. As an example consider the first-order differences method for computing $\{P_k\}_{t=j+1}$. We are computing $\{P_k\}_{t=j+1} = \{P_{k-1}\}_{t=j} + \delta^1_{k-1/k} X_{j-k+1}$ for each product term in the sum for Y_{j+1}, where $\delta^1_{k-1/k}$ may be positive or negative. Irrespective of the sign of $\delta^1_{k-1/k}$, we can get the absolute value of the partial product by computing the product $|\delta^1_{k-1/k}||X_{j-k+1}|$. We can then add it to or subtract it from the term $\{P_{k-1}\}_{t=j}$ depending on the product of the signs of $\delta^1_{k-1/k}$ and X_{j-k+1} to obtain $\{P_k\}_{t=j+1}$. We have no control over the sign of the partial product anyway (irrespective of the sign of the difference $\delta^1_{k-1/k}$) since it depends on the sign of X_{j-k+1}. This technique can also be used for algorithms using greater order differences.

4.1.1.6 Sorted Recursive Differences

One limitation of DCM is that it can be applied only to systems where the envelope generated by the coefficient sequence (and various orders of differences) is a smoothly varying continuous function; thus it was beneficial largely for low-pass FIR filters. We present an improved version of DCM, called the *sorted recursive differences* (SRD) method, which uses recursive sorting of coefficients and various orders of differences to maximize the computational reduction. This recursive sorting is made possible by the transposed direct form of FIR output computation. Thus there are no restrictions on the coefficient sequence to which it is applicable (or the sequences of various orders of differences). The effective word length reduction using the DCM was not the same for each coefficient. Instead of pessimistically using the worst case reduction as mentioned earlier, one can use a simple statistical estimate for the effective reduction in the number of 1's in the coefficients [16].

4.1.1.7 Transposed Direct-Form Computation

In a transposed direct form (TDF) FIR computation, all N product terms involving a particular data are computed before any product term with the next sequential data is computed. The product terms are accumulated in N different registers since they belong to N sequential outputs. The effective throughput remains the same as direct-form computation. Signal flow graphs for direct-form and TDF computation are shown in Figures 4.3 and 4.4, respectively.

One advantage of using TDF computation lies in the fact that it does not matter in which order we compute the product terms involving a particular data so long as we accumulate them in the right registers. Therefore we can sort the coefficients in nondecreasing (or nonincreasing) order before taking first-order differences. It can be shown that this ordering minimizes the sum of the absolute value of first-order differences. Sorting can also be applied to differences. Thus we can generate the second-order differences from the

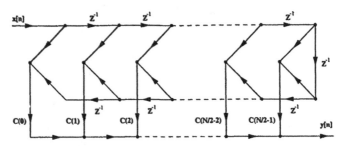

Figure 4.3 Signal flow graph for direct form realization of an even-length FIR system.

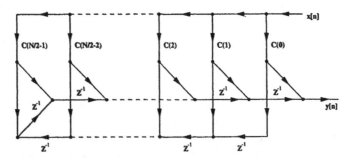

Figure 4.4 Signal flow graph for transposed direct form realization of the same FIR system as in Figure 4.3.

sorted set of first-order differences. This could be recursively done up to any order of differences. The various permutations of the sets of different orders of differences needed for a correct restoration could be hardwired in the control unit. This would enable it to accumulate partial products in appropriate locations so that the correct output is produced. Thus if the DCM is applied in a TDF realization with recursive sorting, we can eliminate the restrictions that were earlier imposed on the coefficients and various orders of differences for the DCM to be viable.

Whereas sorting guarantees that differences of all orders will be nonnegative, the coefficients can be positive or negative. When two consecutive coefficients in the sorted sequence are of opposite signs, the magnitude of the algebraic difference between them is larger than the magnitude of either one. To decrease the range of the coefficients (hence smaller differences), one can use absolute values to compute differences and then manipulate the sign. As an example, consider two consecutive coefficients C_r, C_{r+1} with values $+0.8$ and -0.9. The first-order differences algorithm would compute $C_{r+1}X_j$ as $C_rX_j + \delta^1_{r/r+1}X_j$. Let $\delta^1_{r/r+1}$ be generated as the difference between the absolute values of C_r and C_{r+1}, which in this case is $+0.1$. Then we can compute the partial product $+0.1X_j$, add it to C_rX_j, and complement the sign bit of the sum to obtain the correct value of $C_{r+1}X_j$. The sign-bit manipulation could be handled by the control unit, and if the differences are known in advance, the necessary sequence of sign-bit manipulations could be hardwired in the control unit. As another example, suppose the two terms had values -0.7 and $+0.6$. In this case too we could compute $+0.1X_j$, add it to C_rX_j, and complement the sign bit of the sum to obtain the correct value of $C_{r+1}X_j$. Thus we see that a sign–magnitude number representation is very suitable for SRD and differences can be computed between absolute values of terms, and it is assumed that necessary sign-bit changes are done by the control unit.

The results of simulating four different kinds of linear phase LTI–FIR systems of 51, 101, and 201 tap lengths are presented below. The simulator tracks the exact number of adds and shifts required for direct-form and SRD computations. The metrics are for computation of all $\lceil N/2 \rceil$ product terms with randomly generated data. The coefficients of these systems and the data were generated as floating point numbers with four different mantissa word widths of 8, 12, 16, and 24 bits. Table 4.1 shows the maximum percentage savings in the net number of adds using SRD over DF computation and the order m at which the maximum savings was obtained. Though higher order differences reduce the multiplication complexity, the overhead additions required by SRD increase with increasing orders of differences. Thus there is a point at which the overall number of adds is minimized. This is explicitly shown in Figure 4.6, which plots the net savings in adds as a function of the order of differences for a low-pass FIR filter. It can be seen that there is a maxima at a different order for each of the three tap lengths. It can also be seen that the maximum savings increases with tap length. Table 4.2 shows the

TABLE 4.1   Maximum Savings in Adds Using SRD

FIR System	N	Simulated max S_{ADD}, opt m			
		8 bits	12 bits	16 bits	24 bits
	51	37.9, 2	24.9, 3	23.8, 4	11.0, 1
Low pass	101	56.0, 2	44.3, 3	39.8, 4	30.7, 5
	201	63.5, 2	56.5, 4	52.5, 3	45.1, 5
	51	46.0, 2	31.1, 3	23.5, 4	11.5, 6
Band stop	101	50.4, 2	44.7, 3	36.9, 4	30.2, 4
	201	62.2, 2	55.5, 2	50.1, 3	43.9, 5
	51	43.5, 2	32.7, 3	23.9, 4	12.4, 3
Band pass	101	51.4, 2	43.8, 3	36.9, 3	30.3, 4
	201	62.9, 2	55.8, 3	50.2, 4	43.4, 5
	51	34.9, 2	21.5, 3	23.0, 4	12.4, 1
High pass	101	53.9, 2	43.6, 3	39.6, 3	31.9, 5
	201	62.8, 2	56.2, 4	52.6, 3	45.5, 6

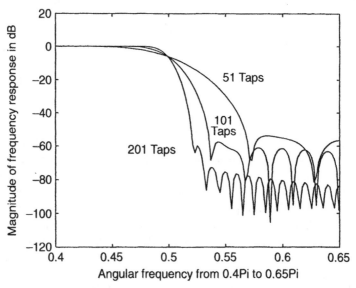

Figure 4.5   Detail of frequency response for the example low-pass filter in Tables 4.1 and 4.2 designed using a Hamming window.

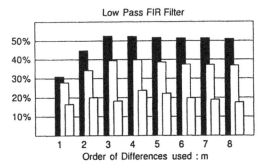

Figure 4.6 Savings in adds for the low-pass filter of Figure 4.5 as a function of the order of differences using 16-bit mantissas. The black bars are for $N = 201$, the grey bars are for $N = 101$, the white bars are for $N = 51$.

TABLE 4.2 Savings in Shifts Using SRD

FIR System	N	Simulated S_{SHIFT} at opt m			
		8 bits	12 bits	16 bits	24 bits
Low pass	51	87.10	82.27	82.81	19.35
	101	95.19	94.27	93.33	83.62
	201	98.64	99.29	94.35	93.51
Band stop	51	89.13	84.83	81.22	74.14
	101	94.19	93.40	92.71	74.16
	201	98.51	89.62	92.86	93.13
Band pass	51	88.17	85.42	81.63	46.71
	101	94.80	93.03	82.51	74.28
	201	98.52	97.99	97.48	92.78
High pass	51	86.05	82.09	84.32	20.96
	101	95.56	94.85	84.60	85.15
	201	98.61	99.46	94.82	96.44

percentage savings in the net number of shifts at the optimum order m at which the net adds are minimized. It is also possible to arrive at an optimum order based on the minimum weighted sum of net adds and shifts.

4.1.2 Power-Constrained Least-Squares Optimization for Adaptive and Nonadaptive Filters

The power constrained least-squares (PCLS) technique can be applied to both nonadaptive digital filters [46]. The basic idea in this technique is to reduce the number of 1's in the representation of the coefficients such that the number of additions in a multiplier can be reduced, thus achieving low-power dissipation. However, due to changes in the coefficients representation, the performance of the filter can change. Hence, changes in the filter coefficients can be allowed within a range such that the change in performance is within the tolerance limit.

Complexity reduction is one of the oldest topics in the classical signal processing literature [47, 48]. However, the goal pursued historically targeted making proposed filter implementation possible for practical state-of-the-art applications. Another motivation has been to reduce the cost of implementation. Higher speed translates to low power using a voltage scaling approach [45]. Further, lower complexity in terms of number of operations directly improves power. A classical measure of complexity has been to compute the total number of add and multiply operations. Further, power consumption is related to such a measure, however, only indirectly. A direct estimate considers number of switching events and a more complex algorithm may consume lower power if it causes less *switching activity* [45]. Hence, a low-power algorithm design approach is one that attempts to reduce the overall switching activity.

Recall that PCLS (Power Constrained Least Squares) coefficients attempt to reduce the switching activity by reducing the number of 1 bits in the coefficients. Since each 1 bit corresponds to a *shift-and-add* (SAA) operation, it consumes energy. Hence, by constraining 1 bits, we can constrain the dissipated power.

4.1.2.1 Constrained Least-Squares Technique for Nonadaptive Filters

In this section, we will present the constrained least-squares (CLS) approach used to compute the modified coefficient vector $\mathbf{k} = [k_0, k_1, \ldots, k_{M-1}]^T$. The original coefficient vector is represented by $\mathbf{c} = c_0, c_1, \ldots, c_{M-1}]^T$. We define the *code class* of a number, α, as the number of 1 bits in its binary representation. Then, the maximum code class allowable per coefficient will be represented by κ. The vector $\mathbf{k}$ obtained using the minimization technique replaces $\mathbf{c}$ when in the actual implementation.

To formulate a CLS problem, we first need to decompose the coefficient vectors into constituent bits. For simplicity we will ignore the sign bits of the coefficient values and concentrate only on the magnitude. Further, we will assume sign–magnitude representation of numbers. Now, our goal is to find a vector $\mathbf{k}$ that is closest to a given coefficient vector $\mathbf{c}$ in a least squares sense, constrained by a maximum number of allowable add operations. Clearly solution to this problem will not invert the sign bit in a component of $\mathbf{k}$ relative to the corresponding component in $\mathbf{c}$, as we can always find another solution yielding a smaller or equal least-squares estimator (LSE) without having the sign bit inverted. Hence, the sign bits of the individual components of the solution will be assumed identical to the sign bit of the corresponding component of the original coefficient vector.

In the sequel, the bit-decomposed ith coefficient c_i, shall be represented as $c_i = [c_{i,0}, c_{i,1}, \ldots, c_{i,N-2}]^T$, where $c_{i,j}$ represents the jth bit of the ith coefficient (note that $k_{i,N-1} = c_{i,N-1}$ is the sign bit). Then, the power constraint can be expressed as $\sum_{i=0}^{M-1}\sum_{j=0}^{N-2}k_{i,j} = M\kappa$. Let us also define a matrix $\mathscr{C} = [c_0, c_1, \ldots, c_{M-1}]^T$, such that the (i, j)th entry of $\mathscr{C}$ represents the jth bit of the ith coefficient. Let $\mathscr{K}$ be defined similarly for entries in **k**. To account for binary weighting of constituent bits of a coefficient, we define a binary weight vector $\mathbf{w} = [2^0, 2^1, \ldots, 2^{N-2}]^T$. Hence, $\mathbf{c} = \mathscr{C}\mathbf{w}$ and $\mathbf{k} = \mathscr{K}\mathbf{w}$. Note that this formulation is correct as sign bit in **k** is already preselected to the correct value. We will now present the CLS solution for two different LSE definitions, which will be referred to as *error definitions I* and *II*, respectively.

Error Definitions for the CLS Problem The first error definition computes LSE ε_I as a sum of squares of the differences between each corresponding bit of c_i and k_i, $0 \le i \le M - 1$, weighted appropriately according to the position of the bit. We define $\mathbf{c_I} = [c_0^T, c_1^T, \ldots, c_{M-1}^T]^T$, $\mathbf{k_I} = [k_0^T, k_1^T, \ldots, k_{M-1}^T]^T$, and $\mathbf{w_I} = [\mathbf{w}^T, \mathbf{w}^T, \ldots, \mathbf{w}^T]^T$ such that $\mathbf{w_I}$ is a column vector of size $(N - 1)M$. The subscript I has been placed to remind the reader that the coefficient vectors are formulated to define ε_I which is given as

$$\varepsilon_I = (\mathbf{c_I} - \mathbf{k_I})^T \mathbf{w_I} w_I^T (\mathbf{c_I} - \mathbf{k_I}) \tag{4.15}$$

Note that the error surface ε_I is convex as $\partial^2 \varepsilon_I / \partial k_{i,j}^2 > 0$ and $\partial^2 \varepsilon_I / \partial k_{i,j} \partial k_{l,m} = 0$ for $i \ne l$, $j \ne m$. It is important to note that the more closely we match $\mathbf{k_I}$ to $\mathbf{c_I}$ by assigning 1's to the positions where they differ, the smaller ε_I we will get. In fact, the optimal solution to power unconstrained least squares problem is simply $\mathbf{k_I} = \mathbf{c_I}$. However, the objective is to identify a strategy using the LSE that can be minimized for a given power constraint.

In the sequel, we will use the term *assignment* to mean that if $k_{i,j} = 0$ and $c_{i,j} = 1$, for some i, j (where $0 \le i \le M - 1$ and $0 \le j \le N - 2$), we assign a 1 to $k_{i,j}$. Hence, assignment means matching a bit in $\mathbf{k_I}$ to the corresponding bit in $\mathbf{c_I}$. Note that an assignment of 1 to a bit in $\mathbf{k_I}$ results in an add operation when an input data is multiplied by $\mathbf{k_I}$.

Error definition II computes LSE ε_{II}, using a more natural definition of error [47]:

$$\varepsilon_{II} = (\mathbf{c} - \mathbf{k})^T (\mathbf{c} - \mathbf{k}) = \mathbf{w}^T (\mathscr{C} - \mathscr{K})^T (\mathscr{C} - \mathscr{K})\mathbf{w} \tag{4.16}$$

Note that this definition is a sum of squares of differences, $c_i - k_i$, for $0 \le i \le M - 1$. We further note that $\partial^2 \varepsilon_{II} / \partial k_{i,j}^2 > 0$, $\partial^2 \varepsilon_{II} / \partial k_{i,j} \partial k_{i,m} > 0$, and $\partial^2 \varepsilon_{II} / \partial k_{i,j} \partial k_{l,m} = 0$ for $i \ne l$. To obtain the CLS solution sum, we must minimize differences between corresponding components of the two vectors in such a way that the given power constraint is not violated.

The CLS solutions that minimize ε_I and ε_{II}, respectively, are integer programming problems that require a branch-and-bound technique to arrive at an integer solution [46]. As these methods are computationally expensive, it provides a motivation to arrive at a solution using simplistic approaches using greedy search. Results on low-power filters suggest that the technique can reduce additions in computation by more than 40%.

4.1.3 Circuit Activity Driven Architectural Transformations

The digital filters, a very important class of DSP circuits, can be represented by state equations on which architectural transforms can be applied. Chatterjee and Roy [53] have considered heuristic transforms based on properties of associativity and commutativity of linear operations on linear time invariant digital circuits. Figure 4.7 shows the data flow graph of an infinite impulse response (IIR) filter. The computation tree for $s_1(t + 1)$, for example, is described by Eq. $s_1(t + 1) = c_1 s_1(t) + c_3 s_2(t) + ku(t)$. The filter can be implemented using either word-parallel or bit-serial arithmetic. In the word-parallel case, each signal (arc) of the data flow graph of Figure 4.7 represents W bits of data comprising a data word. The W bits $b_W, b_{W-1}, \ldots, b_1$ are fed in parallel to the respective adders and multipliers. The delays are designed to hold W bits in parallel. At time $t + 1$, let z out of W bits have different logic values than at time t. The signal activity is defined as the ratio of z over W and is given by $\beta(t) = z/W$. The variable $\beta(t)$ is a random variable for different values of t and represents a stochastic process. The authors define the average activity $\theta(t, t + N)$ of a signal over N consecutive time frames as

$$\theta(t, t + N) = \frac{1}{N} \sum_{i=t}^{i=t+N} \beta(i) \tag{4.17}$$

In case of bit-serial arithmetic, the bit values of the data word are transmitted serially over a single data line over consecutive time steps. Thus

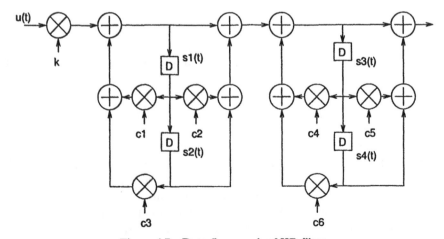

Figure 4.7 Data flow graph of IIR filter.

it is not inter-word differences in bit values, but intra-word bit differences that cause node activity. Experiments over large values of N show that the average activity $\theta(0, N)$ remains constant, showing that the stochastic process is *strict sense stationary*. Average power dissipation in this case is proportional to

$$\text{Cost} = \sum_{i=1}^{i=\gamma} \theta_i(0, N) * C_i \tag{4.18}$$

where θ_i is the average activity at node i (out of γ such nodes) and C_i is the capacitive load on the ith node.

The architectural transforms on the DSP filters are based on the following observations obtained through extensive simulation [53]. Consider a word-parallel computation tree with I inputs $i_1, i_2, \ldots, i_I$ and output $y = \sum_{j=1}^{j=I} \alpha_j b_j$, α_j being rational constants. For simplicity, let $I = 2^l - 1$, where I is the number of levels in a perfectly balanced adder tree, as shown in Figure 4.8. If input values of the tree are mutually independent, then:

- The minimum average value of θ_i over all nodes of a balanced adder tree with I inputs is obtained when (a) $\alpha_1 \geq \alpha_2 \geq \cdots \geq \alpha_I$ or (b) $\alpha_1 \leq \alpha_2 \leq \cdots \leq \alpha_I$.

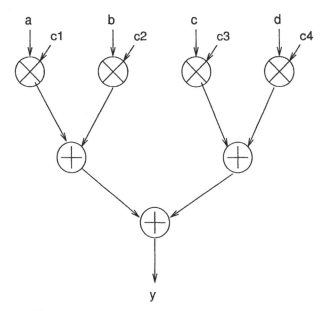

Figure 4.8 Balanced computation tree for $l = 2$.

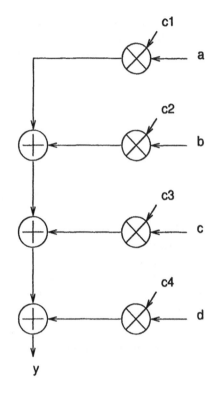

Figure 4.9 Linear array of adders.

For the case of a linear array of adders as shown in Figure 4.9:

- The minimum θ_i over all the nodes of a computation tree is achieved when $\alpha_1 \leq \alpha_2 \leq \cdots \alpha_I$.

The observations above are for mutually independent inputs. Due to reconvergent fanout, signals can become correlated at the internal nodes. However, the transformations seem to apply reasonably well even with such correlated signals. The synthesis algorithm is based on the above observations and on simulation. The given circuit is simulated at the functional level with random, mutually independent circuit input values. The activities at the inputs to all adders are noted. By applying the above two hypotheses, the adder trees are restructured and the average activities recomputed. The above analysis is carried out until there are no further improvements. The procedure forces additions with high activity to be moved closer to the root of a computation tree and vice versa. Note that no assumptions were made regarding the implementation details of the adders or the multipliers. Assuming that the capacitances at the internal nodes are all equal, improvement of up to 23% in power dissipation can be achieved.

4.1.4 Architecture-Driven Voltage Scaling

As noted in Chapter 2, a large improvement in power dissipation can be obtained if the supply voltage is scaled down, as V_{dd} appears as a square term in the expression for average power dissipation. However, one immediate side effect is the increase in circuit delay due to the voltage reduction. The increase in circuit delay can be possibly compensated for by scaling down the device sizes. However, in the submicrometer range the interconnect capacitances do not scale proportionately and can become dominant. Hence, it is worth looking at architectural transformations to compensate for the delay to achieve lower power dissipation by scaling down the supply voltage [56, 80]. One simple way to maintain throughout while reducing the supply voltage is to utilize parallel or pipelined architecture.

Let us consider an example from [56]. A simple data path with an adder and a comparator is shown in Figure 4.10. Let the power dissipation for the data path be given by $P = CV^2 f$, where C is the total effective capacitance being switched per clock cycle. Due to voltage scaling by a factor of 40%, assume that the data path unit works at half the speed. To maintain the throughput if one uses the parallel configuration shown in Figure 4.11, the effective capacitance is almost doubled. In fact, the effective capacitance is more than doubled due to the extra routing required. Assuming that the capacitance is increased by a factor of 2.15, the power dissipation is given by

$$P_{par} = (2.15C)(0.58V)^2(0.5f) \approx 0.36P \qquad (4.19)$$

This method of using parallelism to reduce power has the overhead of more than twice the area and is not suitable for area constrained designs.

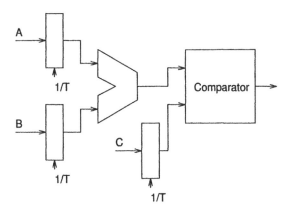

Figure 4.10 Data path operator.

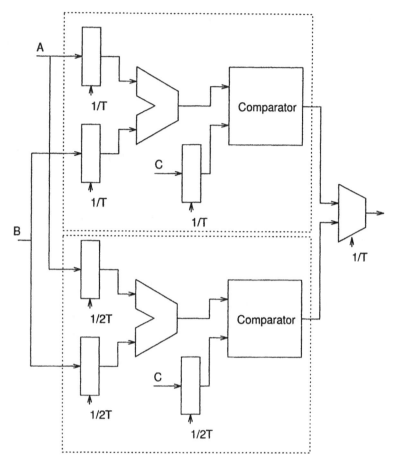

Figure 4.11   Parallel implementation.

Another approach described in [56] is pipelining which has the advantage of smaller area overhead. With the additional pipeline latch (Figure 4.12, the critical path becomes $\max[T_{\text{adder}}, T_{\text{comparator}}]$, allowing the adder and the comparator to operate at a slower speed. If one assumes the two delays to be equal, the supply voltage can again be reduced from 5 to 2.9 V, the voltage at which the delay doubles, with no loss in throughput. Due to the addition of the extra latch, if the effective capacitance increases by a factor of 1.15, then

$$P_{\text{pipe}} = (1.15C)(0.58V)^2(f) \approx 0.39P \qquad (4.20)$$

As an added bonus in pipelining, increasing the levels of pipelining has the effect of reducing logic depth and hence power contributed due to hazards. An obvious extension is to use a combination of pipelining and parallelism to obtain area and power constrained design.

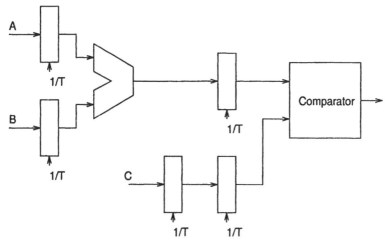

Figure 4.12 Pipelined implementation.

4.1.5 Power Optimization Using Operation Reduction

The easiest way to reduce the weighted capacitance being switched is to reduce the number of operations performed in the data flow graph. Reducing the operations count reduces the total capacitance associated with the system and, hence, can reduce the power dissipation [55]. However, reduction in the number of operations can have adverse effect on critical paths.

Let us consider the implementation of function $X^2 + AX + B$. A straightforward implementation is shown in Figure 4.13a. However, the implementation of Figure 4.13b has one less multiplier. The critical path lengths for both implementations are the same. Hence, the second design is preferable from both power and area points of view. However, operations reduction sometimes increases the critical path length, as shown in Figure 4.14. The figure computes the polynomial given by $X^3 + AX^2 + BX + C$. Figure 4.14a shows

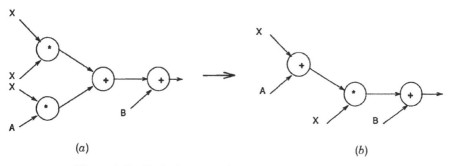

(a) $\qquad\qquad\qquad\qquad\qquad\qquad$ (b)

Figure 4.13 Reducing operations maintaining throughput.

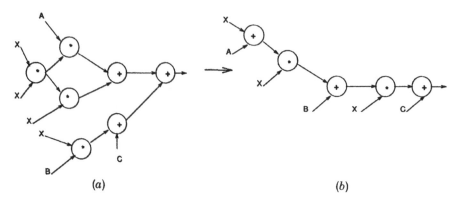

Figure 4.14 Reducing operations with less throughput.

an implementation with four multipliers and three adders while the implementation of Figure 4.14b has two multipliers and three adders. The latter implementation is suitable for area and power; however, the critical path is longer than the former one.

4.1.6 Power Optimization Using Operation Substitution

It is well known that certain operations require more computational energy than others. In DSP circuits, multiplication and addition are the two most important operations performed. Multiplication consumes more energy per computation than addition. Hence, during high-level synthesis, if multiplication can be replaced by addition, one cannot only save area but also achieve improvement in power dissipation [55]. However, such transformations are usually associated with an increase in the critical path of the design. Let us consider the example shown in Figure 4.15. Using the concept of distributivity and common subexpression utilization, the circuit of Figure 4.15a can be transformed into the circuit of Figure 4.15b. The critical path length of the transformed circuit is longer than the critical path of circuit (a); however, it is possible to substitute a multiplication operation by an addition operation, thereby reducing the energy of computation.

Other useful transformations can be noted. The multiplication with constants is widely used in discrete cosine transforms (DCT), filters, and so on. If the multiplication is replaced by shift-and-add operations, then considerable savings in power can be achieved. Experiments on an 11-tap FIR filter shows that the power consumed by the execution unit is lowered by one-eighth of the original power and the power consumed by the registers is also reduced. A small penalty is paid in the controller power. An overall improvement of 62% in power dissipation was seen for the FIR filter [55].

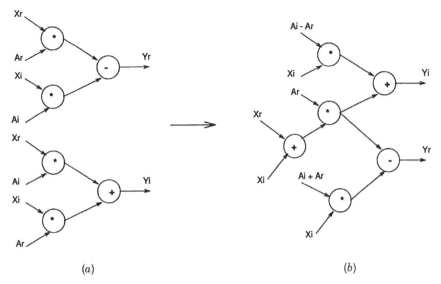

Figure 4.15 Substituting addition for multiplication.

4.1.7 Precomputation-Based Optimization for Low Power

Alidina et al. [82] presented an optimization technique that precomputes the output logic values of the circuit one clock cycle before they are required. The precomputed logic values are used in the following clock cycle to reduce the switching activity at the internal nodes of a circuit. The main problem with this scheme is that the precomputation logic can not only add to the total area but also slow down the clock period. However, for a certain class of circuits, the precomputation scheme can save significant power.

The precomputation architecture is shown in Figure 4.16. The block labeled Comb represents a combinational logic fed by register $R1$ and $R2$ and has one output, as shown in the figure. A set of inputs $x_1, \ldots, x_m$ are fed to register $R1$, while set of inputs $x_j, \ldots, x_n$ are fed to register $R2$. The inputs to register $R1$ are also fed to the logic blocks marked f_1 and f_2, which are the predictor functions given by the relation

$$f_1 = 1 \Rightarrow Z = 1 \qquad f_2 = 1 \Rightarrow Z = 0$$

It can be observed that f_1 and f_2 will never simultaneously evaluate to logic 1. Therefore, during clock cycle t, if either f_1 or f_2 evaluates to 1, then the load-enable line of register $R2$ is turned off and, hence, the outputs of register $R2$ during clock cycle $t + 1$ do not change. However, the outputs of register $R1$ change and, hence, Z evaluates to the correct logic value. It can be noted that only a subset of the input values to the combinational logic is changing. Therefore, the switching activity at the internal nodes of Comb are

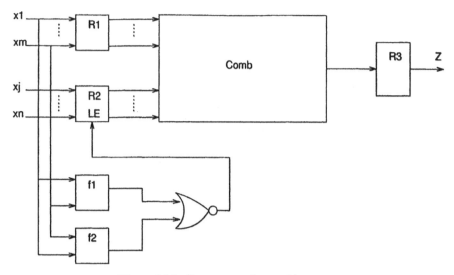

Figure 4.16   Precomputation architecture.

minimized. However, depending on the complexity of the logic functions f_1 and f_2, the switching activity at the internal nodes of f_1 or f_2 can be significant. Hence, for the precomputation scheme to work effectively, the set of inputs fed to register $R2$ should be large, while the complexity of the logic blocks f_1 and f_2 should be small. One would also like to have the signal probability of $f_1 + f_2$ that is, $P(f_1) + P(f_2) - P(f_1)P(f_2)$ to be large for the scheme to work effectively on an average. It has been noted that for some common functions considerable savings in power can be achieved by properly selecting the set of inputs for register $R1$ and $R2$. Let us consider such examples.

Figure 4.17 shows an n-bit comparator circuit using the precomputation configuration of Figure 4.16. The combinational logic implements function $A > B$. It can be easily observed that if the MSBs of A and B [i.e., $A(n - 1)$ and $B(n - 1)$] differ, one can determine whether A is greater than B or not. And if they do differ, it is possible to turn the latch enable of register $R2$ off at time $t + 1$. Any switching of signals on lines $A(n - 2)$ and $B(n - 2)$ through $A(0)$ and $B(0)$ will not be seen by the combinational logic, and hence, the internal nodes of the combinational logic will have very few switching. The parameter Z can still be evaluated accurately using $A(n - 1)$ and $B(n - 1)$. Let us now consider the representation of function f_1 and f_2. They are given by

$$f_1 = A(n - 1) \cdot \overline{B(n - 1)} \qquad f_2 = \overline{A(n - 1)} \cdot B(n - 1)$$

Here, f_1 and f_2 still obey the constraints given earlier. Notice that the combination of f_1, f_2, and the NOR gate of Figure 4.16 can be optimized to

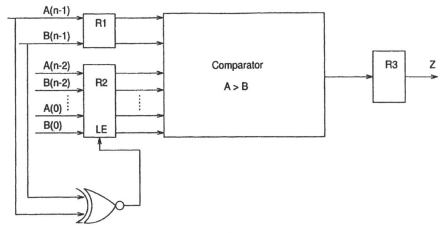

Figure 4.17 n-bit comparator.

one XNOR gate, shown in Figure 4.17. If one assumes a uniform signal probability of 0.5 for the primary inputs, then the signal at the output of the XNOR gate will be logic 1 50% of the time regardless of n. Hence, for large n, one can neglect the power dissipation due to the XNOR gate, and a 50% savings in power can be achieved. Note that $A(n - 2)$ and $B(n - 2)$ can possibly be fed to register $R1$ to try to achieve even larger savings in power. However, care should be taken not to offset the savings in power from the combinational logic by extra dissipation due to the precomputation logic f_1 and f_2.

It is possible to precompute output values that are not required in the succeeding clock cycle but are required two or more cycles later. Let us consider the adder–comparator circuit of Figure 4.18. The function Z is computed in two clock cycles. In this case, two-cycle precomputation can reduce the switching activity by close to 12.5%, as shown below. If functions f_1 and f_2 are given by

$$f_1 = A(n - 1) \cdot B(n - 1) \cdot \overline{C(n - 1)} \cdot \overline{D(n - 1)}$$

$$f_2 = \overline{A(n - 1)} \cdot \overline{B(n - 1)} \cdot C(n - 1) \cdot D(n - 1)$$

where f_1 and f_2 satisfy the constraints. It can be easily seen that signal probability $P(f_1 + f_2)$ is equal to $\frac{2}{16}$, or 0.125. Hence, the registers $A(n - 2:0)$, $B(n - 2:0)$, $C(n - 2:0)$, and $D(n - 2:0)$ can be disabled 12.5% of the time resulting in switching activity reduction.

The precomputation circuits described above work well only for certain logic operations. However, a precomputation scheme can be used for all logic circuits using Shannon's expansion, as shown in Figure 4.19. A logic function

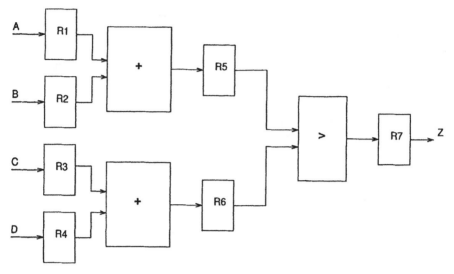

Figure 4.18 Adder–comparator circuit.

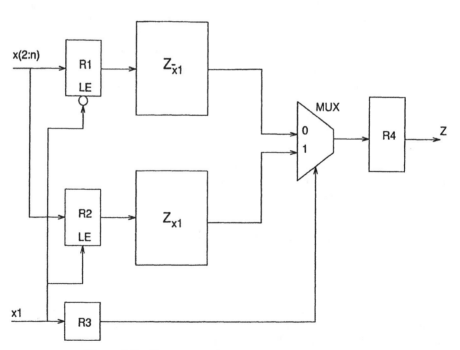

Figure 4.19 Precomputation using Shannon's expansion.

Z with input set $X = \{x_1, \ldots, x_n\}$ can be expressed as follows using Shannon's expansion:

$$Z = x_j Z_{x_j} + \overline{x_j} Z_{\overline{x_j}}$$

where Z_{x_j} and $Z_{\overline{x_j}}$ are the cofactors of Z with respect to x_j. For Figure 4.19, depending on the value of x_j, only one of the cofactors is evaluated while the other is disabled by disabling the load-enable line. Input x_j is also used to select the proper output using the multiplexer. Note that no precomputation logic is required in this case and the method is applicable to all circuits. One of the disadvantages of the architecture is that the number of registers for the inputs is duplicated. The selection of the splitting variable for Shannon's expansion is important for minimization of area and power. One of the heuristics for the selection of the splitting variable is to use the most binate variable present in function Z [84]. A variable is defined as the most binate in a function Z based on the number of times it appears in complemented and uncomplemented form. If such a variable is used for splitting, then the complexities of function Z_{x_j} and $Z_{\overline{x_j}}$ are minimized.

4.2 LOGIC LEVEL OPTIMIZATIONS FOR LOW POWER

When, to begin with, a design is described at the behavior level, first behavioral synthesis and then logic synthesis are performed. Unfortunately for the proponents of behavioral synthesis, it has not received very wide acceptance. Most designs are initially described at the Register Transfer or logic level. Hence logic synthesis tools are often the first tools in the synthesis flow. The data path sections of the designs usually have regular and preoptimized structures. As a result, logic optimization of data path sections is usually limited to binding the logic gates to the cells or gates in the technology library. The control sections, on the other hand, require technology-independent optimizations to achieve small area and delay.

For FSMs found in the control logic, state assignment has to be carried out. State assignment strongly influences the random logic associated with the FSMs. Hence, proper state assignment for low-power dissipation is required. For the random logic, which can be looked upon as a set of Boolean functions to map the inputs into outputs [60], logic synthesis involves determining the common set of subfunctions between the given Boolean equations so that area is optimized. However, it has been observed that minimum area is not always associated with minimum power dissipation for CMOS circuits [79, 58]. Hence, based on signal activity values defined in Chapter 3, the common subexpressions can be selected so that the signal activity of each node times the capacitance associated with it (which is proportional to the average power dissipation due to that node) is minimized over all nodes of the circuit. For custom VLSI designs, the complex CMOS gates obtained using the above procedure can be sized properly to obtain a design optimized for power, area, and delay. However, for library-based designs a technology mapping phase after logic synthesis is required to map the design into the logic elements of the library. For FSM synthesis, there is a state assignment phase that defines the combinational logic. In the following sections, we first consider FSM state assignment and logic synthesis for custom VLSI applications followed by technology mapping for low power.

4.2.1 FSM and Combinational Logic Synthesis

Finite-state machine assignment and combinational logic synthesis have been conventionally targeted at reducing area and critical path delay [85–87]. However, power dissipation during the logic synthesis process has begun to be considered only recently. The synthesis process consists of two parts: state assignment, which determines the combinational logic function, and multi-level optimization of the combinational logic, which tries to minimize area while trying to reduce the sum over all circuit nodes, the product of the circuit activity at a node, and the capacitance at the node.

The state assignment scheme considers the *likelihood of state transitions*—the probability of a state transition (say, from state S_1 to state S_2) when the primary input signal probabilities are given. The state assignment minimizes the total number of transitions occurring at the V inputs (or the present state inputs) of the state machine shown in Figure 4.20. It can be observed that scaled-down supply voltage technologies can still be applied after logic synthesis to further reduce power dissipation. The multilevel logic optimization process is iterative. During each iteration, the best subexpression from among all promising common subexpressions is selected. The

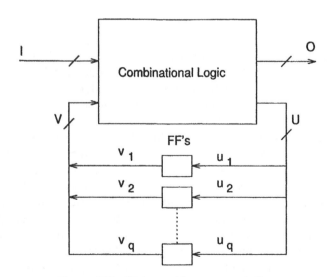

Figure 4.20 State machine representation.

objective function is based on both area and power savings. The selected subexpression is factored out of all affected expressions.

4.2.1.1 Representation of FSMs

Finite-state machines are represented by probabilistic state transition graphs (PSTGs). Probabilistic STGs are directed graphs consisting of a set of nodes S and a set of edges E. Each node $S_i \in S$ represents a state of the FSM. There is a directed edge $S_i - S_j$ between nodes S_i and S_j if there exists input set I, which when applied to the machine at state S_i produces a transition from state S_i to state S_j with output O. Hence, each edge is associated with a label L_{ij} that carries information on the values of primary inputs that caused the transition and the value of the primary outputs corresponding to the state transition. Each edge is also associated with a number p_{ij}, $0 < p_{ij} \le 1$, which denotes the conditional probability of a state transition from state S_i to S_j, given that the state machine is at state S_i and is directly related to the signal probabilities of the primary input nodes. The cardinality of set S, N_S, gives the total number of states in the machine. The number of primary inputs and primary outputs are denoted by N_I and N_O, respectively. Only completely specified FSMs are considered in this analysis. Hence, if there are m outgoing edges from node S_j, each associated with a probability of p_{jm}, then

$$\sum_m p_{jm} = 1$$

If a machine is incompletely specified, then self-loops are introduced at each state S_j corresponding to the don't care input to that state to make the machine completely specified. Figure 4.21 shows a state machine with five

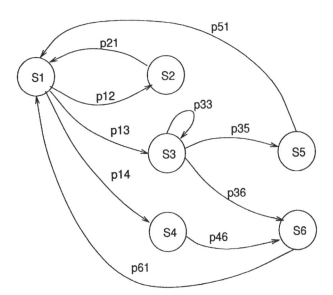

Figure 4.21 State transition diagram.

states where state S_i has three outgoing edges each associated with a transition probability such that $p_{12} + p_{13} + p_{14} = 1$. For node S_3, $p_{33} + p_{35} + p_{36} = 1$.

The state assignment problem involves assigning unique Boolean numbers to different states of an FSM so as to satisfy some given criteria. Given a state assignment, the *Hamming distance* between any two states S_i and S_j is given by

$$H(S_i, S_j) = \mathcal{T}(S_i \oplus S_j)$$

where $\oplus$ represents the exclusive-or operation. The function $\mathcal{T}(x)$ determines the number of 1 in a Boolean representation of x. In other words, $H(S_i, S_j)$ denotes the total number of bits in which states S_i and S_j differ.

Now, the expression for average power dissipation in a CMOS circuit can be written as

$$\text{Power}_{\text{avg}} = \tfrac{1}{2} V_{DD}^2 \sum C_i D(i)$$

summing over all circuit nodes i. The term D_i is the signal activity at node i and is explained in Chapter 3. During the multilevel logic synthesis process, the capacitive load C_i at each node of a circuit is approximated by the fanout factor at that node. If V_{dd} is assumed to be constant and the capacitive load at node i is proportional to the fanout at node i, then average power dissipation is proportional to Φ and is given by

$$\Phi = \sum_i \text{fanout}_i D(i)$$

where fanout_i is the number of fanouts at node i.

4.2.1.2 State Assignment for Finite-State Machines

This section addresses the problem of encoding of states synchronous sequential machines based on input signal probabilities and activities. The state encoding scheme uses the *likelihood of state transition* information. Let us consider the PSTG of Figure 4.21. State S_1 has three outgoing and three incoming edges. Let us assume that p_{12} is much larger than both p_{13} and p_{14}. From the diagram it is clear that $p_{51} = p_{61} = p_{21} = 1$, because there is one outgoing edge from each of the states S_5, S_6, and S_2, and the machine is assumed to be completely specified. Hence, the likelihood of state transition from state S_1 to S_2, or vice versa, is high. Therefore, states S_1 and S_2 should be assigned codes such that there are a minimum number of transitions between these two states, that is, $H(S_1, S_2)$ should be minimum. For example, if state S_1 is assigned a 4-bit code of 0000 and state S_2 is assigned a 4-bit code of 1000, then $H(S_1, S_2) = 1$. When there is a transition from S_1 to S_2, only one flip-flop of Figure 4.21 will undergo transition from a logic value of

0 to 1. However, if S_2 is assigned a code of 1111 instead of 1000, then all the four flip-flops will undergo a transition from 0 to 1.

Let p_{ij} and L_{ij} respectively denote the conditional state transition probability (given that the FSM is in state S_i, p_{ij} is the probability that the next state is S_j) and the label associated with an edge $S_i - S_j$ of a PSTG with N_I primary inputs each having a signal probability P_x, $x \le N_I$. Each primary input x in L_{ij}, which causes the state transition to occur, can assume a logic value $value(x)$, from the set $\{1, 0, - \}$, where $-$ represents the "don't care" condition. Here, P_x is the probability that $value(x) = 1$. Hence, $(1 - P_x)$ denotes the probability that $value(x) = 0$. The condition $value(x) = -$ obviously occurs with a probability of 1. Assuming that the inputs are mutually independent, the state transition probability p_{ij} is given by

$$p_{ij} = \prod_{x \in \text{inputs in } L_{ij}} W_x$$

where

$$W_x = \begin{cases} P_x & \text{if value}(x) = 1 \\ 1 - P_x & \text{if value}(x) = 0 \\ 1 & \text{if value}(x) = - \end{cases}$$

The signal probabilities and the activities of the primary inputs can be obtained by system level simulation of the circuit over a large number of clock periods and noting the signal values and transitions at the boundary pins.

The state assignment algorithm tries to minimize the number of transitions or the transition activity at the present state inputs to the state machine. The basic assumption is that if the transitions at the present state inputs are minimized, then the combinational portion of the state machine can be more efficiently synthesized for low-power dissipation. The higher the input transition activities at the input to a combinational circuit, the higher is the rate at which the internal nodes of the circuit will switch, dissipating more power.

If there are N_S states, then the minimum number of flip-flops required for coding is $\lceil \log_2 N_S \rceil$. It should be noted that if one-hot coding [86] with N_S flip-flops is used, the Hamming distance between any two states is always 2, and hence, an optimum assignment to minimize the number of switching at the present state inputs might not be obtainable. Besides, one-hot coding also increases the number of present state inputs to the combinational logic of Figure 4.20. The average number of switching at the present state inputs to a state machine can be minimized if the state assignment scheme is such that the following objective function is minimized:

$$\gamma = \sum_{\text{over all edges}} p_{ij} H(S_i, S_j)$$

The above function represents the summation of all Hamming distances between two adjacent states weighted by the state transition probability. The higher the state transition probability between two states, the lower should be the Hamming distance between those two states.

Due to the complex nature of the objective function γ, simulated annealing can be used to solve the problem. It starts with a random assignment of states with the prescribed number of bits. Two types of moves are allowed during annealing: interchanging the codes of two states or assigning an unassigned code to the state that is randomly picked for exchange. The move is accepted if the new assignment decreases γ. If the move increases the value of the objective function γ, the move is accepted with a probability of $e^{-|\delta(\gamma)|/\text{Temp}}$, where $|\delta(\gamma)|$ is the absolute value of the change in the objective function, and Temp denotes the annealing temperature. When Temp is high, moves that increase γ are accepted with a higher probability so that the solution does not get stuck at any local minima. As the annealing temperature decreases, the probability of accepting such a move decreases.

Figure 4.22 shows a state machine that produces an output 1 whenever a sequence of five 1's appears, else it outputs a 0. The machine was implemented using three D-type flip-flops using the two assignment schemes shown in the table. The input signal probability is assumed to be 0.5, and hence, each edge of the state transition graph has a state transition probability of 0.5. For coding 1, $\gamma = 10$, whereas for coding 2, $\gamma = 5.5$. Both machines

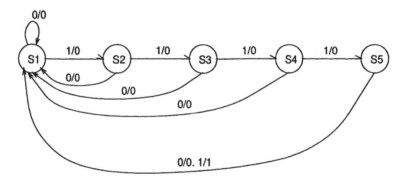

	Coding 1	Coding 2
S1	010	000
S2	101	100
S3	000	111
S4	111	010
S5	100	011

Figure 4.22   State assignment.

can be implemented using 34 transistors and 3 flip-flops. Circuit simulation using SPICE with random inputs shows that the time average power dissipation with the first encoding is about 15% more than the second one.

4.2.1.3 Signal Probabilities and Activities at Present State Inputs

State assignment determines the functionality of combinational logic. The combinational logic is represented by $\mathscr{F}(I,V)$, where I is the set of primary inputs and V represents the present state inputs (refer to Figure 4.20). The signal probabilities and transition activities are given for each input $i_k \in I$. Given the combinational logic $\mathscr{F}(I,V)$, the signal probabilities and activities for the V inputs have to be determined in order to synthesize multilevel combinational logic based on power dissipation measures. The V inputs are the same as U outputs but are delayed by a clock period. Hence, for a steady-state stationary process, the signal probabilities and the activities of V inputs are equal to the corresponding values for the U outputs. After state assignment, the state machine is simulated with different inputs to determine the signal probabilities and activities at the present state inputs. Otherwise, one can use the analysis shown in Chapter 3 to determine the signal distribution at the V inputs using probabilistic technique.

The simulation proceeds as follows. Primary input signals are randomly generated such that the signal probabilities and transition densities conform to the given distribution. The state machine is simulated to determine the percentage of time bit v_j of the state machine has a logical value of 1. Similarly, the number of transitions occurring at bit v_j of the machine is also determined. The number of transitions divided by the total number of simulations gives transition activity at input $D(v_j)$. The unit of $D(v_j)$ is transitions per clock period. The simulations can be carried out very fast because the state transition diagram is only simulated.

4.2.1.4 Multilevel Logic Representation

Multilevel logic can be described by a set $\mathscr{F}$ of completely specified Boolean functions. Each Boolean function $f \in \mathscr{F}$, maps one or more input and intermediate signals to an output or a new intermediate signal. A circuit is represented as a *Boolean network*. Each *node* has a Boolean *variable* and a Boolean *expression* associated with it. There is a directed edge to a node g from a node f if the expression associated with node g contains in it the variable associated with f in either true or complemented form. A circuit is also viewed as a *set of gates*. Each gate has one or more input pins and (generally) one output pin. Several pins are electrically tied together by a signal. Each signal connects to the output pin of exactly one gate, called the driver gate.

4.2.1.5 *Power Dissipation Driven Multilevel Logic Optimization*

A complicated two-level Boolean function is often implemented to have additional levels between inputs and outputs. Multilevel optimization of a set of such Boolean functions involves creating new intermediate signals and/or removing some of the existing ones to achieve reduction in area and to meet other design constraints such as performance. The global area optimization process of MIS [85] turned out to be very well suited for extensions to consider the impact on power dissipation of the circuit. The input to the optimization process is a set of Boolean functions. A procedure called *kernel* finds some or all *cube-free* multiple or single-cube divisors of each of the functions and retains those divisors that are common to two or more functions. The best few common divisors are factored out, and the affected functions are simplified by a second procedure called *substitution*. This process is repeated until no common divisors can be found. The "good"ness of a divisor is measured by the magnitude or area saving it brings about. Roy and Prasad [59] have extended this to also take power saving into account.

4.2.1.6 *Factoring and Power Saving*

In this section the effect of factoring out a common subexpression from several expressions is considered. Let $g = g(u_1, u_2, \ldots, u_K)$, $K \geq 1$, be a common subexpression of functions $f_1, f_2, \ldots, f_L$, $L \geq 2$. Let $v_1, v_2, \ldots, v_M$, $M \geq 0$, be the nodes internal to g. Each input u_k to g is either primary input or the output of a node in the circuit. Figure 4.23 is a pictorial representation of the circuit. When g is factored out of $f_1, f_2, \ldots, f_L$, the signal probabilities and activities at all the nodes of the Boolean network remain unchanged. However, the capacitances at the output of the driver gates of $u_1, u_2, \ldots, u_K$ change. Each such gate now drives $L - 1$ fewer gates than before. This results in a reduction in the power dissipation, which is given by

$$(L - 1)V_{dd}^2 C_0 \sum_{k=1}^{K} n_{u_k} D(u_k)$$

Here $D(x)$ is the activity at node x, n_{u_i} is the number of gates belonging to node g and driven by signal u_k (there are gates not belonging to g that are also driven by u_k), and C_0 is the load capacitance due to a fanout equal to one gate.

The driver gate of the newly created node g drives exactly as many gates as the driver gates of all its copies (which existed prior to factorization) taken together, so there is no change in this component of the total power dissipation. Since there is only one copy of g in place of L, there are $L - 1$ fewer copies of internal nodes $v_1, v_2, \ldots, v_M$ switching and dissipating power.

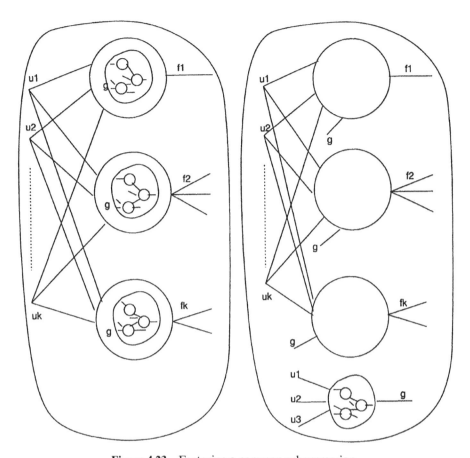

Figure 4.23   Factoring a common subexpression.

The saving in power is given by

$$(L - 1)V_{dd}^2 C_0 \sum_{m=1}^{M} n_{v_m} D(v_m)$$

Here n_{v_m} is the number of gates driven by signal v_m.

The total power saving on factoring out g is the sum of the above two components and is given by

$$\Delta W(g) = (L - 1)V_{dd}^2 C_0 \left(\sum_{k=1}^{K} n_{u_k} D(u_k) + \sum_{m=1}^{M} n_{v_m} D(v_m) \right) \quad (4.21)$$

The magnitude of the saving $\Delta W(g)$ depends on the transition activities at the boundary and internal nodes of g.

The area saving $\Delta A(g)$ due to a divisor g is found as in [85]. Let $T(g)$ be the number of literals in the factored form of g. Then,

$$\Delta A(g) = (L - 1)[T(g) - 1] - 1$$

The net saving is given by

$$S(g) = \alpha_W \cdot \frac{\nabla W(g)}{W_T} + \alpha_A \cdot \frac{\Delta A(g)}{A_T} \qquad (4.22)$$

Here W_T and A_T are the area and average power dissipation of the input Boolean network, and α_W and α_A are weight factors: $0 \le \alpha_A, \alpha_W \le 1$, and $\alpha_W + \alpha_A = 1.0$.

4.2.1.7 The Optimization Procedure

At the beginning of the optimization procedure, signal probabilities and activities for each internal and output node are computed. Each time a common divisor $g = g(u_1, u_2, \ldots, u_K)$ is factored out, the $P(u_k = 1)$ and $D(u_k)$, $1 \le k \le K$, are known but $P(v_m = 1)$ and $D(v_m)$, $1 \le m \le M$, are not. The latter are computed when $\Delta W(g)$ is being evaluated and are retained. Thus, once again, $P(s = 1)$ and $D(s)$ for each node s are known.

The parameter N_0 is used to control the number of kernel intersections (cube-free divisors common to two or more functions) that are substituted into all the functions before the set of kernel intersections is recomputed. Recomputing after a single substitution is wasteful, as only some of the functions have changed. On the other hand, with each substitution, some of the kernel intersections become invalid.

Algorithm: Power Dissipation Driven Multilevel Logic Optimization
   Inputs: Boolean network F, input signal probability $P(x_i = 1)$ and transition activity $D(x_i)$ for each primary input x_i, N_0.
   Output: Optimized Boolean network F', $P(s = 1)$ and $D(s)$ for each node in the optimized network.
   Step 0: Compute $P(s = 1)$ and $D(s)$ for each node s in F.
   Step 1: Repeat steps 2–4.
   Step 2: $G' = \bigcup_{f \in F} K(f)$, where $K(f)$ = set of all divisors of f. The set of kernels (cube-free divisors) is computed for each function. G' is the union of all the sets of kernels.
   Step 3: $G = \{g \mid g(\in G') \wedge (g \in K(f_i)) \wedge (g \in K(f_j)) \wedge (i \ne j)\}$. The set of kernel intersections, G, is the set of those kernels that apply to more than one function.
   Step 4: Repeat steps 5–7 N_0 times.
   Step 5: Find g, p_g, d_g such that

$$(g \in G) \wedge (\forall h \in G)[S(g) \ge S(h)] \wedge [p_g = P(g = 1)] \wedge [d_g = D(g)]$$

If $\Delta A(g) < 0$, terminate procedure. The kernel intersection g brings about the largest net saving. The signal probability and transition activity of the output signal of g are remembered. If the area component of net saving is negative, there are no more multiple-cube divisors common to two or more functions and so we stop.

Step 6: For all f such that $f \in F \wedge g \in K(f)$, substitute variable g in f in place of the subexpression $g(u_1, u_2, \ldots, u_K)$. Each function, which has the expression g as one of its kernels, has the new variable g substituted into it in place of the expression.

Step 7: $F = F \cup \{g\}, G = G - \{g\}$. Here, $P(g = 1) = p_g, D(g) = d_g$. The new function g is added to the set of functions F. The newly added node is assigned signal probability and activity values from step 5.

An Example

Let us consider a small circuit to illustrate the application of the above procedure.

Let $F = \{f_1, f_2\}$ be a two-output circuit given by

$$f_1 = ad + bcd, ae + f_2 = a + bc + dh + eh$$

The signal probabilities and the transition activities at the primary inputs are assumed to be

$$P(a) = P(b) = P(c) = P(d) = P(e) = P(h) = 0.5$$
$$D(a) = 0.1, D(b) = 0.6, D(c) = 3.6, D(d) = 21.6,$$
$$D(e) = 129.6, D(h) = 3.6$$

Since F is a small circuit, we recompute the set of kernel intersections after every substitution, that is, $N_0 = 1$.

Figure 4.24 shows the circuit F as an interconnection of logic gates. The area and power dissipation of the unoptimized circuit are

$$A_T(F) = 6 + 6 = 12 \qquad W_T(F) = 503.35 \text{ units}$$

The sets of kernels for f_1 and f_2 are computed:

$$K(f_1) = \{a + bc, d + e\} \qquad K(f_2) = \{a + bc, d + e\}$$

The union of the sets of kernels of all the functions, G', is computed as

$$G' = \{a + bc, d + e\}$$

The set of kernel intersections, G, that is, those kernels that apply to two or more functions, is computed as

$$G = \{a + bc, d + e\}$$

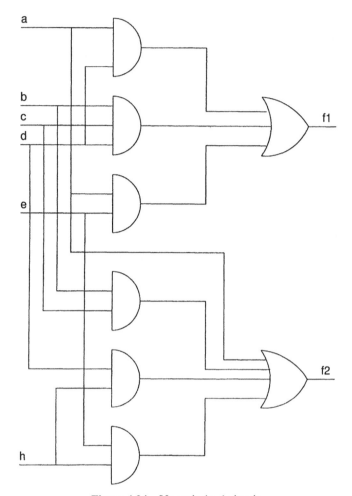

Figure 4.24   Unoptimized circuit.

Let us first consider $\alpha_A = 1.0$, $\alpha_W = 0$ (area optimization only). The net saving due to each of the kernel intersections, $g \in G$, is determined and the kernel intersection corresponding to the largest net saving is selected:

$$g = a + bc, \Delta A(g) = 1, \quad \Delta W(g) = 6.4, \quad P(g) = 0.625,$$
$$D(g) = 1.125, \quad S(g) = 0.083$$
$$g = d + e, \Delta A(g) = 0, \quad \Delta W(g) = 151.2, \quad P(g) = 0.75,$$
$$D(g) = 75.6, \quad S(g) = 0$$

Hence, $g = a + bc$ is selected. It is substituted into functions in F and added to F to give F^*:

$$F^* = \{f_1, f_2, f_3\}$$

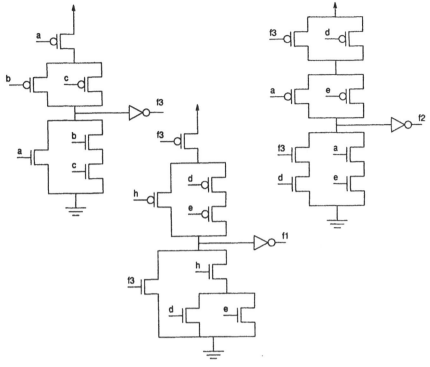

Figure 4.25   Complex gate implementation of the circuit optimized for area alone.

where

$$f_3 = a + bc$$
$$f_1 = f_3d + ae$$
$$f_2 = f_3 + dh + eh$$

The total area and power dissipation of circuit F^* are

$$A_T(F^*) = 3 + 4 + 4 = 11 \qquad W_T(F^*) = 476.5 \text{ units}$$

No more kernel intersections can be found and the procedure terminates. The complex logic gate implementation of the optimized circuit is shown in Figure 4.25. It requires 28 transistors.

Next we consider $\alpha_A = 0$, $\alpha_W = 1.0$, which causes optimization for low-power dissipation. Once again each of the kernel intersections $g \in G$ is evaluated and the best is selected:

$$g = a + bc, P(g) = 0.625, \quad D(g) = 1.125, \quad \Delta A(g) = 1,$$
$$\Delta W(g) = 6.4, \quad S(g) = 0.013$$
$$g = d + e, \Delta A(g) = 0, \quad \Delta W(g) = 151.2, \quad P(g) = 0.75,$$
$$D(g) = 75.6, \quad s(g) = 0.3$$

This time $g = d + e$ is selected. It is substituted into functions in F and added to F to give F^{**}:

$$F^{**} = \{f_1, f_2, f_3\}$$

where

$$f_3 = d + e$$
$$f_1 = f_3 a + bcd$$
$$f_2 = a + bc + f_3 h$$

The total area and power dissipation of circuit F^{**} are

$$A_T(F^{**}) = 2 + 5 + 5 = 12 \qquad W_T(F^{**}) = 423.12 \text{ units}$$

No more kernel intersections can be found. The complex logic gate implementation of the optimized circuit is shown in Figure 4.26. It requires 30 transistors.

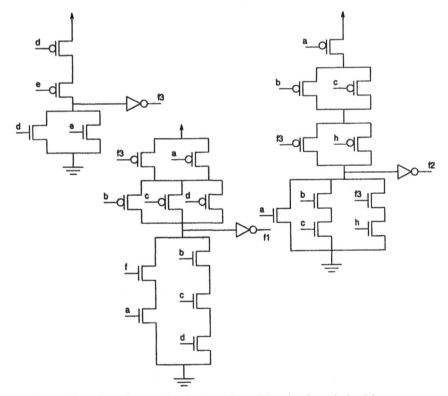

Figure 4.26 Complex gate implementation of the circuit optimized for power.

Hence using $\alpha_A = 0$, $\alpha_W = 1.0$, gives us a large area (12 literals vs. 11 literals or 30 transistors vs. 28 transistors) but a smaller power dissipation (423.12 vs. 476.12).

4.2.1.8 Results

The synthesis problem is broken up into two parts: the state assignment problem, where the objective function γ is minimized so as to reduce the activities at the present state inputs V, and the multilevel combinational logic synthesis process based on power dissipation measure and area. The state assignment and the subsequent logic synthesis process can get greatly affected if the primary input signal probabilities and activities are altered. Results on MCNC benchmarks show that more than 25% improvement in power dissipation with about 5% increase in area can be achieved compared to circuits optimized for area only.

4.2.2 Technology Mapping

The generic technology mapping problem can be stated as follows: Given a Boolean network potentially optimized by technology independent optimization procedures and a target library, find a binding of nodes in the network to gates in the library such that some given cost is optimized [61]. To facilitate mapping, the logic-synthesized circuit is decomposed into a set of base functions, such as two-input NAND/NOR gates. The technology mapping algorithms identify possible gate matching at nodes, such that the circuit graph is covered with the elements from the library with minimum cost. The tree covering problem can be used for technology mapping, as shown by Keutzer [62]. The circuit is decomposed into trees, and the tree covering problem is used on all the trees separately. There exists efficient tree covering algorithms using dynamic programming. Given a match g at a node in the subject graph, area costs of the best matches at the inputs to g are independent of each other. In case of the minimum cost for area, the following cost function is used:

$$\text{Area}(g) + \sum_{n_i \in \text{inputs } (g)} \text{MinArea}(n_i) \qquad (4.23)$$

For the above cost function, the dynamic programming algorithm would traverse the tree once from the leaves to the root, visiting each node exactly once.

Lin and de Man [61] have considered power dissipation in the cost function for technology mapping. The power dissipation model that was used is similar to the one used in logic synthesis. The formula to estimate power dissipation is given by

$$\text{Power} = \sum_{i=1}^{i=n} \frac{1}{2} V_{dd}^2 a_i C_i f \qquad (4.24)$$

where f is the clock frequency and C_i is the capacitive load associated with node i. The switching activity factor a_i is defined as $2P_i(1 - P_i)$ and defines the probability of a transition (logic 0 to 1 or 1 to 0). Here P_i denotes the probability that the signal at node i assumes a value logic 1 and has the same significance as described in Chapter 3. Glitches can be ignored if a zero-delay model is used for estimating power dissipation.

The key to the approach of lowering power dissipation during technology mapping is to estimate the partial power consumption of intermediate solutions during the dynamic programming procedure. Hence, the cost function for power dissipation during tree mapping can be formulated as

$$\text{power}(g, n) = \text{power}(g) + \sum_{n_i \in \text{inputs } (g)} \text{MinPower}(n_i) \qquad (4.25)$$

where $\text{power}(g, n)$ is the power cost of choosing gate g to match at node n. The term $\sum_{n_i \in \text{inputs}(g)} \text{MinPower}(n_i)$ is the sum of minimum power costs for the corresponding subtrees rooted at the input pins of g. The term $\text{power}(g)$ indicates the power contribution of gate g to implement node n and is given by

$$\text{power}(g) = Ka_i C_i \qquad (4.26)$$

The term K is a constant and is equal to $0.5fV_{dd}^2$, a_i is the switching probability at node i, and C_i is the equivalent capacitive load. Hence, the contribution of $\text{power}(g)$ cannot be determined until the fanouts of g have been mapped. However, the *best match* at node i can be determined from the term $\sum_{n_i \in \text{inputs}(g)} \text{MinPower}(n_i)$ because $Ka_i C_i$ is the same for all matches at the node. The tree mapping problem for area–power optimization under delay constraints can now be formulated as follows:

$$\text{Minimize} \quad wP + (1 - w)R \quad \text{such that } T \leq T_{\max}$$

where P and R represent the total power dissipation and area, respectively, and w is the weight factor that reflects the relative importance of power versus area and is a number between 0 and 1. The parameter T measures the worst-case delay through the circuit and must satisfy a given timing constraint $T_{\max}$. However, it should be noted that the delay cost of a match cannot be determined until matches have been obtained for its fanout nodes since the delay cost is determined by the fanout factor. Hence, a set of partial solutions is computed and stored with each node in the tree during the forward pass of the algorithm [61]. The partial solutions are represented by a set of triplets (t, p, r) representing timing, power, and area. The set of solutions is referred to as a *trade-off curve*. Separate *trade-off curves* for each possible fanout load is used by rounding the possible load values to a coarser grid. The point (t, p, r) on a curve represents the arrival time at the output of the node, the partial power cost, and the partial gate area cost for a match under load value C_i corresponding to the curve. A partial solution (t_1, p_1, r_1) is said to dominate another partial solution (t_2, p_2, r_2) if and only if one of the following holds:

- $t_1 \leq t_2$, $p_1 \leq p_2$ and $r_1 < r_2$
- $t_1 \leq t_2$, $p_1 < p_2$ and $r_1 \leq r_2$
- $t_1 < t_2$, $p_1 \leq p_2$ and $r_1 \leq r_2$

At each node, partial solutions that are dominated by other solutions under the same load values are removed from the solution set. To further prune the number of partial solutions, two additional removal rules are used:

- $t_1 \leq t_2, wp_1 + (1 - w)r_1 < wp_2 + (1 - w)r_2 + \varepsilon$
- $t_1 < t_2, wp_1 + (1 - w)r_1 \leq wp_2 + (1 - w)r_2 + \varepsilon$

where ε is a user-defined threshold.

The overall topological pass proceeds from the leaves to the root of the tree. Let us consider mapping at an intermediate node n. Let match(n) be the set of matches. For each gate $g \in$ match(n), it is assumed that the inputs to g have already been processed. For the sake of explaining, let us consider that there are two input nodes n_a and n_b for gate g. For each partial solution x for the *trade-off curve* in n_a, the partial solutions for the corresponding *trade-off curve* in n_b whose arrival times are less than or equal to the arrivals of x are determined. Based on that, the *trade-off curves* for the matching gates are computed.

After the forward pass, a reverse pass from the root of the tree is performed to construct the solution. Starting from the root, the minimum-cost solution according to the condition for area, power, and delay trade-off that satisfies the required time at the root is selected. The matching gate, g, and the corresponding set of inputs are identified. Similar calculations are done for all the fanin nodes of gate g. The reverse pass continues until all the leaf nodes have been reached.

Experimental results show that average power consumption can be reduced by 22% while maintaining similar delay performance. However, in some cases area increases can be as high as 39%, which demonstrates once again that area and power minimization objectives do not necessarily correlate.

4.3 CIRCUIT LEVEL

The circuit design style, circuit optimization, and transistor sizing can also be used to further minimize power dissipation. In this section, we will first consider circuit optimization techniques such as transistor input reordering and transistor sizing. The effect circuit design style on power dissipation is considered in Chapter 5.

4.3.1 Circuit Level Transforms

Most performance-driven circuit optimization algorithms are based on iterative improvement. Any such algorithm has four components:

1. a method for determining which gates in the circuit to examine next,
2. a set of transformations to apply to the gates being examined,
3. methods for computing the overall improvement due to transformations, and
4. a method for updating the circuit after each transformation.

In the case of an algorithm for power minimization that has to work with more than one transformation, the methods referred to in 1 and 4 should be independent of the individual transformations. Some of the transformations referred to in 2 should be able to influence the power consumption of the circuit. Methods referred to in 3 should be able to evaluate the effect of transformations on the power consumption of the circuit.

For our analysis we will only consider *synchronous* circuits composed of *gates* from a library. Each gate has one or more input pins and one output pin. Several pins are electrically tied together by a signal. Each signal connects to the output pin of exactly one gate, called the driver gate.

4.3.1.1 Gate Delay Model

The delay characteristics of all the gates in the library are known and are in the form of a pair $\{T_{i,j}^i(G), R_{i,j}(G)\}$ for every pair of an input terminal I_i and an output terminal O_j of every gate G. Here $T_{i,j}^i(G)$ is the fanout load independent delay and is called the *intrinsic delay*. The superscript i is for "intrinsic," $R_{i,j}(G)$ is the additional delay per unit fanout load. Each input terminal I_i of each gate G has a capacitance $C_i(G)$ associated with it. The total propagation delay through a gate G from a given input terminal I_i to a given output terminal O_j is given by

$$T_{i,j}^i(G) + R_{i,j}(G)C_j(G)$$

where $C_j(G)$ is the total fanout load capacitance at O_j. Separate delays are associated with rising and falling transitions.

4.3.1.2 Switching Event Probabilities

The signal probability p_y of the logic signal $y(t)$ at node y is the probability that $y(t) = 1$ at a given instant of time t. Then $1 - p_y$ is the probability that it is a 0. Let us assume, without any loss of generality, that all signal transitions occur at the leading edge of the clock. Let y be the node in the circuit that has been monitored for a large number N of clock cycles. Let the node be observed at a point in the clock cycle that is separated from the leading edge of the clock by a long enough interval to allow the logic level at the node to reach its stable value. Then it follows that we would have observed a 1 during $p_y N$ clock cycles and a 0 during $(1 - p_y)N$ clock cycles.

We normalize all activities by dividing them by the transition activity of the clock ($= 2f$, where f is the clock frequency). The normalized activity d_y, a real number between 0 and 1, of the node y is the probability that if we selected a clock cycle at random, there would be a transition at the node at the beginning of the clock cycle. Then $1 - d_y$ is the probability that there would be no transition. We denote the probability of a rising or a falling transition at node y by $p_y^{\updownarrow}$. Hence,

$$p_y^{\updownarrow} = d_y \tag{4.27}$$

Since a rising transition at any node (including an internal node) has the same probability of occurring as a falling transition (because a rising transition has to be followed by a falling transition and vice versa),

$$p_y^{\uparrow} = p_y^{\downarrow} = \tfrac{1}{2} p_y^{\updownarrow} \tag{4.28}$$

where $p_y^{\uparrow}$ is the probability of a rising transition at node y and $p_y^{\downarrow}$ is the probability of a falling transition at node y.

4.3.2 CMOS Gates

As submicrometer feature sizes become commonplace and the push to shrink feature sizes further continues, the changing electrical properties of the circuit structures affect their usefulness. An example is CMOS gates with series-connected transistors. Because the delay scales linearly with number of transistors in series, large NAND/NOR/complex gates are used infrequently. However, Sakurai et al. [65] report a result that should encourage more extensive use of NAND/NOR/complex gates. They show that the ratio of the delay of NAND/NOR to the delay of the inverter becomes smaller in the submicrometer region. This is because the V_{ds} and V_{gs} of each transistor in a series connection of several transistors are smaller than those of a transistor in an inverter. The smaller voltages make carrier velocity saturation less severe of a problem.

Another reason for infrequent use of complex gates in particular is the lack of broad availability of good technology mapping tools. In the very recent past a number of technology mapping algorithms have been reported in the literature that are able to explicitly minimize delay [66, 67].

If and when larger NAND/NOR/complex gates begin to be used, it will become even more important to order the series-connected transistors in them properly. For example, in the case of a four-input CMOS NAND gate in the Texas Instruments BICMOS gate-array library, for a load of 13 inverters (ignoring interconnect), the delay varies by 20% and power dissipation by 10% between the good and the bad transistor order. For this reason we have selected transistor reordering or gate input reordering as the sample transformation to use with our optimization algorithm. It is to be noted that this transformation has the nice feature that it has negligible impact on area.

4.3.2.1 *Power Consumption of CMOS Gates*

Consider the CMOS gate in Figure 4.27. Input pattern $x_1 = \uparrow, x_2 = 0, x_3 = 1$ causes a falling transition at the output and so does the input pattern $x_1 = 0, x_2 = \uparrow, x_3 = 1$. But in case of the first pattern, the capacitance being discharged equals $C_y + C_{z_1}$, which is more than C_y, the capacitance being discharged in the second pattern. This simple example points out two things:

1. The gate presents a variable capacitance to the power/ground rails. The magnitude of this capacitance depends on the logic values at the inputs to the gate.
2. Given two signals A and B that are connected to the two equivalent inputs x_1 and x_2 of the gate in Figure 4.27, if A has much longer signal activity than B then A should be connected to x_2 and B to x_1. This results in lower power consumption.

Capacitances at the internal nodes of a CMOS gate charge and discharge, resulting in additional power consumption. Some of these, but not all, happen at the same time instants as the charging and discharging of the capacitance at the gate output. For example, when the input to the gate in Figure 4.27 is $(x_1 = \downarrow, x_2 = 1, x_3 = 1)$, there may be a rising transition at z_1 though y remains at 0. On the other hand, when the input is $x_1 = \uparrow, x_2 = 0, x_3 = 1$, there is a falling transition at both z_1 and y. Therefore, internal

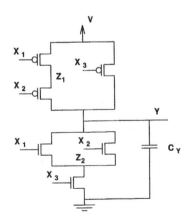

Figure 4.27   CMOS gate $y = \overline{(x_1 + x_2)x_3}$.

nodes of a gate need to be treated as circuit nodes distinct from (though related to) the output node of the gate. Hence, the average power dissipation is given by

$$W_{av} = V_{dd}^2 f \left(C_y d_y + \sum_i C_{z_i} d_{z_i} \right) \tag{4.29}$$

where the z_i's are the internal nodes of the gate and d_{z_i} is the normalized transition activity of the "signal" at node z_i. It should be noted that the internal node voltage savings is given by $(V_{dd} - V_{th})$ and is not reflected in Eq. (4.29).

In [68] the authors show that determining $p_{z_i}^{\updownarrow} (= d_{z_i})$ is NP-hard. The difficulty in determining $p_{z_i}^{\updownarrow}$ arises from the fact that for a rising transition to occur at an internal node at the beginning of a clock cycle, (i) the logic signal at the node must have had value 0 in the immediately previous clock cycle and (ii) the conduction state of at least one path from node to V_{dd} must have changed from off to on at the beginning of the clock cycle. The probability of a conducting path existing between any two nodes of a gate can be determined under the assumption that inputs to the gate are mutually independent. But determining the logic value at an internal node is difficult because an internal node may have been isolated from both V_{dd} and ground during the immediately previous clock cycle and possibly in all of the zero or more clock cycles immediately preceding that. In effect, it is required to keep track of the past transitions at the node potentially going back to $-\infty$—an impossible task. The problem can be resolved by assuming that whenever the state of an internal node cannot be determined, a transition occurs. This is done by using, in place of $p_{z_i}^{\uparrow}$, the probability that the number of conducting paths from z_i to V_{dd} changes from zero to a number larger than zero. The latter probability is an upper limit on $p_{z_i}^{\uparrow}$. This is well suited for the problem of determining the worst-case peak supply current. In this analysis it is required that to make a comparison between power dissipated in the same gate for two different input orderings, the errors need to be small and comparable in the two cases.

A more accurate upper limit on $p_{z_i}^{\uparrow}$ can be obtained from the observation that for a pair of complementary transitions to occur at an internal node, the number of conducting paths from the node to V_{dd} must change from zero to a number larger than zero followed by a similar change in that to ground. If the number of conducting paths from a node to V_{dd} changes from zero to a number larger than zero once every 40 clock cycles but that to ground only once every 100 clock cycles, then the node cannot have more than two transitions every 100 clock cycles. Therefore,

$$p_{z_i}^{\updownarrow} = \begin{cases} p_{z_i, V_{dd}}^{\uparrow} & \text{if } p_{z_i, V_{dd}}^{\uparrow} \leq p_{z_i, V_{ss}}^{\uparrow} \\ p_{z_i, V_{ss}}^{\uparrow} & \text{otherwise} \end{cases}$$

where $p_{z_i, z_j}^\uparrow$ is the probability that the number of conducting paths from z_i to z_j changes from zero to greater than zero.

The $p_{z_i, z_j}^\uparrow$'s can be determined by applying a reduction procedure to a graph obtained from the schematic of the gate. To each node of the gate there corresponds a vertex in the graph. Hence there is a vertex for ground, V_{dd}, y, and each of the z_i's. To each transistor with its drain and source connected to two nodes, there corresponds an edge between the corresponding vertices. Each edge e has two labels p_e and $p_e^\uparrow$, where p_e denotes the probability that edge e is on or conducting, and $p_e^\uparrow$ denotes the probability that edge e turns from off to on. Note p_e equals p_x if e corresponds to an nMOS transistor with logic signal x_i connected to its gate and equals $1 - p_{x_i}$ otherwise. Transition probability $p_e^\uparrow$ is simply $d_{x_i}/2$ in both cases. These graphs are a restricted class of graphs called series–parallel graphs [69].

The required path conduction probabilities can now be determined using an operation termed graph reduction [70]. Let e_p (e_s) denote the single edge equivalent to the parallel (series) combination of two edges e_1 and e_2; then

$$p_{e_p} = p_{e_1} + p_{e_2} - p_{e_1} p_{e_2} \tag{4.30}$$

$$p_{e_s} = p_{e_1} p_{e_2} \tag{4.31}$$

$$p_{e_p}^\uparrow = p_{e_1}^\uparrow (1 - p_{e_2}) + p_{e_2}^\uparrow (1 - p_{e_1}) - p_{e_1}^\uparrow p_{e_2}^\uparrow \tag{4.32}$$

$$p_{e_s}^\uparrow = p_{e_1}^\uparrow p_{e_2} + p_{e_2}^\uparrow p_{e_1} - p_{e_1}^\uparrow p_{e_2}^\uparrow \tag{4.33}$$

Let the path from node x_i to node x_j be reduced to edge e_r; then

$$p_{x_i, x_j}^\uparrow = p_{e_r}^\uparrow$$

4.3.2.2 Characterization of Gates in the Library

For any given n, the total number of functions with n inputs is 2^{2^n}, a very large number even for small n. Most typical semicustom libraries will have gates corresponding to a very small subset of these and usually limited to $n \leq 6$. In that case, it is efficient to analyze each of these in advance. Subsequently the results of the analysis are used when processing a circuit containing these gates.

The characterization process derives symbolic expressions for $p_{z_i, V_{dd}}^\uparrow$ and $p_{z_i, V_{ss}}^\uparrow$ for each internal node z_i of each gate in the library. The symbols in these expressions are the p_{x_i}'s and $p_{x_i}^\uparrow$'s, where x_i's denote the input signals to the gate.

In addition, the node capacitances C_{z_i} for each internal node z_i of each gate are also computed.

4.3.2.3 Transistor Reordering

Delay Since the load capacitance at the output of every gate in the circuit is known, given a gate, the propagation delays from each of its inputs to its output is known. As a result of timing analysis, the total delay for the longest paths through each input of the gate is also known. When the gate is a

NAND or a NOR gate, to reorder its inputs to reduce delay, the latest arriving signal is connected to the input with the smallest delay, and so on. When the gate is a complex gate, the set of input terminals is divided into *permutable* sets. For example, for the complex gate in Figure 4.27 the result is $\{\{x_1, x_2\}\{x_3\}\}$ and only input signals connected to x_1 and x_2 may be interchanged. The set of input signals is also analogously divided. Now each subset of input terminals and the corresponding subset of input signals is taken and reordering is carried out. In full-custom-design styles, it is possible to swap a transistor or a parallel connection of transistors with another transistor or a parallel connection of transistors connected in series with the first. But in semi-custom-design styles, one is restricted to using the gates available in the library without being able to modify their internal connections. Hence in semi-custom-design styles, gate input reordering is a more appropriate name for this transformation than transistor reordering.

Power Consumption   Because of the complex dependence of the power consumption of a gate on signal transition probabilities of its inputs, it is not possible to tell with small computational effort which input order is the best. Hence both exhaustive enumeration method as well as a heuristic method can be used. As most gates only have a small number of transistors in series, exhaustive enumeration is viable. All possible orderings are enumerated and the power consumption in each is computed.

The heuristic we present of computing, for each signal, the probability of the event that it will be switching and all other signals in the same permutable set will have the "on"value (i.e., a 1 for nMOS and a 0 for pMOS). Then the signal with the largest value is connected to the input closest to the output terminal. To understand the reasoning behind the heuristic, consider the case when the signal with the largest value is connected to the input closest to V_{dd}/ground. Each time the signal has an off-to-on transition with the other transistors in series in the on state, all the internal node capacitances discharge/charge. When the signal has an on-to-off transition with the other transistors in series in the on state, internal node capacitances charge/discharge, causing power to be consumed.

Optimization Algorithm   The algorithm is based on breadth-first traversals of the circuit. A traversal going from primary inputs and register outputs to primary outputs and register input is referred to as a *forward* traversal and the one in reverse direction is called a backward traversal. Each forward traversal allows us to compute for each gate G the delay of the longest path from a primary input or register output to this gate output, which we denote by $T^f(G)$. Similarly the backward traversal allows us to compute or recompute for each gate the delay of the longest path from this gate output to a primary output or a register input, which we denote by $T^b(G)$. Obviously the total length of the longest path through a gate G is given by $T^t(G) = T^f(G) + T^b(G)$. Both traversals require time, which is a linear function of the size of the circuit [71].

First we try to meet the performance goal. To do this, let us begin by performing a forward traversal and then a backward traversal. Hence when a gate G is reached during the backward traversal, $T^t(G)$ is known as $T^f(G)$ was computed earlier and $T^b(G)$ has just been computed. If $T^t(G)$ is smaller than the specified delay, nothing is done. If $T^t(G)$ is greater than the specified delay, we try to reorder the inputs of G to reduce $T^t(G)$. If we succeed in doing this, the delay through G would have changed and the $T^f(G')$'s for each gate G' in the fanout cone of G and $T^b(G'')$'s for each gate G'' in the fanin cone of G become invalid. But we do not need either of them during the current traversal! The current backward traversal has already finished with gates G' in the fanout cone of G and is in the process of computing $T^b(G'')$'s for each gate G'' in the fanin cone of G. When this backward traversal is completed, there exist valid $T^b(G)$'s for each gate G, but if any reordering took place, then not all $T^f(G)$'s are valid. If no reorderings took place then either the performance goal has been met or the performance of the circuit cannot be improved further. In both cases we proceed to power minimization.

If some reorderings took place, then we perform a forward traversal next. Once again, when a gate G is reached, $T^t(G)$ is known as $T^b(G)$ was computed during the backward traversal just completed and $T^f(G)$ has just been computed. Just as during the backward traversal described above, we continue the forward traversal, trying to reorder gate inputs when $T^t(G)$ is larger than specified.

The alternating forward and backward traversals are continued until during a traversal no reorderings take place. When that happens, either the performance goal has been met or the performance of the circuit cannot be improved further. In both cases we proceed to power minimization.

Power minimization is also carried out using alternating forward and backward traversals. But now each gate is evaluated differently. We determine the increase in delay for the input order xorresponding to the least estimated power dissipation. If the increase in delay is less than the available slack, the inputs are reordered. The available slack is defined as the difference between two delay values. The first delay value is the larger of the specified maximum acceptable delay and the delay of the longest path in the circuit. The second delay value is the delay of the longest path through the gate.

Results on the MCNC synthesis benchmark show power dissipation improvement of 7–8% without any increase in critical path delays. However, it should be noted that due to input reordering there is very little area penalty. The penalty is due to the change in routing complexity.

4.3.3 Transistor Sizing

Independent of supply voltage scaling or the logic family being used to implement logic operations, transistor sizing plays an important role in power dissipation minimization. Datta, Nag, and Roy have considered transistor sizing for low-power and high-performance static CMOS circuits [72]. The transistors on the critical path(s) of a logic circuit are sized to obtain a better power and delay performance. To improve the switching speed and the output transition characteristics of a particular circuit block on the critical path, one may seek to increase the widths of the transistors in the block. This results in an increased current drive and better output transition time. Faster input/output transition times imply both faster switching speeds and lower rush-through current, hence smaller short-circuit power dissipation. Thus, we can target lower delay and smaller short-circuit power dissipation at the same time because their requirements are not conflicting. A sluggish input ramp not only causes the transistors of a gate to switch late but also causes a substantial amount of power dissipation. To see this, let us consider a simple inverter at the transistor level. A rising step input turns the pMOS transistor off and the nMOS transistor on at the same time (instantaneously), and so there is no short-circuit current. But when the input is a slow ramp, as is often the case in real life, for some time along the rising (or falling) slope, both the pMOS and the nMOS transistors are on, opening a direct path from *power* to *ground*, leading to a considerable amount of short-circuit current. This short-circuit current gives rise to a power dissipation called the short-circuit or the rush-through power. It is to be noted, however, that even though the delay of a particular block and the rush-through power of the succeeding block are reduced, an increase in the transistor widths of the block also increases the capacitive loading of the preceding block (on the same path) and may adversely affect its delay and switching power. More-over, an increased current drive for the present block, coupled with a slower transition time at the output of the preceding block (due to an increased loading), may result in a slow input transition for the present block, thereby increasing its short-circuit power dissipation. Thus, the issues regarding the delay and the power dissipation are fairly interlinked. The optimization technique attempts to size the transistors such that the resulting solution is a satisfactory trade-off between them. Since the area calculation is straightfor-ward and directly related to the size of the transistors, we incorporate an area metric in our optimization scheme so that an area constraint can also be taken into account while optimizing a circuit for power and delay.

The algorithm minimizes the delay, the area, and the power dissipation (or a combination thereof) of a circuit by optimizing the sizes of the gates on the N most critical paths of the circuit. The critical paths are obtained from a timing analyzer. In the course of the optimization process, there may arise a situation where an alteration in the size of a component can reduce the delay of a specific critical path at the expense of increasing the delay of some other path(s). In such a case, the change is effected only if the resultant maximum delay of all the paths after the size change is less than the maximum delay before the change. The global picture of the circuit is taken into account, and hence, no circuit changes are made that actually worsen the overall perfor-mance. Thus, it is possible to optimize the power performance of a circuit, given the delay constraints.

Accurate analytical models are used in the formulation of the optimization cost function and use closed-form equations for calculating the gate delay and the power dissipation. The delay and the power approximations for inverter circuits are based on Sakurai's α-power law MOSFET model [73], which accounts for the velocity saturation effect. This model has been found to be in reasonable agreement with SPICE [88] for typical circuit sizes and loading. For static gates other than the inverter, each gate is mapped to an equivalent inverter having an equivalent output capacitance [74]. The size of the inverter and the value of the modeling capacitor depend upon the gate type, the gate size, the switching input, the input transition, and the process technology.

The optimization technique is based on simulated annealing. Given a particular combinational logic circuit, the sizes of the transistors are optimized on the critical paths of interest in order to minimize the delay, the power dissipation, and/or the area of the circuit. It is to be noted that the closed-form expressions used to predict the delay and the power of each block on a certain critical path are functions not only of the parameters (the transistor widths) of the particular block but also of those of the preceding and the succeeding blocks on the path. The output transition time of the preceding block and the input capacitance of the succeeding block determine, respectively, the input transition time and the output loading of the current block. The input transition time and the transistor sizes characterize the short-circuit power, whereas the loading capacitance determines the switching power dissipation. This interdependence of blocks makes it almost impossible to derive (and hence optimize) a single closed-form expression connecting the delay and the power dissipation of all the circuit blocks. Added to this is the need to consider a global picture of the circuit during the optimization. As a result, the objective function to be minimized takes a complicated form, and simulated annealing seemed to be a prudent choice for optimizing such an objective function.

The optimization algorithm can be tailored to yield suitable trade-offs between delay, power, and area, depending upon which parameter is more critical for the design under consideration. This is achieved by suitably choosing the weights for these parameters. Each weight is a number in the closed interval [0, 1] and determines the relative sensitivity of the optimization cost function to a change in the corresponding parameter value. It is this concept of relative weights that enables a reduction in the power dissipation of a circuit, given a delay and/or area constraint.

4.4 SUMMARY AND FUTURE DIRECTIONS

In this chapter, we have described automatic synthesis of digital logic circuits for low-power dissipation. We first considered behavioral or architectural level synthesis, which probably has the most potential in minimizing power dissipation by exploring the entire design space efficiently. Most of the research for behavioral synthesis has been in the area of signal processing, in which the systems can be represented by state variable equations. We also considered supply voltage directed architecture synthesis. Scaling down of supply voltage can considerably reduce power dissipation at the cost of increased delay. Hence, parallel or pipelined processing can possibly be used along with supply voltage reduction to achieve the same performance with lower power dissipation but increased area. Next we considered logic synthesis for low-power dissipation. In the technology-independent phase we considered state assignment and combination logic synthesis, which is based on the MIS algorithm. In the technology mapping phase the circuit graph (DAG) is broken up into a large number of trees so that a dynamic programming-based tree matching algorithm can be used to cover the trees. The cost function for power dissipation is based on the switching probability at each node of the circuit graph. Finally we considered circuit level optimization for low power by transistor resizing and input signal reordering to the equivalent inputs of complex CMOS logic gates.

With the increasing use of portable electronic systems, research is required to minimize power dissipation at all possible levels in VLSI design. One expects lower improvements in power as we go from the behavioral down to the transistor level. More formal methods need to be developed at the behavioral or data flow level to explore the design space efficiently. Along with this, efficient means of estimating power dissipation at that level is required. Scheduling and allocation algorithms [75] should be extended to incorporate power dissipation. Some of the recent scheduling techniques using multiple supply voltages to achieve low power under performance constraints have been covered in Chapter 5.

REFERENCES

[1] A. Chatterjee, R. K. Roy, and M. A. d'Abreu, "Greedy Hardware Optimization for Linear Digital Systems Using Number Splitting and Repeated Factorization," *Sixth International Conference VLSI Design*, India, Jan. 1993, pp. 154–159.

[2] A. G. Dempster and M. D. Macleod, "Use of Minimum-Adder Multiplier Blocks in FIR Digital Filters," *IEEE Trans. Circuits Syst. II—Analog Digital Signal Process.*, vol. 42, no. 9, pp. 569–577, 1995.

[3] A. P. Chandrakashan and R. Brodersen, *Low Power Digital CMOS Design*, Kluwer Academic, 1996.

[4] D. M. Kodek and K. Steiglitz, "Filter-Length Word-Length Tradeoffs in FIR Digital Filter Design," *IEEE Trans. Acoustics, Speech, Signal Proc.*, vol. ASSP-28, no. 6, pp. 739–744, 1980.

[5] H. J. Shin et al., "Custom Design of CMOS Low-Power High-Performance Digital Signal-Processing Macro for Hard-Disk-Drive Applications," *IBM J. Res. Dev.*, vol. 39, nos. 1–2, pp. 83–90, 1995.

[6] H. Samueli, "An Improved Search Algorithm for the Design of Multiplierless FIR Filters with Powers-of-Two Coefficients," *IEEE Trans. Circuits Syst.*, vol. 36, no. 7, pp. 1044–1047, 1989.

[7] J. Anderson, S. Seth, and K. Roy, "A Coarse-Grained FPGA Architecture for High-Performance FIR Filtering," *Sixth International Symposium on Field Programmable Gate Arrays*, Monterrey, CA, 1998.

[8] J. Rabaey, "Reconfigurable Computing: The Solution to Low Power Programmable DSP," *ICASSP*, Munich, April 1997.

[9] J. W. Adams and A. N. Willson, Jr., "A New Approach to FIR Digital Filters with Fewer Multipliers and Reduced Sensitivity," *IEEE Trans. Circuits Syst.* vol. CAS-30, no. 5, pp. 277–283, 1983.

[10] J. W. Adams and A. N. Willson, Jr., "Some Efficient Digital Prefilter Structures," *IEEE Trans. Circuits Syst.*, vol. CAS-31, no. 3, pp. 260–265, 1984.

[11] L. R. Rabiner and R. E. Crochiere, "A Novel Implementation for Narrow-Band FIR Digital Filters," *IEEE Trans. on Acoustics, Speech, Signal Proc.* vol. ASSP-23, no. 5, pp. 457–464, 1975.

[12] M. A. Soderstrand and K. Al-Marayati, "VLSI Implementation of Very-High Order FIR Filters," *IEEE Int. Symp. Circuits Syst.*, vol. 2, pp. 1436–1439, 1995.

[13] M. Mehendale, S. D. Sherlekar, and G. Venkatesh, "Coefficient Optimization for Low Power Realization of FIR Filters," *IEEE Workshop on VLSI Signal Processing*, Japan, 1995.

[14] M. Yagyu, A. Nishihara, and N. Fujii, "Fast FIR Digital Filter Structures Using Minimal Number of Adders and Its Application to Filter Design," *IEICE Trans. Fund.*, vol. E79-A, no. 8, pp. 1120–1128, 1996.

[15] N. Sankarayya, K. Roy, and D. Bhattacharya, "Algorithms for Low-Power and High Speed FIR Filter Realization Using Differential Coefficients," *IEEE Trans. Circuits Syst. II—Analog Digital Signal Proc.*, vol. 44, no. 6, pp. 488–497, 1997.

[16] N. Sankarayya, K. Roy, and D. Bhattacharya, "Optimizing Computations for Reducing Energy Dissipation in Realization of High Speed LTI-FIR Systems," *ACM/IEEE International Conference on Computer-Aided Design*, Santa Clara, 1997, pp. 120–125.

[17] O. Monkewich and W. Steenaart, "Stored Product Digital Filtering with Non-Linear Quantization," *IEEE International Symposium Circuits and Systems*, 1976, pp. 157–160.

[18] P. W. Wong and R. M. Gray, "FIR Filters with Sigma-Delta Modulation Encoding," *IEEE Trans. Acoustics, Speech, Signal Proc.*, vol. 38, no. 6, pp. 979–989, 1990.

[19] P. P. Vaidyanathan, "Efficient and Multiplierless Design of FIR Filters with Very Sharp Cutoff Via Maximally Flat Building Blocks," *IEEE Trans. Circuits Syst.* vol. CAS-32, no. 3, pp. 236–244, 1985.

[20] R. Hartley, "Optimization of CSD Multipliers for Filter Design," *IEEE International Symposium on Circuits and Systems*, vol. 4, 1991, pp. 1992–1995.

[21] S. R. Powell and P. M. Chau, "Efficient Narrow-Band FIR and IFIR Filters Based on Powers-of-Two Sigma-Delta Coefficient Truncation," *IEEE Trans. Circuits Syst. II—Analog, Digital Signal Proc.*, vol. 41, no. 8, pp. 497–505, 1994.

[22] T. Saramaki, "Design of FIR Filters as a Tapped Cascaded Interconnection of Identical Subfilters," *IEEE Trans. Circuits Syst.*, vol. CAS-34, no. 9, pp. 1011–1029, 1988.

[23] T. Saramaki, Y. Neuvo, and S. K. Mitra, "Design of Computationally Efficient Interpolated FIR Filters," *IEEE Trans. Circuits Syst.*, vol. 35, no. 1, pp. 70–87, 1988.

[24] Y. C. Lim and S. R. Parker, "FIR Filter Design Over a Discrete Powers-of-Two Coefficient Space," *IEEE Trans. Acoustics, Speech, Signal Proc.*, vol. ASSP-31, no. 3, pp. 583–590, 1983.

[25] Y. C. Lim, "Predictive Coding for FIR Filter Word-Length Reduction," *IEEE Trans. Circuits Syst.*, vol. CAS-32, no. 4, pp. 365–371, 1985.

[26] Y. C. Lim, "Frequency-Response Masking Approach for the Synthesis of Sharp Linear Phase Digital Filters," *IEEE Trans. Circuits Syst.*, vol. CAS-33, no. 4, pp. 357–364, 1986.

[27] A. P. Chandrakashan, S. Sheng, and R. Brodersen, "Low Power CMOS Digital Design," *IEEE Trans. Solid-State Circuits*, vol. 27, no. 4, pp. 473–483, 1992.

[28] A. P. Chandrakashan et al., "Optimizing Power Using Transformations," *IEEE Trans. on Computer-Aided Design Integrated Circuits and Systems*, vol. 14, no. 1, pp. 12–31. 1995.

[29] K. Roy and S. Prasad, "Circuit Activity Based Logic Synthesis for Low Power Reliable Operations," *IEEE Trans. VLSI Syst.* pp. 503–513, 1993.

[30] S. Prasad and K. Roy, "Circuit Optimization for Minimization of Power Consumption Under Delay Constraint," *International Workshop on Low Power Design*, Napa Valley, 1994.

[31] T.-L. Chou, K. Roy, and S. Prasad, "Estimation of Circuit Activity Considering Signal Correlations and Simultaneous Switching," *IEEE International Conference on Computer-Aided Design*, 1994.

[32] B. P. Brandt and B. A. Wooley, "A Low-Power Area-Efficient Digital Filter for Decimation and Interpolation," *IEEE J. Solid State Circuits*, pp. 679–686, June 1994.

[33] C. Gebotys, "ILP Model for Simultaneous Scheduling and Partitioning for Low Power System Mapping," Dept. of ECE, VLSI Group Technical Report, April 1995.

[34] M. Mehendale, S. D. Sherlekar, and G. Venkatesh, "Low Power Realization of FIR Filters Using Multirate Architectures," *International Conference on VLSI Design*, India, 1996.

[35] M. Mehendale, S. D. Sherlekar, and G. Venkatesh, "Techniques for Low Power Realization of FIR Filters," *Asia and South Pacific Design Automation Conference*, Japan, 1995.

[36] M. Mehendale, S. D. Sherlekar, and G. Venkatesh, "Coefficient Optimization for Low Power Realization of FIR Filters," *IEEE Workshop on VLSI Signal Processing*, Japan, 1995.

[37] M. Mehendale, S. D. Sherlekar, and G. Venkatesh, "Synthesis of Multiplierless FIR Filters with Minimum Number of Additions," *IEEE International Conference on Computer-Aided Design*, Nov. 1995, pp. 668–671.

[38] M. Mehendale, S. D. Sherlekar, and G. Venkatesh, "Optimized Code Generation of Multiplication-Free Linear Transforms," paper presented at the ACM/IEEE Design Automation Conference, Las Vegas, June 1996.

[39] A. Chatterjee and R. K. Roy, "Synthesis of Low Power Linear DSP Circuits Using Activity Metrics," *Seventh International Conference on VLSI Design*, Jan. 1994, pp. 265–270.

[40] T. Burd, "Low-Power CMOS Library Design Methodology," M.S. Thesis, University of California, Berkeley, 1994.

[41] N. Weste and K. Eshraghian, *Principles of CMOS VLSI Design*, Addison-Wesley, Reading, MA, 1984.

[42] A. V. Oppenheim and R. W. Schafer, *Discrete-Time Signal Processing*, Prentice-Hall, Englewood Cliffs, NJ, 1989.

[43] R. J. Higgins, *Digital Signal Processing in VLSI*, Prentice-Hall, Englewood Cliffs, NJ, 1990.

[44] A. Bellaoar and M. I. Elmasry, *Low-Power Digital VLSI Design*, Kluwer Academic, 1995.

[45] J. M. Rabaey, *Digital Integrated Circuits: A Design Perspective*, Prentice-Hall, Englewood Cliffs, NJ, 1996.

[46] K. Muhammad and K. Roy, "On Complexity Reduction of FIR Digital Filters Using Constrained Least Squares Solution," in *Proc. 1997 IEEE International Conference on Computer Design (ICCD)*, pp. 196–201, Oct. 1997.

[47] S. Haykin, *Adaptive Filter Theory*, Prentice-Hall, Englewood Cliffs, NJ, 1996.

[48] L. L. Scharf, *Statistical Signal Processing: Detection, Estimation, and Time Series Analysis*, Addison-Wesley, Reading, MA, 1991.

[49] E. K. P. Chong, *An Introduction to Optimization*, Wiley, New York, 1996.

[50] C. Mead and L. Conway, *Introduction to VLSI Systems*, Addison-Wesley, Menlo Park, CA, 1980.

[51] H. B. Bakoglu, *Circuits, Interconnections, and Packaging for VLSI*, Addison-Wesley, Menlo Park, CA, 1990.

[52] F. N. Najm, "Transition Density, a Stochastic Measure of Activity in Digital Circuits," *ACM/IEEE Design Automation Conference*, 1991.

[53] A. Chatterjee and R. Roy, "Synthesis of Low Power Linear DSP Circuits Using Activity Metrics," *Seventh International Conference on VLSI Design*, 1994, pp. 265–270.

[54] A. Chandrakasen, M. Potkonjak, J. Rabaey, and R. Brodersen, "HYPER-LP: A System for Power Minimization Using Architectural Transforms," *International Conference on Computer-Aided Design*, 1992, pp. 300–303.

[55] A. Chandrakasan, M. Potkonjak, R. Mehra, J. Rabaey, and R. Bordersen, "Optimizing Power Using Transformations," *IEEE Trans. Computer-Aided Design*, pp. 12–31, Jan. 1995.

[56] A. Chandrakasen, S. Sheng, and R. Brodersen, "Low-Power CMOS Digital Design," *IEEE J. Solid State Circuits*, pp. 473–484, Apr. 1992.

[57] R. Brodersen, A. Chandrakasen, and S. Sheng, "Technologies for Personal Communications," *VSLI Circuits Symposium*, 1992, pp. 5–9.

[58] K. Roy and S. Prasad, "Circuit Activity Based Logic Synthesis for Power Reliable Operations," *IEEE Trans. VLSI Syst.*, pp. 503–513, Dec. 1993.

[59] K. Roy and S. Prasad, "SYCLOP: Synthesis of CMOS Logic for Low Power Applications," *IEEE International Conference on Computer Design*, 1992, pp. 464–467.

[60] S. Prasad and K. Roy, "Circuit Activity Driven Multilevel Logic Optimization for Low Power Reliable Operations," *European Design Automation Conference*, 1993, pp. 368–372.

[61] B. Lin and H. de Man, "Low-Power Driven Technology Mapping Under Timing Constraints," *International Workshop on Logic Synthesis*, 1993, pp. 9a-1–9a-16.

[62] K. Keutzer, "DAGON: Technology Binding and Local Optimization," *ACM/IEEE Design Automation Conference*, 1987, pp. 341–347.

[63] C. Tsui, M. Pedram, and A. Despain, "Technology Decomposition and Mapping Targeting Low Power Dissipation," *ACM/IEEE Design Automation Conference*, 1993, pp. 68–73.

[64] V. Tiwari, P. Ashar, and S. Malik, "Technology Mapping for Low Power," *ACM/IEEE Design Automation Conference*, 1993, pp. 74–79.

[65] T. Sakurai and A. R. Newton, "Delay Analysis of Series-Connected MOSFET Circuits," *IEEE J. Solid-State Circuits*, vol. 26, no. 2, pp. 122–131, 1991.

[66] K. Chaudhary and M. Pedram, "A Near Optimal Algorithm for Technology Mapping Minimizing Area Under Delay Constraints," *Proc. 29th ACM/IEEE Design Automation Conf.*, pp. 492–498, 1992.

[67] H. J. Touati, C. W. Moon, R. K. Brayton, and A. Wang, "Performance-Oriented Technology Mapping," *Proc. 6th MIT Conf., Advanced Res. VLSI*, E. J. Dally, Ed., pp. 79–97, 1990.

[68] F. N. Najm and I. Hajj, "The Complexity of Test Generation at Transistor Level," Report No. UILU-ENG-87-2280, Coordinated Science Lab., University of Illinois at Urbana Champaign, Dec. 1987.

[69] F. Harary, "Combinatorial Problems in Graphical Enumeration," in *Applied Combinatorial Mathematics*, E. F. Beckenbach, Ed., Wiley, New York, 1984.

[70] F. Najm, R. Burch, P. Yang, and I. Hajj, "Probabilistic Simulation for Reliability Analysis of CMOS VLSI Circuits," *IEEE Trans. Computer-Aided Design*, vol. 9, no. 4, pp. 439–450, 1990.

[71] R. B. Hitchcock, G. L. Smith, and D. D. Cheng, "Timing Analysis of Computer Hardware," *IBM J. Res. Dev.*, vol. 26, no. 1, pp. 100–105, 1982.

[72] S. Data, S. Nag, and K. Roy, "ASAP: A Transistor Sizing Tool for Area, Delay, and Power Optimization of CMOS Circuits, *International Conference on Circuits and Systems*, 1994.

[73] T. Sakurai and R. Newton, "Alpha-Power Law MOSFET Model and Its Applications to CMOS Inverter Delay and Other Formulas," *Proc. IEEE J. Solid-State Circuits*, vol. 25, pp. 584–593, 1990.

[74] H. Chen and S. Dutta, "A Timing Model for Static CMOS Gates," *Proc. IEEE International Conference on CAD*, pp. 72–75, 1989.

[75] D. Gajski, N. Dutt, A. Wu, and S. Lin, *High-Level Synthesis: Introduction to Chip and System Design*, Kluwer Academic, Boston, 1992.

[76] A. Ghosh, S. Devdas, K. Keutzer, and J. White, "Estimation of Average Switching Activity in Combinational and Sequential Circuits," *ACM/IEEE Design Automation Conference*, 1992.

[77] R. Iyer, D. Rossetti, and M. Hsueh, "Measurement and Modeling of Computer Reliability as Affected by System Activity," *ACM Trans. Computer Systems*, vol. 4, no. 3, pp. 214–237, 1986.

[78] T. Lengauer and K. Melhorn, "On the Complexity of VLSI Computations," in *Proceedings of CMU Conference on VLSI*, Computer Science Press, 1981, pp. 89–99.

[79] G. Kissin, "Measuring Energy Consumption in VLSI Circuits: A Foundation," *14th Annual ACM Symposium on Theory of Computing*, 1982, pp. 99–104.

[80] R. W. Brodersen, A. Chandrakashan, and S. Sheng, "Technologies for Personal Communications," *1991 Symposium on VLSI Circuits*, Tokyo, pp. 5–9.

[81] A. P. Chandrakashan, S. Sheng, and R. Brodersen, "Low Power CMOS Digital Design," *IEEE Trans. Solid-State Circuits*, vol. 27, no. 4, pp. 473–483, 1992.

[82] M. Alidina, J. Monteiro, S. Davadas, A. Ghosh, and M. Papaefthymiou, "Precomputation-Based Sequential Logic Optimization for Low Power," *IEEE International Conference on Computer-Aided Design*, 1994, pp. 74–81.

[83] K. Yano et al., "A 3.8-ns CMOS 16 × 16 Multiplier Using Complementary Pass Transistor Logic," *IEEE J. Solid-State Circuits*, 388–395, 1990.

[84] R. Bryaton et al., "Fast Recursive Boolean Function Manipulation," *International Conference on Circuits and Systems*, 1982, pp. 58–62.

[85] R. Brayton, R. Rudell, A. Sangiovanni-Vincentelli, and A. Wang, "MIS: A Multiple-Level Logic Optimization System," *IEEE Trans. Computer-Aided Design*, pp. 1062–1081, 1987.

[86] G. De Micheli, R. Brayton, and A. Sangiovanni-Vincentelli, "Optimal State Assignment of Finite State Machines," *IEEE Trans. Computer-Aided Design*, pp. 269–284, 1985.

[87] S. Devadas, H-K T. Ma, A. Newton, and A. Sangiovanni-Vincentelli, "MUSTANG: State Assignment of Finite State Machines Targeting Multi-Level Logic Implementations," *IEEE Trans. Computer-Aided Design* pp. 1290–1300, 1988.

[88] L. Nagel, "SPICE2: A Computer Program to Simulate Semiconductor Circuits," Memo ERL-M520, University of California, Berkeley, May 9, 1975.

[89] D. Pickens, "Power Simulator for CMOS Circuits," Texas Instruments, Dallas, TX, unpublished report, 1993.

CHAPTER 5

DESIGN AND TEST OF LOW-VOLTAGE CMOS CIRCUITS

5.1 INTRODUCTION

We have noted earlier that power dissipation in CMOS circuits consists of dynamic and static components. Since dynamic power is proportional to V_{dd}^2 and static power is proportional to V_{dd}, lowering the supply voltage is one of the most effective ways to reduce power dissipation in CMOS circuits. With the scaling of supply voltage and device dimensions, the transistor threshold voltage also has to be scaled to achieve the required performance. Unfortunately, such scaling does not come for free and can increase the leakage current. Due to the exponential nature of leakage current in the subthreshold regime of the transistor, leakage current can no longer be ignored. In fact, leakage current can be a major component of the total power dissipation in CMOS circuits. Figure 5.1 plots power dissipation of an inverter with respect to supply voltage and transistor threshold. The plot clearly shows that the power dissipation can exponentially increase when the threshold of the transistors is very low. The figure also suggests that there is an optimum value of supply voltage and transistor threshold to obtain the minimum power dissipation for iso-performance. Therefore, techniques to reduce the leakage current should be considered in low-voltage CMOS circuit design. Hence, a major portion of the chapter is devoted to leakage control techniques.

It can be observed that, for a large design, the different sections of the design can possibly run at different speeds. Since delay through a logic block is inversely proportional to the supply voltage and the power dissipation is proportional to the square of the supply voltage, the slower parts of the

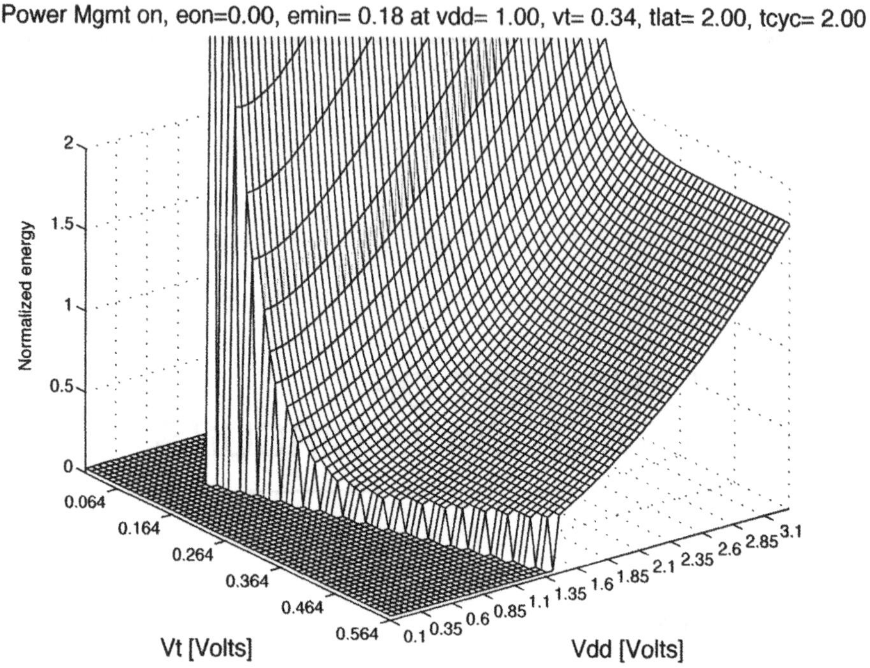

Power Mgmt on, eon=0.00, emin= 0.18 at vdd= 1.00, vt= 0.34, tlat= 2.00, tcyc= 2.00

Figure 5.1   Power dissipation as a function of supply and threshold (V_{th}) voltage.

circuit can possibly run at a lower voltage than the parts of the circuit that are more critical in terms of speed. This will considerably reduce the power dissipation without any performance penalty. In this chapter we consider multiple-supply design style and discuss the pros and cons of using multiple supply voltage.

Circuit design style also plays an important role in determining the performance, power dissipation, and supply threshold voltage scalability (noise considerations). For example, fully complementary designs are usually robust and dissipate lower power but cannot achieve the performance of domino or other precharged logic styles. However, fully complementary designs are probably more scalable than domino circuits. Hence, it is essential to consider different circuit styles when low voltage designs are considered. We will devote a section in this chapter to consider different logic styles, their scalability, and their impact on the design of low-power, high-performance circuits.

High leakage current can also make current testing techniques invalid. Let us consider quiescent power supply current (I_{DDQ}) monitoring technique to detect the defects and analyze failures in CMOS ICs. The basic idea behind such testing technique is that the quiescent current in static CMOS circuits during standby mode is small. In the presence of faults (such as bridging faults) the I_{DDQ} can be large. Any large enough deviation of quiescent current from nominal value can be considered due to the presence of faults [8]. However, if the intrinsic leakage of a design is high, a defect may not produce a large percentage difference in quiescent current compared to the nonfaulty value. Hence, modified I_{DDQ} test techniques are required to achieve high defect coverage. We will present some recent techniques to achieve high defect coverage using modified I_{DDQ}.

This chapter is organized as follows. We will first consider different logic styles and their impact on power dissipation. That discussion will be followed by leakage mechanisms in deep submicrometer transistors and how such leakage can influence circuit design and test issues. Finally, we will consider the pros and cons of using multiple supply voltages and its impact on low-power design.

5.2 CIRCUIT DESIGN STYLE

Circuit performance and power dissipation depend largely on the circuit design style. Circuit design styles can be broadly classified as *nonclocked* and *clocked* types. The clocked design styles are usually faster but consume more power. Let us analyze some representative logic families from the different logic styles to determine the power dissipation, performance, and scalability issues. Our aim in this analysis is not to be exhaustive but to give our readers the feel for the importance of design styles from a power dissipation point of view. A volume of literature exists on how these logic families operate [3, 4]. A designer mainly chooses the style of design based on circuit speed, circuit size, power dissipation, wiring complexity, noise margin, and ease of design.

5.2.1 Nonclocked Logic

In this section we will analyze some of the representative logic families, like fully complementary logic, pass transistor logic, and cascode voltage switch logic.

5.2.1.1 Fully Complementary Logic

Fully complementary CMOS (or conventional CMOS) logic has excellent properties in many areas, such as ease of design, low-power dissipation, low sensitivity to noise and process variations, and scalability. However, the logic family suffers from lower performance, especially for large fanin gates. Let us analyze the power dissipation and performance of fully complementary CMOS in greater detail.

- Dynamic power consists of both switching and short-circuit components, and due to different delays through different paths of the circuit, glitches or spurious transitions can occur. As we have noticed earlier, such spurious transitions can be a significant component of total power in completely unbalanced designs.

- During the stand-by mode of operation, power dissipation is only due to leakage current through the transistors, and for designs having comparatively large V_{th}, this component of power dissipation can be neglected.
- Since fully complementary designs have high noise margin, the design style is more scalable than dynamic logic circuits. Hence, transistor threshold voltages can be lower than dynamic circuits. The lower threshold can possibly help in achieving better performance for ultra-low-voltage designs.
- The circuit performance sometimes suffers due to large PMOS transistors resulting in large input capacitances and weak output driving capability caused by series transistors. Performance also suffers due to the presence of short-circuit current in standard CMOS designs.

5.2.1.2 NMOS and Pseudo-nMOS Logic

Pseudo-nMOS logic is a ratioed logic that uses pMOS transistor load with the gate connected to V_{SS}. This logic style is very similar to NMOS logic with depletion load with the gate connected to the source (since $V_{th} < 0$, the load is always on). To keep the low noise margin high, it is important to have the resistance of the load much larger than the resistance of the pull-down network. The logic family is called *ratioed* because a careful ratioing of the pull-up and the pull-down network is required for proper operation. Figure 5.2 shows the schematic of a depletion load NMOS and a pseudo-nMOS logic. The salient features of the logic family are as follows:

- Reduced complexity of logic and hence, lower capacitance, and faster speed.

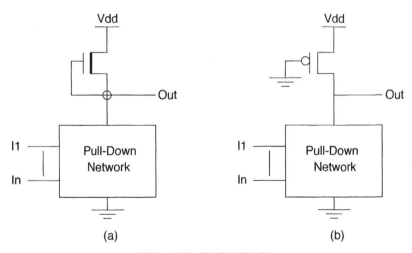

Figure 5.2 Ratioed logic.

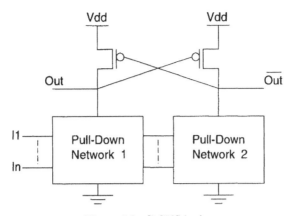

Figure 5.3 DCVS logic.

- The NMOS logic, with depletion load has the source of the load connected to the output, and hence as the output voltage changes, the threshold of the transistor also changes due to the body effect. The drive current also changes due to such V_{th} variation. Therefore, pseudo-NMOS is a better choice because the load device is not body affected.
- A ratioed design style is better suited for large fanin gates such as NOR. Each input connects to a single transistor, presenting a smaller load to the preceding gate.
- The power dissipation in this logic family is usually higher than conventional CMOS design, because static current always flows through the logic gate whenever the pull-down network is on.

5.2.1.3 Differential Cascade Voltage Switch (DCVS) Logic

The DCVS logic is shown in Figure 5.3. Here a differential output signal is available, and a careful analysis indicates that the logic family eliminates any static power dissipation that is present in ratioed logic style. However, DCVS logic still has the speed advantage of ratioed logic. The two pull-down networks implement complementary functions (Out and $\overline{Out}$), and hence, the logic style has larger area or switched capacitance. The switching power dissipation of this logic family is high. However, it should be noted that it is sometimes possible to share some logic in the two pull-down networks to reduce the area. The complementary outputs can also help eliminate inverters, which might otherwise be required.

5.2.1.4 Pass-Transistor Logic

Pass-transistor logic (PTL) uses pass-transistor networks to implement different logic functions. A combination of AND (pass transistors or switches in series implement AND functions) and OR (pass transistors or switches in parallel) functions implements different functional networks. Figure 5.4 shows

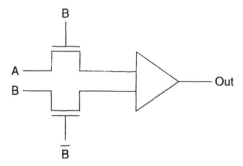

Figure 5.4   Pass-transistor logic style.

the implementation of an AND gate. The PTL has the advantage of being fast, and complex logic gates can be implemented with minimal number of transistors. However, the nMOS pass transistors do not transmit a good 1 ($V_{dd} - V_{th}$), and hence, level restorers may be required at the output of logic gates. Therefore, the designs can have reduced noise margin. Second, static power dissipation is possible when the nonrestored logic values are applied to standard CMOS logic.

It is possible to implement the above logic style with zero-threshold transistors and differential input and output signals. Such a logic family is called complementary pass-transistor logic (CPL). Figure 5.5 shows the implementation of simple AND/NAND and XOR/XNOR gates. It can be noted that very simple implementation of XOR functions is possible making this kind of logic suitable for adders and multipliers. The main advantages of this design style are that the designs can be modular while differential output signals are available. Differential outputs can help eliminate inverters and can make the designs faster. The logic family has reduced internal capacitances and is suitable for power efficient implementation coupled with the

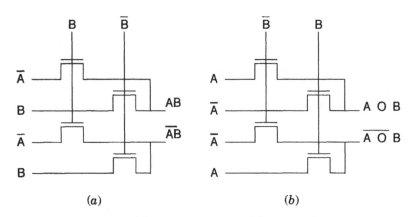

Figure 5.5   CPL (*a*) NAND/AND and (*b*) XOR/XNOR gates.

fact that reduced voltage swing is possible in this logic family. Recent comparison of CPL and standard CMOS logic [14] shows that for a 32-bit adder (buffered parallel prefix adder) the power-delay product of CPL is about 10% better than standard CMOS logic. Earlier comparisons predicted CPL to have considerably better power–delay product than standard CMOS [1].

5.2.2 Clocked Logic Family

The clocked logic families that we will discuss are usually faster than the static logic families described above. However, the performance usually comes at the expense of higher power dissipation. First we will consider the domino logic family. The second logic family that we consider, called differential current switch logic (DCSL), has better power–delay product than domino circuits due to low internal node voltage swing and the possibility of having large fanin logic gates.

5.2.2.1 Domino Logic

Figure 5.6 shows a two-input domino NAND gate. When the clock signal is low, the output of the logic gate is precharged to logic 1. When the clock is high, the output is conditionally discharged. The inverter at the output helps in cascading similar stages. The pMOS keeper transistor (small size) at the output helps keep the output at logic 1 and is especially useful when the

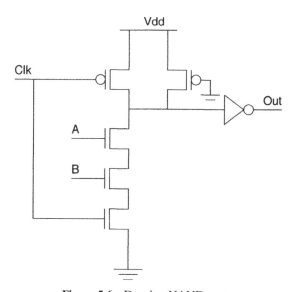

Figure 5.6 Domino NAND gate.

transistors are leaky. The salient features of the logic family are as follows:

- Domino circuits only implement noninverting logic gates, and hence, it is not easy to use the domino design style.
- Domino design style is better suited for large fanin gates such as NOR. Each input connects to single transistor, presenting smaller load to the preceding gate. Hence, it is possible to achieve high performance.
- The output of a gate is sometimes precharged only to discharge in the evaluation phase. Therefore, the signal activity at the output can be high. Increased signal activity along with the extra load that the clock line has to drive is primarily responsible for high power dissipation in domino compared to static CMOS circuits (see Chapter 3 on power estimation).
- Spurious transitions are not possible in domino circuits since the output can only make a 1 to 0 transition.
- It has been shown that domino circuits are not as scalable as standard CMOS due to lower noise immunity in domino compared to standard CMOS circuits. Hence, the domino circuits will usually require a higher transistor threshold for proper operation [15].

5.2.2.2 Differential Current Switch Logic

Differential current switch logic (DCSL) belongs to the class of clocked differential cascode voltage switch logic circuits. It is a generic methodology that applies to clocked DCVS gates to reduce internal node voltage swings. The circuit topology of precharged high DCSL is shown in Figure 5.7 [16].

The precharged high DCSL (DCSL1) consists of an nMOS evaluation tree, a cross-coupled inverter pair (T2, T3, T6, and T7), and precharge transistors (T1 and T4). The presence of transistors T5 and T8 is what differentiates this gate from other DCVS logic gates. Operation of the gate starts with CLK low and nodes Q and $\overline{Q}$ being charged high. Gate evaluation begins with stable inputs to the nMOS tree and CLK going high. The CLK high switches T9, T10, and T11 on while Q and $\overline{Q}$ being high ensures that T5, T6, T7, and T8 are switched on. Here Q and $\overline{Q}$ discharge towards ground through T6, T7, and T10. The discharge of Q and $\overline{Q}$ is not symmetrical because the nMOS tree assures that one of the outputs, say Q, has a stronger path to ground. This causes Q to fall faster than $\overline{Q}$. The cross-coupled inverter functions as a sense amplifier and boosts the output voltage differential in the right direction. Once the inverter switch threshold is crossed by Q, $\overline{Q}$ swings high. The low going transition of Q disconnects the nMOS tree from $\overline{Q}$ by progressively cutting off transistor T8. Hence the rising node $\overline{Q}$ is isolated from the nMOS evaluation tree. This action limits the charge up of internal nodes in the nMOS tree. This is unlike other DCVS circuits where the nMOS pull-down tree is never disconnected from the output. The DCVS

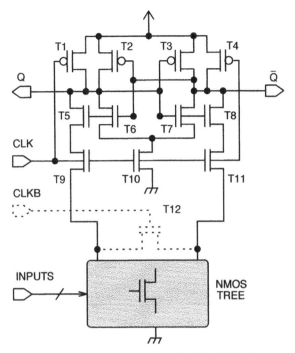

Figure 5.7 Precharge high DCSL (DCSL1).

circuits charge the internal nodes of the nMOS tree up to $V_{CC} - V_{tn}$ (V_{tn} is the threshold voltage of nMOS transistor), where V_{CC} is the supply voltage (5 V) and V_{tn} is the threshold voltage of the nMOS device (of the order of 1 V). In contrast, DCSL charges internal nodes to much smaller voltages. SPICE simulations show that internal node voltage swings for DCSL are of the order of 1 V with a supply voltage of 5 V. The gate comes to rest in a state with Q low and $\overline{Q}$ high. Transistor T12 may be required to prevent charge buildup on internal nodes of the gate.

On completion of the evaluation, the fact that the high output ($\overline{Q}$ in the previous case) is disconnected from the nMOS tree assures us that further changes in inputs do not propagate to the output. This is unlike most CMOS logic styles, where changes in the inputs of clocked logic cause DC through currents or the gate output being destroyed. Strict adherence to the design constraint of building DCVS nMOS trees is no longer required. Gate inputs may cause paths to ground in both halves of the nMOS tree. However, assuring that one of the paths has a stronger pull-down than the other allows the DCSL gate to evaluate its inputs. On completion of the evaluation, no static current paths from V_{CC} to ground exist.

The circuit has strongly driven outputs in all phases of gate operation. A completion signal may be generated by taking a NAND of the two outputs. The addition of transistors T5 and T8 increases the capacitance of the output

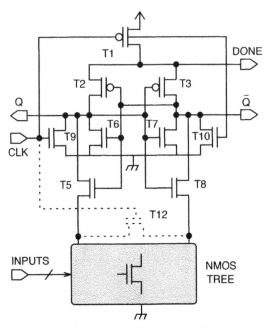

Figure 5.8 Precharge low DCSL (DCSL2).

nodes. However, decoupling the nMOS tree from gate outputs reduces the effective internal node capacitance seen at the gate output. The gate becomes increasingly better than SSDL [9] in terms of power consumed and propagation delay as the nMOS tree height rises.

The previous discussion considers DCSL logic with outputs precharged high. Figure 5.8 shows a DCSL gate (DCSL2) with outputs charged low. The gate imposes a lower clock load and has the advantage of generating a completion signal (DONE).

DCSL2 is similar to EDCL [10]. The circuit is in the precharged state with Q and $\bar{Q}$ low when CLK is high. The circuit evaluates with CLK going high. Since the gate outputs are initially low, the nMOS evaluation tree is disconnected from the gate outputs by T5 and T8. The circuit enters the evaluate mode with the CLK going low. Since evaluation starts only after the outputs have crossed V_{tn}, the gate propagation delay degrades.

Further refinement of this circuit topology gives DCSL3, shown in Figure 5.9. The circuit has better speed characteristics than the circuit shown in Figure 5.8. The improvement is obtained by removing the pull-down transistors T9 and T10 in Figure 5.8 and replacing them by the single transistor T9 in Figure 5.9. Gate operation starts with CLK high, T9 on and nodes Q and $\bar{Q}$ equalized. In addition, this causes transistors T5, T6, T7, and T8 to have their gate drains connected. Hence Q and $\bar{Q}$ discharge to a voltage that is V_{tn} or lower, through transistors T6 and T7. Limiting the discharge voltage to V_{tn}

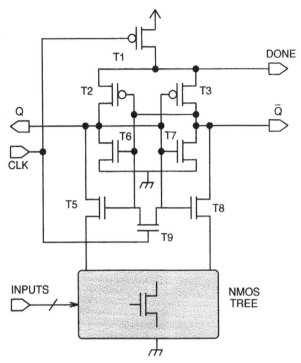

Figure 5.9   An improved precharge low DCSL (DCSL3).

lowers the power dissipation. Additionally, performance as compared to DCSL2 improves marginally because of the smaller voltage swing needed at the outputs. However, unlike DCSL1 and DCSL2, the outputs are not always at 0 or 5 V (supply voltage). Evaluation starts with a low going CLK. Transistor T1 switches on, and charges the output nodes high through T2 and T3, which are on because of the existing low-voltage precharge at the outputs. The output node, which has a path to ground through the nMOS tree, is prevented from rising. The other node charges up, leading to the inverter loop taking on the correct state.

To carry out a comparison of similar clocked circuit structures, the layout of a parity generator circuit was carried out using the MOSIS 1.2-μm SCMOS process design rules. SPICE simulations of the extracted circuits were carried out. Results obtained with various nMOS tree heights (fanin) for a parity generator circuit are shown in Figures 5.10 and 5.11. SPICE simulations are carried out using the level 3 SPICE model at 27°C. The following circuits were compared: SSDL, ECDL, dynamic cascode voltage switch logic (CVSL), and the DCSL circuits shown in Figures 5.7–5.9.

Figure 5.10 shows the energy consumption per clock cycle for the various logic styles. The graph does not include the power expended in clocking the circuits. The power advantage of DCSL is greater than a factor of 2 as

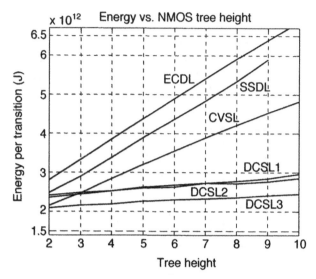

Figure 5.10   Energy consumption.

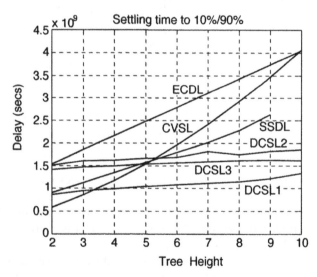

Figure 5.11   Comparison of CLK-to-Q (90%/10%) delay.

compared to similar DCVS logic with tree heights greater than 8. Of the three DCSL topologies, DCSL3 is better in terms of power dissipation. The power dissipation of DCSL shows a weaker dependence on the nMOS tree height. The power advantage is primarily because of the very low internal voltage swings. Internal node voltages in standard CMOS gates (supply voltage of 5 V) is of the order of 4 V when a transistor threshold of 1 V is used. Internal node voltage swings for the DCSL circuit in Figures 5.7 and

5.8 are of the order of 1 V and 0.3 V, respectively. Such low-voltage swing at the internal nodes produces a large savings in energy dissipation.

The CLK-to-Q delay of the various logic families as a function of tree height is compared in Figures 5.11 and 5.12. In general, DCSL exhibits superior settling times by a factor of 2 at tree heights greater than 8. Unlike CVSL, SSDL, and EDCL, in which increasing nMOS tree height progres-

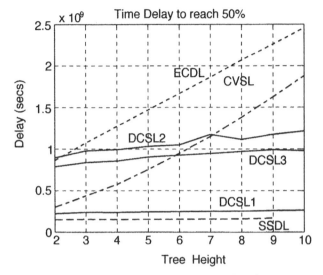

Figure 5.12 Comparison of CLK-to-Q (50%) delay.

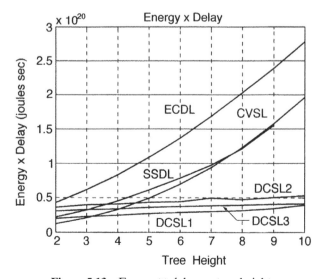

Figure 5.13 Energy $\times$ delay vs. tree height.

sively worsens gate propagation delays, DCSL delays are substantially independent of nMOS tree heights.

Figure 5.13 plots the energy–delay product for the various logic families and clearly shows that the performance and power advantage of DCSL increases with increasing gate heights. This is because the gate delay and energy consumption are weakly dependent on the nMOS tree height, unlike the other logic families. The DCSL logic favors implementing high-complexity gates.

A low-voltage DCSL is also proposed in [17] that is able to operate at much lower supply voltage and a comparatively high transistor threshold.

5.3   LEAKAGE CURRENT IN DEEP SUBMICROMETER TRANSISTORS

We have seen earlier that the scaling of supply voltage and transistor threshold has a large impact on the leakage current. Transistor off-state current is the drain current when the gate–source voltage is zero. The off-state leakage in long-channel devices is dominated by drain–well and well–substrate reverse-bias pn junctions. For short-channel transistors, the off-state current is influenced by threshold voltage, channel physical dimensions, channel/surface doping profile, drain/source junction depths, gate oxide thickness, the supply voltage (V_{dd}), the drain, and the gate voltages. We will first delineate the components of transistor leakage that contribute to the total off-state current and then present some experimental data from [39] that will help in the design of circuits with "leaky transistors."

5.3.1   Transistor Leakage Mechanisms

For long-channel transistors, the leakage current is dominated by the pn-junction leakage and the weak-inversion current. Other leakage mechanisms are peculiar to the small geometries themselves. Figure 5.14 shows the summary of leakage mechanism in a short-channel transistor.

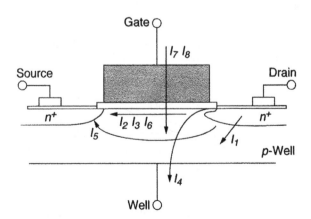

Figure 5.14   Leakage mechanism in short-channel transistors.

5.3.1.1 pn Reverse-Bias Current (I_1)

A reverse-bias pn-junction leakage (I_1) has two main components: one is the minority-carrier drift near the edge of the depletion region and the other is due to electron–hole pair generation in the depletion region of the reverse-bias junction [6]. If both n and p regions are heavily doped, Zener tunneling may also be present. For an MOS transistor, additional leakage can occur between the drain and well junction from gated diode device action (overlap and vicinity of gate to drain to well pn junctions) or carrier generation in drain-to-well depletion regions with influences of the gate on these current components [7]. The pn reverse-bias leakage (IREV) is a function of junction area and doping concentration [6, 8]. The IREV for pure diode structures is usually a minimal contributor to total transistor I_{off}.

5.3.1.2 Weak Inversion (I_2)

Weak-inversion current between source and drain in a MOS transistor occurs when gate voltage is below the threshold voltage V_{th}. The weak-inversion region is shown in Figure 5.15 as the linear portion of the curve (plotted on a log scale for a 3.5μ technology). The channel has virtually no horizontal electric field, and hence, the carriers move by diffusion similar to charge transport across the base of bipolar transistors. The exponential relation between driving voltage on the gate and the drain current is a straight line in a log plot. Weak inversion typically dominates modern device off-state leakage due to the low V_{th} that is used.

Table 5.1 shows transistor $\log(I_D)$ versus V_G subthreshold slopes (S) for the different technologies listed in Table 5.2. For the test transistors, S was about 80 mV/decade at room temperature. And S has not increased as

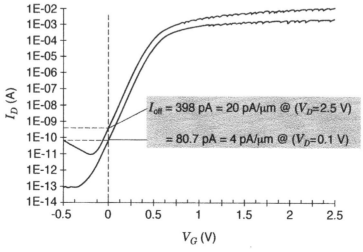

Figure 5.15 I_D vs. V_G.

TABLE 5.1   Subthreshold Slopes of Various Technologies at Room Temperature

Technology (μm)	Doping	S_t (mV/decade)
0.8-μm, 5-V CMOS	LDD	86
0.6-μm, 5-V CMOS	LDD	80
0.35-μm, 3.3-V BiCMOS	LDD	80
0.35-μm, 2.5-V CMOS	HDD	78
0.25-μm, 1.8-V CMOS	HDD[a]	85

[a] From [16].

TABLE 5.2   Comparison of Process Technologies

Technology (μm)	V_{DD} (V)	T_{ox} (Å)	V_T (V)	L_{eff} (μm)	I_{off} (pA/μm)
1.0	5	200	n/a	0.80	4.1×10^{-4}
0.8	5	150	0.60	0.55	5.8×10^{-2}
0.6	3.3	80	0.58	0.35	0.15
0.35	2.5	60	0.47	0.25	8.9
0.25	1.8	45	0.43	0.15	24
0.18	1.6	30	0.40	0.10	86

technologies advance mainly because t_{ox} (gate oxide thickness) has been scaled and substrate doping profiles have improved. Here, S is a function of gate oxide thickness and the surface doping adjust implants. An S of 100 mV/decade is an indication that the device technology is leaky and unsuitable for high-volume manufacturing. Lower S values indicate better control of SCE (short-channel effect) and a lower I_{off} for a given threshold voltage.

5.3.1.3   Drain-Induced Barrier-Lowering Effect (I_3)

Drain-induced barrier lowering (DIBL) occurs when the depletion region of the drain interacts with the source near the channel surface to lower the source potential barrier. The source then injects carriers into the channel surface without the gate playing a role. Thus DIBL is enhanced at higher drain voltage and shorter L_{eff}. Surface DIBL typically happens before deep bulk punchthrough. Ideally, DIBL does not change the subthreshold slope S but does lower V_{th}. Higher surface and channel doping and shallow source/drain junction depths reduce the DIBL leakage current mechanism [11, 12].

Figure 5.16 illustrates the DIBL effect as it moves the curve up and to the right as V_D increases. DIBL can be measured at constant V_G as the change in I_D for a change in V_D. For V_D between 0.1 and 2.7 V, I_D changed 1.68 decades, giving a DIBL of 1.55 mV/decade change of I_D (for the technology reported in [39]).

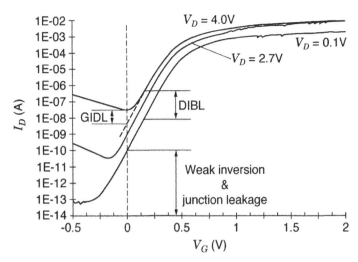

Figure 5.16 An n-channel I_D vs. V_G showing DIBL, GIDL, weak inversion, and pn-junction reverse-bias leakage components.

5.3.1.4 Gate-Induced Drain Leakage (I_4)

Gate induced drain leakage (GIDL) current arises in the high electric field under the gate/drain overlap region causing deep depletion [12]. GIDL occurs at a low V_G and high V_D bias and generates carriers into the substrate and drain from surface traps or band-to-band tunneling. It is localized along the channel width between the gate and drain. The GIDL current is seen as the "hook" in the waveform of Figure 5.16 that shows increasing current for negative values of V_G. Thinner t_{ox}, higher V_{DD}, and LDD (lightly doped drain) structures enhance the electric-field-dependent GIDL. GIDL is a major obstacle in I_{off} reduction.

The GIDL current was isolated by measuring source current $\log(I_s)$ versus V_G [39]. It is seen as the dotted line extension of the $V_D = 4.0$ V curve in Figure 5.16. The GIDL contribution to I_{off} is small at 2.7 V, but as the drain voltage rises to 4.0 V (close to burn-in voltage), the off-state current on the $V_D = 4.0$ V curve increases from 6 nA (at the dotted line intersection with $V_G = 0$ V) to 42 nA for a GIDL of 36 nA.

5.3.1.5 Punchthrough (I_5)

Punchthrough occurs when the drain and source depletion region approach each other and electrically "touch" deep in the channel. Punchthrough is a space-charge condition that allows channel current to exist deep in the subgate region, causing the gate to lose control of the subgate channel region. Punchthrough current varies quadratically with drain voltage and S increases, reflecting the increase in drain leakage [6, p. 134]. Punchthrough is regarded as a subsurface version of DIBL.

5.3.1.6 *Narrow-Width Effect* (I_6)

Transistor V_{th} in non-trench-isolated technologies increases for geometric gate widths on the order of 0.5 μm. An opposite and more complex effect is seen for trench-isolated technologies that show decrease in V_{th} for effective channel widths on the order of $W \leq 0.5$ μm [13].

5.3.1.7 *Gate Oxide Tunneling* (I_7)

Gate oxide tunneling current I_{ox} from high electric field (E_{ox}) can cause direct tunneling through the gate or Fowler–Nordheim (FN) tunneling through the oxide bands [Eq. (5.1)] [6]. Fowler–Nordheim tunneling typically lies at a higher field strength than found at product use conditions and will probably remain so and has a constant slope for $E_{ox} > 6.5$ MV/cm (Figure 5.17). Figure 5.17 also shows significant direct tunneling at lower E_{ox} for thin oxide:

$$I_{ox} = AE_{ox}^2 \, e^{-B/E_{ox}} \tag{5.1}$$

Oxide tunneling current is presently not an issue but could surpass weak inversion and DIBL as a dominant leakage mechanism in the future as oxides get thinner.

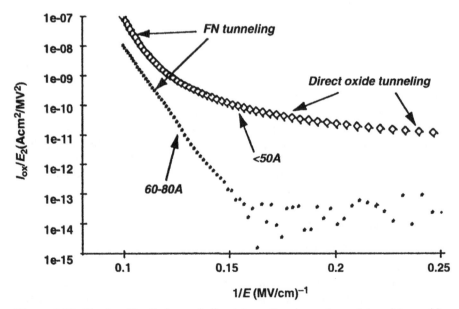

Figure 5.17 Fowler–Nordheim and direct tunneling in *n*-channel transistor oxide. The 60–80-Å curve shows dominance of FN tunneling while the < 50-Å curve shows FN at high E_{ox} but significant direct tunneling at low electric fields.

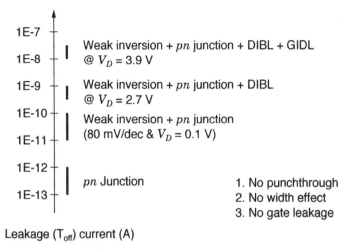

Figure 5.18   Leakage summary.

5.3.1.8   *Hot-Carrier Injection* (I_8)

Short-channel transistors are more susceptible to injection of hot carriers (holes and electrons) into the oxide. These charges are a reliability risk and are measurable as gate and substrate currents. While past and present transistor technologies have controlled this component, it increases in amplitude as L_{eff} is reduced unless V_{DD} is scaled accordingly.

Figure 5.18 summarizes relative contributions of all components of intrinsic leakage for a typical 0.35-μm CMOS technology. We can see that, for a typical drain voltage of 2.7 V (consistent with typical power supply voltage of the technology), DIBL is the dominant component of leakage. At elevated burn-in voltage of 3.9 V, GIDL dominates. Finally, at low V_D, weak inversion is the primary leakage mechanism.

5.3.2   Leakage Current Estimation

The leakage power of a CMOS circuit is determined by the leakage current through each transistor, which has two main sources: reversed-biased diode junction leakage current and subthreshold leakage current. Diode junction leakage is very small and can be ignored [3]. Subthreshold leakage exponentially increases with the reduction of threshold voltage [19], making it critical for low-voltage circuit design. Therefore, let us first focus on subthreshold leakage power estimation.

In order to estimate leakage power accurately, a general transistor model [20] that considers subzero gate-to-source voltage (V_{GS}) for nMOS and superzero V_{GS} for pMOS (occurs when multiple series-connected transistors are turned off), body effect, and DIBL is used. The following analysis is done for nMOSFETs but is equally applicable to pMOSFETs.

From the BSIM2 MOS transistor model [22–24], the subthreshold current of a MOSFET can be modeled as

$$I_{\text{sub}} = A \exp\left(\frac{q}{n'kT}(V_G - V_S - V_{\text{th}_0} - \gamma'V_S + \eta V_{DS})\right)\left[1 - \exp\left(\frac{-qV_{DS}}{kT}\right)\right]$$

(5.2)

where

$$A = \mu_0 C_{\text{ox}} \frac{W_{\text{eff}}}{L_{\text{eff}}}\left(\frac{kT}{q}\right)^2 e^{1.8}$$

Here C_{ox} is the gate oxide capacitance per unit area, μ_0 is the zero-bias mobility, n' is the subthreshold swing coefficient of the transistor, and V_{th_0} is the zero-bias threshold voltage. The body effect for small values of V_S is very nearly linear. It is represented by the term $\gamma'V_S$, where γ' is the linearized body effect coefficient. Here η is the DIBL coefficient.

If transistors are connected in parallel and are both turned off (such as in the pull-down network of a NOR gate), then the values of V_{DS} and V_S are the same for each transistor. The leakage contribution of each transistor can be calculated separately and added together. However, things become more complicated if they are in series. Consider the pull-down network of an N-input NAND gate. Without loss of generality, we consider the case where all N nMOS transistors are turned off. The quiescent subthreshold through each transistor must be identical, given that other leakage components are negligible. So we equate the current of the first (top) and second transistor. Equation (5.3) can be obtained by solving for V_{DS_2} in terms of V_{dd} (we assume that $V_{S_1} \ll V_{dd}$):

$$V_{DS_2} = \frac{n'kT}{q(1 + 2\eta + \gamma')}\ln\left(\frac{A_1}{A_2}e^{q\eta V_{dd}/n'kT} + 1\right)$$

(5.3)

$$V_{DS_i} = \frac{n'kT}{q(1 + \gamma')}\ln\left(1 + \frac{A_{i-1}}{A_i}(1 - e^{-(q/kT)V_{DS_{i-1}}})\right)$$

(5.4)

One can similarly equate the current through the $(i - 1)$th and ith transistors, solving for V_{DS_i} in terms of $V_{DS_{i-1}}$. This results in Eq. (5.4) (a more detailed derivation of Eqs. (5.3) and (5.4) can be found in [20]). Equation (5.4) can be used iteratively to find V_{DS_i} for each transistor, starting with the third transistor in the stack. Finally, the voltage offset at the source of each transistor is given by $V_{S_i} = \sum_{j=i+1}^{N} V_{DS_j}$, and V_{DS_1} can be determined by $V_{dd} - V_{S_1}$. Now Eq. (5.2) can be used to calculate the quiescent leakage (I_{DS_q}) for any transistor in the stack, which should be the same for each transistor. Finally, the total leakage power can be determined as the sum of $V_{DS} \times I_{DS_q}$

over all transistors:

$$P_{\text{leak}} = \sum_i I_{DS_{q_i}} V_{DS_{q_i}} \tag{5.5}$$

The general method of computing leakage power for a large circuit involves the following steps. Given a particular set of circuit input values, determine which pull-up and pull-down networks are turned off. Within each network, consider transistors that are turned on to be short circuits. Transistors that are parallel to a transistor that is turned on can be eliminated from the leakage calculation. Given the resulting simplified network, estimate V_{DS} for the remaining transistors using Eqs. (5.3) and (5.4). Finally, the magnitude of leakage current and resulting leakage power can be computed.

The above method is very suitable for leakage power estimation during the standby mode. In the active mode, the time required for the leakage current in transistor stacks to converge to its final value is determined by the internal node capacitance, input conditions, and subthreshold leakage current [21]. Subthreshold leakage current strongly depends on the threshold voltage and temperature. If the internal node capacitance is small and active temperature is high, the given method can also be used to estimate active leakage power of low-V_{th} circuits, especially at low switching activities. Considering the fact that standby leakage current depends on input signal levels, the average leakage power can be evaluated with random patterns applied to primary inputs.

From Eqs. (5.2) and (5.5), we can get the sensitivity of standby leakage power with respect to V_{th} by

$$\frac{\partial P_s}{\partial V_{\text{th}}} = \sum_i \frac{\partial I_{DS}(i)}{\partial V_{\text{th}}} V_{DS}(i) = -\frac{q}{n'kT} P_s \tag{5.6}$$

Clearly, such sensitivity is proportional to leakage power itself.

5.3.2.1 *Proper Input Selection for Stand-By Mode*

The leakage estimation technique uses accurate modeling of transistor stacks to estimate leakage. The analysis clearly suggests that vectors at the input to logic gates have a large impact on the leakage current. A genetic algorithm (GA) based technique to estimate the standby leakage power in CMOS circuits has been developed in [18]. The supply voltage V_{dd} and the zero bias threshold voltage V_{th0} used in this experiment are 1.0 and 0.2 V, respectively. The subthreshold swing coefficient is 1.5, which corresponds to the subthreshold slope of 89.8 mV/decade. The parameters η and γ' are 0.05 and 0.24 for nMOS and 0.047 and 0.11 for pMOS, respectively. For simplicity, all transistors are assumed to have the same channel length of 0.3 μm while the channel widths for pMOSFETs and nMOSFETs are assumed to be 3.6 and 1.8 μm, respectively. However, the method is not limited to such assumptions.

TABLE 5.3   Minimum and Maximum Leakage Current

Circuit Chosen	Minimum Leakage Power (μW)			Maximum Leakage Power (μW)		
	HSPICE	GA	Diff (%)	HSPICE	GA	Diff (%)
Three-input NAND	0.022	0.021	4.5	0.492	0.485	1.4
Full adder	1.818	1.909	5.0	2.220	2.281	2.7
2-Bit multiplier	1.842	1.894	2.8	3.049	3.055	0.2

Table 5.3 shows the minimum and maximum leakage currents for some benchmark circuits. The table clearly suggests that by applying proper input vectors to a circuit during the stand-by mode of operation, it is possible to reduce the static current considerably. It should be noted that, for stand-by leakage management, using proper input selection requires that the states of the flip-flops be saved. States can be saved using modified flip-flops [25].

5.4 DEEP SUBMICROMETER DEVICE DESIGN ISSUES

Scaling down feature size is an important issue for high-performance and high-density VLSI circuits. However, some second order effects become serious for short-channel devices [6, 41], which have to be considered in device and circuit designs. Actually, the short-channel-effect (SCE) reduction, which includes threshold voltage roll-off and DIBL, has become a major challenge in deep submicrometer devices and circuits [42].

5.4.1 Short-Channel Threshold Voltage Roll-Off

Figure 5.19 shows a schematic of a MOS transistor. Here L is the channel length and X_j is the source and drain junction depth. The surface potentials

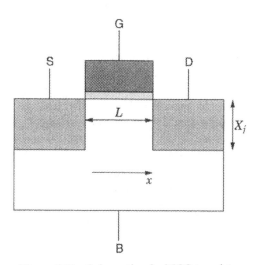

Figure 5.19   Schematic of a MOS transistor.

of short channel and long-channel devices before strong inversion are shown in Figure 5.20. The drain, source, and body voltages are all zero. For a long channel transistor, the barrier (Φ_{b1}) is a constant. Hence, the threshold voltage is not sensitive to the channel length variation. As for a short-channel device, the barrier (Φ_{b2}) is reduced along with the scaling of channel length. Therefore, the smaller the channel length, the lower the threshold voltage. The relationship between threshold voltage and transistor channel length is shown in Figure 5.21. In order to make the device work properly, dV_{th}/dL cannot be too large. This will determine the minimum channel length (L_{min}).

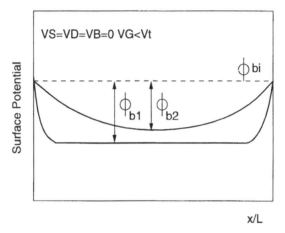

Figure 5.20 Surface potentials of short- and long-channel devices ($V_D = 0$ V).

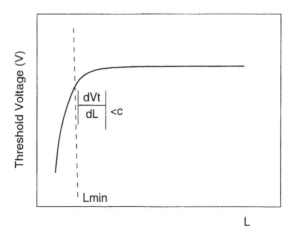

Figure 5.21 Threshold voltage vs. channel length.

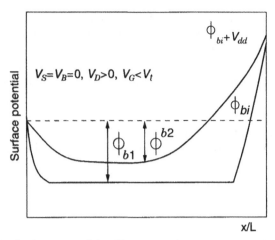

Figure 5.22 Surface potentials of long- and short-channel devices ($V_D > 0$).

5.4.2 Drain-Induced Barrier Lowering

Figure 5.22 shows the surface potentials of long-channel and short-channel devices at a large drain voltage (V_D). For a long-channel device, the barrier is not sensitive to V_D. However, the barrier of a short-channel device will reduce along with the increase of drain voltage, which will cause a higher subthreshold current and lower threshold voltage, as noted earlier.

5.5 KEY TO MINIMIZING SCE

In order to minimize a short-channel effect, a sufficient large aspect ratio (AR) of the device is required. AR is defined as

$$AR = \frac{\text{dimension}_{\text{lateral}}}{\text{dimension}_{\text{vertical}}} \tag{5.7}$$

For a MOSFET, AR can be expressed as

$$AR = \frac{L}{\left[t_{\text{ox}}(\epsilon_{\text{si}}/\epsilon_{\text{ox}})\right]^{1/3} d^{1/3} X_j^{1/3}} \tag{5.8}$$

Where ϵ_{si} and ϵ_{ox} are silicon and oxide permittivities and L, t_{ox}, d, and X_j are channel length, gate oxide thickness, depletion depth, and junction depth, respectively. From Eq. (5.8), we can see that reducing t_{ox}, d, and X_j will reduce the SCE of a MOSFET.

In order to minimize SCE, a modified MOSFET structure can be used. Figure 5.23 shows the low-impurity channel shallow-junction MOSFET [40].

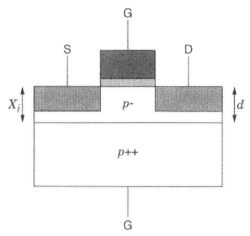

Figure 5.23 Schematic of a low-impurity channel shallow-junction bulk MOSFET.

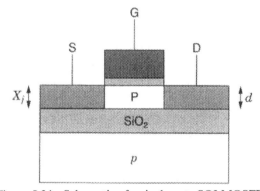

Figure 5.24 Schematic of a single-gate SOI MOSFET.

The small SCE of this transistor is because of the smaller depletion depth and junction depth. The AR of a single gate silicon-on-insulator (SOI) MOSFET is shown in Figure 5.24. Since $d = X_j$, the AR is given by

$$AR = \frac{L}{\left[t_{ox}(\epsilon_{si}/\epsilon_{ox})\right]^{1/3} d^{2/3}} \tag{5.9}$$

Figure 5.25 shows a double-gate silicon-on-insulator (DGSOI) MOSFET. The AR of DGSOI MOSFET is

$$AR = \frac{L}{\left[T_{ox}(\epsilon_{si}/\epsilon_{ox})\right]^{1/3} (d/2)^{1/3} (X_j/2)^{1/3}} \tag{5.10}$$

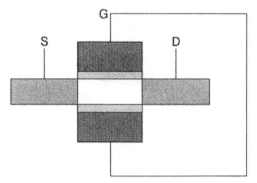

Figure 5.25   Schematic of a DGSOI MOSFET.

It is clear that the effective junction depth and the depletion width are reduced to half of that of a bulk MOSFET. Therefore, the SCE of a DGSOI MOSFET is much smaller than a bulk silicon MOSFET, which makes it a good candidate for deep submicrometer applications [43].

Short-channel threshold voltage roll-off and DIBL are two short-channel effects, which will complicate the transistor operation and degrade device performance. In order to minimize SCE, a sufficient large aspect ratio is required and novel MOSFET structures can be used.

5.6   LOW-VOLTAGE CIRCUIT DESIGN TECHNIQUES

The subthreshold current can be expressed as

$$I_{\text{sub}} = k \exp\left(\frac{V_{gs} - V_{\text{th}}}{S/\ln 10}\right)\left(1 - \exp\left(-\frac{V_{ds}}{V_T}\right)\right) \tag{5.11}$$

where k is a function of the technology, V_{th} is the threshold voltage, S is the subthreshold swing and V_T is the thermal voltage (KT/q). The subthreshold leakage is independent of V_{ds} if V_{ds} is approximately larger than 0.1 V [1].

From the above equation, we know that the subthreshold current can be reduced by lowering V_{gs} and increasing S and V_{th}. Based on the above discussions, a few low-voltage circuit design techniques are described below.

5.6.1   Reverse V_{gs}

The principle to reverse the V_{gs} of MOSFETs (in the case of nMOSFETs) in the standby mode of operation [2] can be best understood by considering the transistor characteristic shown in Figure 5.26. As V_{th} is scaled from V_{th_1} to V_{th_2}, the drain current at $V_{gs} = 0$ (leakage current in the standby mode)

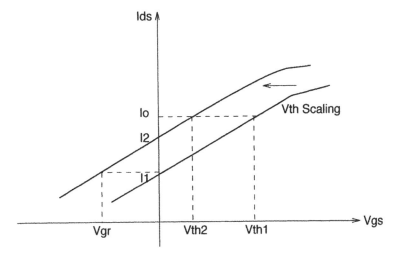

Figure 5.26 Principle of subthreshold reduction.

increases from I_1 to I_2. However, if V_{gs} can be reversed to V_{gr} in the standby mode of operation, the leakage current in the standby mode could be the same as that before scaling.

5.6.2 Steeper Subthreshold Swing

A steeper subthreshold swing helps achieve low leakage. The subthreshold swing S of a MOSFET can be expressed as follows [41]:

$$S = \frac{KT \ln 10}{q}\left(1 + \frac{C_d}{C_{ox}}\right) \qquad (5.12)$$

where K is the Boltzmann constant, T is the absolute temperature, q is the elementary charge, and C_d and C_{ox} are the capacitance of the depletion layer and gate oxide. One way of reducing the subthreshold swing is to operate the circuit at liquid nitrogen temperature, however, that would increase the cost. Besides, for portable applications, room temperature operation is essential. The SOI device structure (Figure 5.27) has the steepest subthreshold swing. For SOI transistors, C_d/C_{ox} is close to zero as the depletion capacitance is negligible. The SOI circuit also has better performance than the bulk silicon circuit because of the smaller parasitic capacitance. Figure 5.28 shows the subthreshold characteristics of bulk silicon and SOI nMOSFETs. For the same drain current at $V_{gs} = 0$, the V_{th} of the SOI device is less than that of the bulk silicon device. In other words, the leakage current of the SOI device in the standby mode is much lower than that of the bulk silicon device for the same threshold voltage.

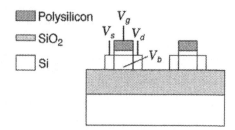

Figure 5.27   SOI structure.

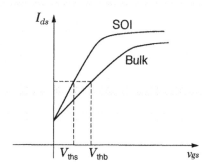

Figure 5.28   Subthreshold characteristics of SOI and bulk silicon devices.

5.6.3   Multiple Threshold Voltage

It is clear from the above discussions that threshold voltage is one of the most important parameters for device and circuit design. For the active mode, the low V_{th} is preferred because of the higher performance. However, for the standby mode of operation, high V_{th} is useful for reduction of leakage power. Hence, if different threshold voltages could be used during the different modes of operation, large improvements in performance are possible without sacrificing the speed. Different threshold voltages can be developed by multiple V_{th} implantation during the fabrication, by changing the substrate and source bias (body effect), by controlling the back gate of double-gate SOI devices (Figure 5.29) [31], and so on. Let us consider some multiple-threshold techniques to achieve low leakage under performance constraint. In particular, we will consider the following techniques: SATS

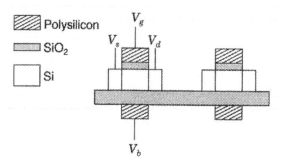

Figure 5.29   Double-gate SOI structure.

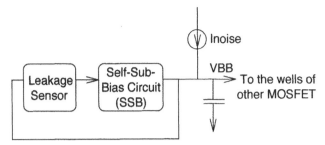

Figure 5.30   Block diagram of self-adjusting threshold-voltage scheme (SATS).

(self-adjusting threshold voltage scheme) [32], MTCMOS (multithreshold-voltage CMOS) [33], DTMOS (dynamic threshold voltage MOSFET) [34, 35], double gate dynamic threshold control SOI (DGDT-SOI) [26], and multiple-threshold CMOS based on path criticality [30].

- *SATS* Figure 5.30 shows a block diagram of the self-adjusting threshold-voltage scheme (SATS). A leakage sensor senses a representative MOSFET and outputs a control signal to self-sub-bias (SSB) circuit. Consider an *n*MOSFET transistor. When the leakage current is higher than a certain value, the SSB will be triggered and will reduce the substrate bias of all the other *n*MOSFETs, which in turn will increase the threshold voltage and reduce the leakage current. For *p*MOSFETs, a similar technique can be used.
- *MTCMOS* Multithreshold CMOS uses both high- and low-threshold-voltage MOSFETs in a single chip and a sleep control scheme is introduced for efficient power management (see Figure 5.31). In the

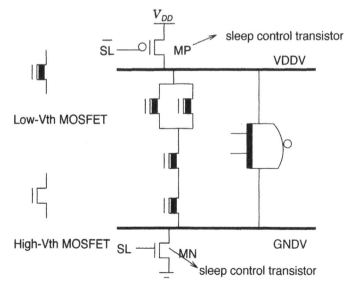

Figure 5.31   Schematic of MTCMOS.

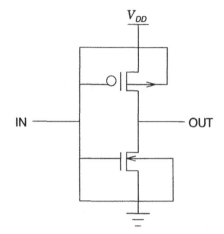

Figure 5.32 Schematic of DTMOS inverter.

active mode, SL is set high and the sleep control transistors (MP and MN) are turned on. The on resistances of the sleep control transistors are small. Hence VDDV and GNDV function as real power and ground lines. In the standby mode, SL is set low, MN and MP are turned off and VDDV and GNDV are floating. The leakage current is suppressed by the high-V_{th} MOSFETs—MN and MP. The technique is simple and achieves a large improvement in leakage current. However, one of the main disadvantages is the use of sleep control transistors, which can affect performance. It should also be noted that the sleep control transistors can be large and, hence, the capacitance being switched for turning on or off those transistors can be large.

- *DTMOS* For DTMOS, the gate and substrate of the transistors are tied together. Figure 5.32 shows the schematic of DTCMOS inverter. Because of the "body effect," the threshold voltage of MOSFETs can be changed dynamically during the different mode of operation. When IN is "low," pMOS is "on" with low V_{th} and nMOS is "off" with normal V_{th} (high V_{th}). In the active mode, the circuit switches from low to high with a higher speed because of the low-V_{th} pMOS. In the standby mode, the static leakage current is decided by the subthreshold current of the high-V_{th} nMOS and is smaller. When IN is "high," the situation is just the opposite. The supply voltage of DTMOS is limited by the diode built-in potential. The pn-junction diode between source and body should not be forward biased. So this technique is only suitable for ultra-low-voltage (0.6-V and below) circuits.

- *DGDT-SOI* Thin-film fully depleted (FD) SOI MOSFETs have nearly ideal subthreshold slope and small parasitic capacitance, which makes it attractive in low-voltage, high-performance applications. However, it is difficult to control the threshold voltage of FD SOI MOSFETs. Considering that the back gate can be used to control the front-gate

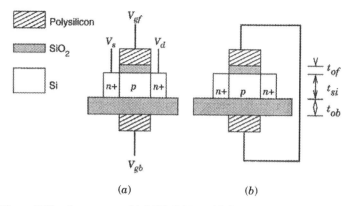

Figure 5.33   Structures of (*a*) FD SOI and (*b*) DGDT SOI MOSFET.

threshold voltage [27–30] and reduce the sensitivity of threshold voltage to the thin silicon film, DTMOSs can be achieved by double-gate FD SOI devices. The DGDT voltage SOI MOSFETs combine the advantages of DTMOSs and FD SOI MOSFETs without the limitation of the supply voltage.

Figure 5.33 shows the structures of the FDSOI MOSFET and the DGDT SOI MOSFET. The back-gate oxide of DGDT SOI MOSFETs is thick enough to make the threshold voltage of the back gate larger than the supply voltage. Since the front-gate and back-gate surface potentials are strongly coupled to each other, the front gate can be regarded as a conducting gate while the back gate acts as a controlling gate for the front gate. Figure 5.34 shows the $I-V$ characteristics of FD SOI MOSFETs and DGDT SOI MOSFETs. The design presumes 0.5-μm channel lengths and dual polysilicon gates. The front-gate oxide thickness (t_{of}), silicon layer thickness (t_{si}), and back-gate oxide thickness (t_{ob}) are 7, 50, and 20 nm, respectively. The body doping densities of nMOSFETs and pMOSFETs are 2.5×10^{17} cm^{-3} and 3.2×10^{17} cm^{-3}, respectively. For FD nMOSFETs (pMOSFETs), the threshold voltage can vary from 0.13 V (-0.13 V) to 0.35 V (-0.36 V) as the back-gate-to-source bias is changed from 1 V (-1 V) to 0 V. The DGDT SOI MOSFETs show better subthreshold characteristics than FD SOI MOSFETs and the threshold voltage can be altered dynamically to suit the operating state of the circuit. A high threshold voltage in the standby mode gives low leakage current (I_{off}), while a low threshold allows for higher current drives (I_{on}) in the active mode of operation. Figure 5.35 shows the variation in the range of threshold voltage for different t_{si} and t_{ob}. The thinner the silicon layer thickness, the smaller is the threshold voltage. This results in a higher drive current, but also larger leakage current. Thinning t_{ob} can improve the controllability of the back gate to the front gate, which increases the threshold voltage variation range.

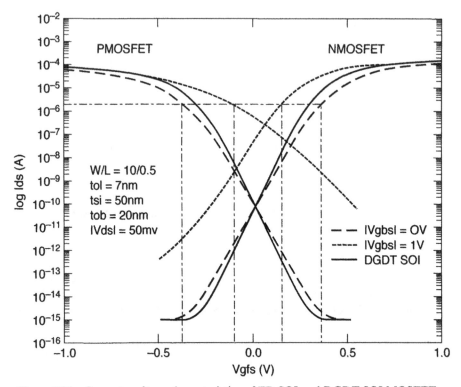

Figure 5.34   Current–voltage characteristics of FD SOI and DGDT SOI MOSFETs.

Even though the back-gate capacitance will also increase, it is much smaller than the front-gate capacitance If the back-gate oxide thickness is compatible with the front-gate oxide thickness, the back channel may conduct. It can improve the drive current, but the leakage current and gate capacitance will strongly increase. Table 5.4 shows the comparison of I_{on}/I_{off} for different FD SOI MOSFETs and DGDT SOI MOSFETs. For typical FD SOI pMOSFETs, the negative back-gate-to-source bias (V_{gbsp}) lowers the threshold voltage, and makes the leakage current too high to be used in low-voltage circuits. The DGDT SOI MOSFET shows symmetric characteristics and the best I_{on}/I_{off}.

Figure 5.36 shows the DGDT SOI inverter voltage transfer character-istics (VTC) for different supply voltages. Good noise margin can be seen even when the supply voltage is scaled down to 0.15 V. A full adder (Figure 5.37) was simulated for DGDT SOI and "modified" FDSOI (V_{gbsp} = 0 V) structures with a supply voltage of 1 V. The propagation delays are 0.625 ns and 0.75 ns while the average power consumptions are 11.5 mW and 10.5 mW for DGDT SOI and "modified" FDSOI full adders, respectively. Simulation results indicate that DGDT voltage SOI MOSFETs are very attractive for low-power, high-performance designs.

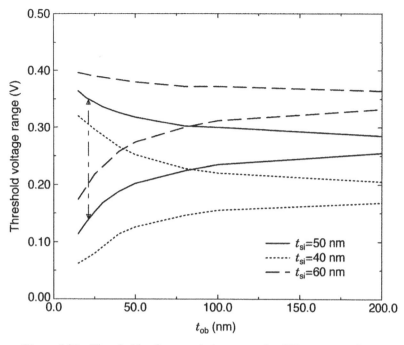

Figure 5.35 Threshold voltage variation range for different t_{si} and t_{ob}.

TABLE 5.4 Comparison of I_{on} / I_{off} for Different FD SOI MOSFETs and DGDT SOI MOSFETs

	V_{th}(V)	I_{on}/w at 1 V (10^{-5} A/μm)	I_{off}/w at 1 V (10^{-11} A/μm)	I_{on}/I_{off} at 1 V (10^6)
FDSOI NMOS	0.35	6.12	1.33	4.6
FDSOI PMOS				
($V_{gbs} = -1$ V)	-0.13	-5.54	-6800	0.0008
FDSOI PMOS				
($V_{gb} = 0$ V)	-0.36	-3.46	-0.735	4.7
DGDT SOI NMOS	0.13–0.35	10.02	1.33	7.53
DGST SOI PMOS	-0.13–0.36	-5.55	-0.735	7.55

It is possible to combine some of the above techniques to achieve even better performance. Let us consider some such techniques.

- *SSI CMOS* The switched source impedance (SSI) CMOS circuit [36] is shown in Figure 5.38. A switched impedance is set at the source of transistor MN, which consists of a resistor (R_S) and a switch (S_S). Here, S_S is turned on during the active mode and turned off during the

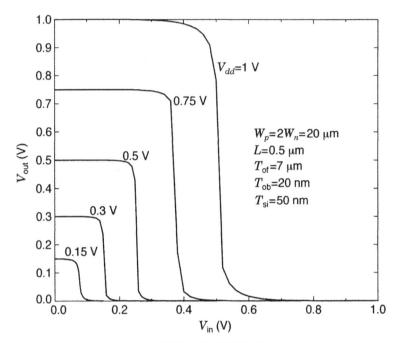

$$W_p = 2W_n = 20 \ \mu m$$
$$L = 0.5 \ \mu m$$
$$T_{of} = 7 \ \mu m$$
$$T_{ob} = 20 \ nm$$
$$T_{si} = 50 \ nm$$

Figure 5.36   VTC of DGDT SOI inverter.

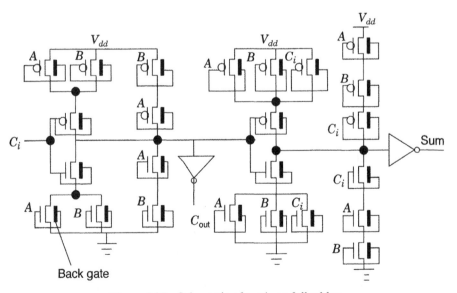

Figure 5.37   Schematic of a mirror full adder.

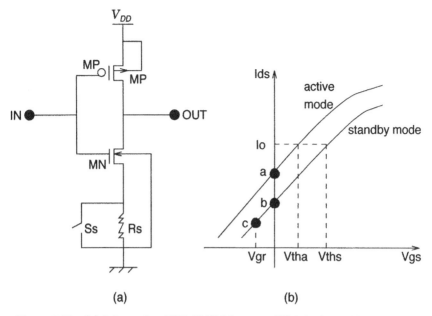

Figure 5.38   (*a*) Schematic of SSI CMOS inverter. (*b*) Principle of SSI CMOS.

standby mode. For nMOSFETs, in the standby mode, V_{gs} is reversed to V_{gr} because of the resistor, and hence, the threshold voltage becomes larger due to the increased bias of the source. From the active to the standby mode, the state of nMOSFETs changes from state a to state c. Similar switched impedance can be set at the source of MP. The reader can very easily see the similarity of this logic with MTCMOS. The high-V_{th} MOSFETs in MTCMOSs are equivalent to the switched impedances in SSI CMOS.

- *DTMOSs on SOI Structure* Although DTMOS can be developed in bulk silicon technology, the advantages of this logic family are more evident

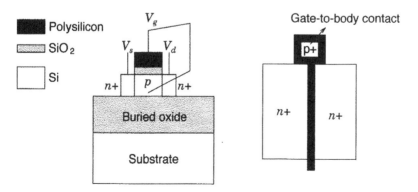

Figure 5.39   Schematics of SOI DTNMOS structure and layout.

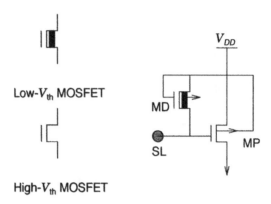

Figure 5.40   Sleep control scheme of MTCMOS SIMOX circuit.

using SOI technology. Figure 5.39 shows the schematic of DTMOS on SOI structure. The structure is simpler and has better characteristics. It has been shown in [35] that excellent DC inverter characteristics down to 0.2 V and good ring oscillator performance down to 0.3 V can be achieved using this method.

- *DTMOS Pass-Gate Circuit on SOI Structure* Pass-transistor logic families, such as CPL and DPL, are very promising for high-performance and low-power applications. By connecting the body of the SOI pass gate to the input gate, a DTMOS pass-gate SOI logic can be obtained. A 0.5 V full adder was developed using this method [37] that has a V_{th} of 0.4 V at 0 V body-bias (standby mode) and 0.17 V at 0.5 V body bias (active mode). The delay was reduced to one third of that of the conventional SOI CPL circuit.

- *MTCMOS with DTMOS on SOI Structure* The 0.5-V MTCMOS circuit of Figure 5.40 has been developed on SIMOX/SOI material [38]. The sleep control scheme is achieved by a modified DTMOS (see Figure 5.40). The body of high V_{th} is connected to the gates through a reverse-biased MOS diode (MD) whose size is about one-tenth of the size of a high-V_{th} MOSFET (MP). So a wider supply voltage range can be applied and the noise problem for DTMOSs can be reduced.

5.6.4 Multiple Threshold CMOS Based on Path Criticality

We noted in earlier sections that multiple threshold voltage can reduce leakage power dissipation. Computer-aided design techniques can also be used to reduce leakage current using logic gates having different threshold transistors. Consider a logic circuit. A higher threshold voltage can be assigned to some transistors on noncritical paths so as to reduce leakage current, while the performance is maintained due to the low-threshold transistors in the critical path(s). Recently, a dual-V_{th} MOSFET process was developed [44], which makes the implementation of dual-V_{th} logic circuits more feasible.

Due to complexity of logic circuits, not all the transistors on noncritical paths can be assigned a high threshold voltage; otherwise, the critical path may change, thereby increasing the critical path delay. An algorithm developed in [30] initially assigns low threshold voltage to all transistors of the design. The threshold voltage is selected based on the supply voltage, critical path delay, and noise considerations. The algorithm then selectively assigns higher threshold voltage in the noncritical paths of the design to reduce leakage current. It can be observed that the leakage power dissipation during both the standby and active modes of operation can be reduced.

The method to reduce leakage power using dual-threshold-voltage transistors was implemented using NAND gates, NOR gates, and inverters. A channel length of 0.3 μm was used. The channel widths for pMOSFETs and nMOSFETs were considered to be same (and were 3.6 and 1.8 μm), respectively.

Figure 5.41 gives an example circuit to show how the algorithm works. Figure 5.41a is the original single-V_{th} circuit, where the supply voltage is 1 V and the threshold voltage is 0.2 V. Figure 5.41b–d show the dual-V_{th} circuits with high threshold voltages of 0.25, 0.395, and 0.46 V, respectively. The low V_{th} is $0.2V_{dd}$. Note that the critical paths and critical delay are maintained after the assignment.

Figure 5.42 shows the standby leakage power of the above example circuit with different high thresholds (V_{th_2}). Here, V_{th_2} varies from 0.2 to 0.5 V ($V_{th_2} = 0.2$ V represents the single low-threshold circuit). The squares represent the leakage power obtained by the estimation technique [30] while the circles denote the leakage power obtained by HSPICE.

For a CMOS digital circuit, total power dissipation includes dynamic and static components in the active mode. Ignoring power dissipation due to the short-circuit current, total active power dissipation can be expressed as follows:

$$P_T = P_{dyn} + P_{static}$$
$$= \sum_i \alpha_i C_i V_i V_{dd} f_{clk} + I_{static} V_{dd} \qquad (5.13)$$

where α_i is the switching activity (the probability of switching), C_i are the load and parasitic capacitances, f_{clk} is the operating frequency, and V_i is the voltage swing which equals V_{dd} at the output node and $V_{dd} - V_{th}$ at internal nodes. The summation is taken over all nodes of the circuit. The term I_{static} is the leakage current through the circuit.

Consider the example circuit (Figure 5.41). Figure 5.43 shows the HSPICE simulation results of the total active power dissipations of single-V_{th} and dual-V_{th} circuits at different frequencies. The circuits were simulated at 1 V

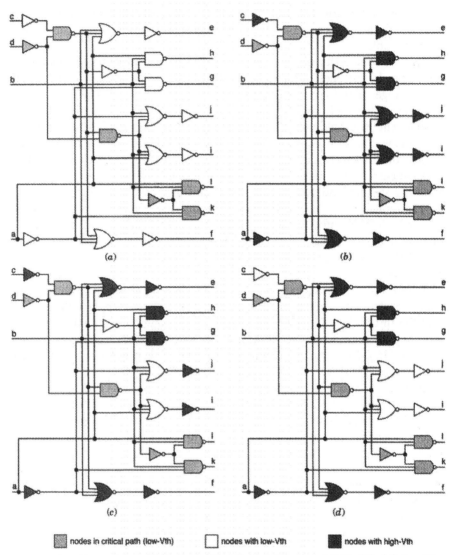

▨ nodes in critical path (low-Vth)   ☐ nodes with low-Vth   ■ nodes with high-Vth

Figure 5.41 Example circuit: (a) $1 - V_{\text{th}}$ $V_{dd} = 1$ V, $v_{\text{th}} = 0.2$ V; (b)–(d) $2 - V_{\text{th}}$, $V_{dd} = 1$ V, $V_{\text{th}_1} = 0.2$ V; (b) $V_{\text{th}_2} = 0.25$ V; (c) $V_{\text{th}_2} = 0.395$ V; (d) $V_{\text{th}_2} = 0.46$ V.

supply voltage and at 110°C. The threshold voltage of the single-V_{th} circuit was 0.2 V. The low and high threshold voltages of the dual-V_{th} circuit were 0.2 and 0.396 V, respectively. At low frequency, the active power savings of the dual-V_{th} circuit, mainly because of the static power reduction, is about 50%. For high-frequency circuits, the active power dissipation is dominated by dynamic consumption. In addition to leakage power saving, the dynamic power is reduced due to the reduction of internal node voltage swing for high

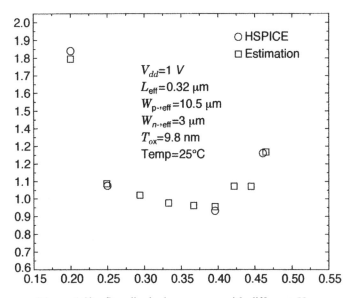

Figure 5.42   Standby leakage power with different V_{th_2}.

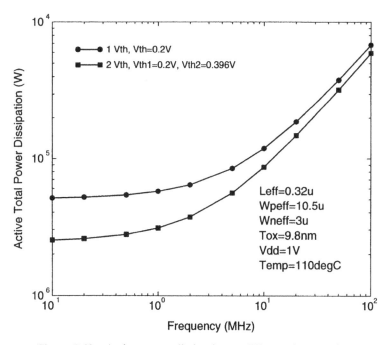

Figure 5.43   Active power dissipation at different frequencies.

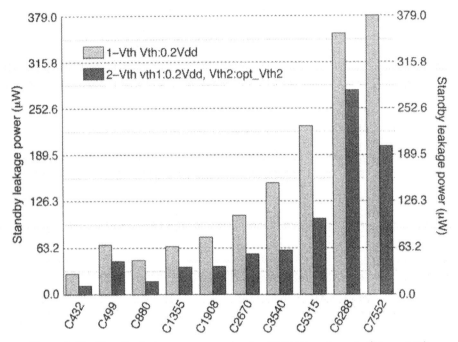

Figure 5.44 Standby leakage power saving for ISCAS benchmarks ($V_{dd} = 1$ V).

threshold gates. In this example, the total power saving can be around 13% at 100 MHz frequency.

Figure 5.44 shows the optimal high threshold and static power saving for ISCAS benchmark circuits (after technology mapping). The percentages of high-threshold-voltage transistors and gates over total transistors and gates for different benchmarks are illustrated in Figure 5.45. A V_{dd} of 1 V was used for all experiments with V_{th_1} of 0.2 V. Results indicate that the percentage of high threshold voltage transistors can be more than 60% and standby leakage power can be reduced by around 50% for most circuits. Even though the optimal high threshold voltage varies for different circuits, most of them are between $0.3V_{dd}$ and $0.4V_{dd}$.

5.7 TESTING DEEP SUBMICROMETER ICs WITH ELEVATED INTRINSIC LEAKAGE

The high leakage current in deep submicron, short-channel transistors is threatening well-established quiescent current (I_{DDQ}) based testing techniques [39]. The deep submicron device properties can be applied to a test application that combines I_{DDQ} and ICs maximum operating frequency (F_{max}) to establish a two-parameter test limit for distinguishing intrinsic and extrinsic (defect) leakages in microprocessors and ICs with high background leakage.

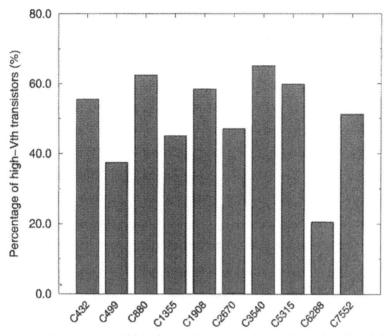

Figure 5.45 Percentage of high threshold gates and transistors for ISCAS benchmarks ($V_{dd} = 1$ V).

A common IC design practice increases the maximum operating frequency (F_{max}) for a given CMOS technology by lithographically reducing transistor channel lengths and L_{eff} to less than a given process target. This increases transistor intrinsic off-state leakage, but the shortened L_{eff} provides higher IC operating clock frequencies. As transistor I_{off} increases, the standby current (I_{SB}) and I_{DDQ} increase.

A cumulative distribution of I_{SB} is shown for a population of experimental data on a microprocessor with intrinsically short channel lengths (Figure 5.46). Curve (b) shows die that are intrinsically leakier than those in curve (a) because wafer (b) has die with smaller L_{eff} transistors. The I_{SB} (I_{DDQ}) values are large due to measurement at 85°C for a population of die with very short channel lengths. The shorter L_{eff} on the die increased I_{SB}, causing the observed tail. The I_{SB} measured at the IC level is a summation of the leakage for all transistors in the IC and is dominated by the transistors with minimum channel length.

An important observation is that the microprocessors in distribution Figure 5.46b are markedly faster than those in (a). Two key observations: (1) no punchthrough existed and (2) no extrinsic defect leakage in the distribution was observed (based on controlled experiment with effective channel length).

The critical parameter that affects device speed and off-state leakage is the transistor effective length, not the geometric length. Several properties

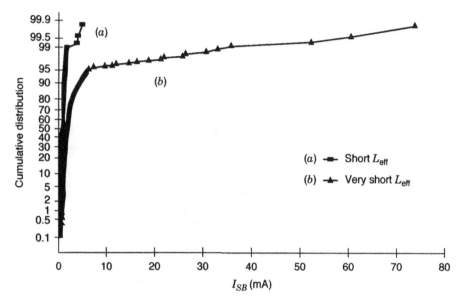

Figure 5.46 Intrinsic cumulative distribution for I_{SB} at 85°C for ICs with (*a*) short L_{gate} and (*b*) very short L_{gate}.

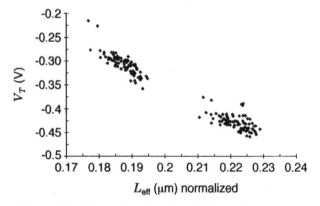

Figure 5.47 V_{th} vs. L_{eff} for short *p*-channel transistors.

change as the intrinsic L_{eff} becomes smaller. First, DIBL increases, allowing more charge to enter the channel from the source and threshold voltage goes down. Figure 5.47 shows this V_{th} effect measured on a distribution of intrinsically short *p*-channel transistors. The test transistors were measured in the wafer scribe lines. The data in Figures 5.47–5.51 were modified for publication (absolute values are not necessary to show these effects).

A related effect is the increase in I_{off} as L_{eff} becomes smaller (Figure 5.48). Here, I_{off} has a log-linear response to L_{eff} while V_{th} has a linear response to L_{eff}.

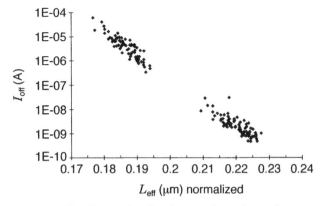

Figure 5.48 I_{off} vs. L_{eff} for short p-channel transistors.

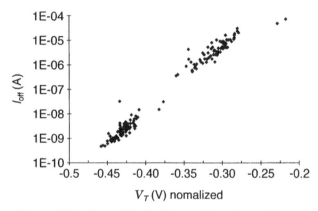

Figure 5.49 I_{off} vs. V_{th} for short p-channel transistors.

The log-linear relation between I_{off} and V_{th} is shown in Figure 5.49. This exponential relation shows orders-of-magnitude reduction in I_{off} for smaller magnitude changes in V_{th}.

Shorter channel lengths reduce the transit time of carriers moving from source to drain so that the device and circuit are faster. A lower V_{th} allows stronger drive to transistors with shorter channel length, but the reduced V_{th} increases I_{off}. If the same gate voltage is applied to a distribution of L_{eff} transistors, then shorter transistors will show more drain current response (g_m) or $I_D(sat)$. Drain current in the saturated state is an indicator of how fast a transistor can charge and discharge load capacitance during logic transition.

Figure 5.50 shows these relations with $I_D(sat)$ versus $1/L_{eff}$ measurements. Transistors with shorter L_{eff} will charge and discharge load capacitances faster than long transistors. Equation (5.14) is the saturation current

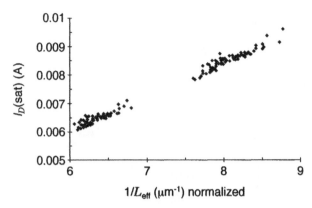

Figure 5.50 $I_D(\text{sat})$ vs. $1/L_{\text{eff}}$ for p-channel transistors.

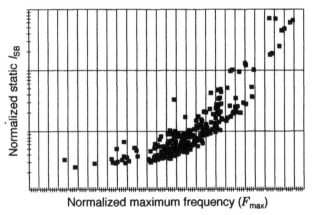

Figure 5.51 I_{SB} vs. F_{max} for 32-bit microprocessor.

equation for a MOSFET and shows $I_D(\text{sat})$ as a linear function of $1/L_{\text{eff}}$ [6]. Here K is the conduction constant:

$$I_D(\text{sat}) = \frac{W}{L_{\text{eff}}} K \left(V_{gs} - V_{\text{th}}\right)^2 \tag{5.14}$$

Importantly, Figure 5.50 supports evidence that the ICs in Figure 5.46 are not in punchthrough; the ICs from the distribution tail are leakier due to weak-inversion and DIBL effects. Therefore, there is a clear correlation between I_{DDQ} (I_{SB}) and the maximum operating frequency (F_{max}) of a microprocessor as both are functions of L_{eff}.

Figure 5.51 summarizes the previous figures, showing how F_{max} and I_{SB} track each other. The parameters I_{SB} and F_{max} are fundamentally related as

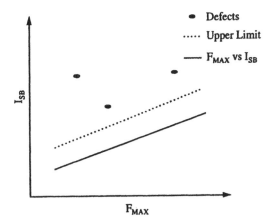

Figure 5.52   Test application.

both are functions of channel length. This relation can be used in testing. An adjustable I_{SB} limit can be set based upon the parameters I_{SB} and F_{max} to establish a two-parameter test limit that distinguishes fast and slow die from those that are defective. The concept allows for improved signal-to-noise ratio for defect detection for high-performance ICs with high background leakage levels.

Figure 5.52 further illustrates the test application. We have assumed a linear dependency for simplicity. Intrinsic values of I_{SB} can be distinguished from extrinsic (defect-driven) I_{SB} values and a limit set up to reject the defective ICs. The I_{SB} limit moves up as F_{max} increases.

The test technique shows that I_{DDQ} testing can be effective testing methodology even for deep submicron designs provided it is coupled with the F_{max} parameter.

5.8   MULTIPLE SUPPLY VOLTAGES

Since power dissipation decreases quadratically with the scaling of supply voltage, while the delay is proportional to $V_{dd}/(V_{dd} - V_{th})^2$, it is possible to use high supply voltage in the critical paths of a design to achieve the required performance while the off-critical paths of the design use lower supply voltage to achieve low-power dissipation [45–49]. In [46], dual supplies have been used at the gate level of the design. The concept is very similar to the multiple-threshold design shown in Section 5.6.4. Whenever a logic gate operating at a low voltage is driving a logic gate operating at a higher voltage, level converters are required. Such a level converter may not be required when a logic gate operating at a higher voltage drives logic operating at a lower voltage. A DCVS type of level converter is shown in Figure 5.53 [50].

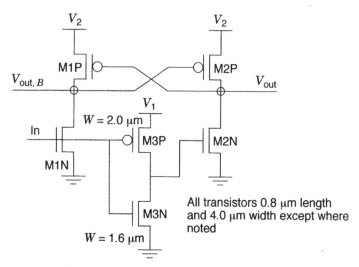

Figure 5.53 A DCVS voltage level converter.

A model is needed that could accurately indicate the power dissipation and propagation delay of the DCVS level converter as a function of the input logic supply voltage V_1, output logic supply voltage V_2, and load capacitance.

Multiple supply voltages have also been used for digital signal processing data path designs [45, 47–49]. During the scheduling of data flow graphs, when each operation of the data path is scheduled to meet the timing under resource constraints, different supply voltages are assigned to different functional blocks such as adders and multipliers. The basic concept is to slow down the operations using lower supply voltages if the operations are not in the critical path. Level converters are required for driving functional blocks operating at higher voltages with blocks operating at lower voltages. Figure 5.54a shows an example data flow graph. The scheduled data flow graph with multiple supplies is shown in Figure 5.54b. It is clear that multiple supply voltage helps in meeting the latency constraint while the functional blocks on noncritical paths and operating at lower supplies help reduce the power dissipation. In [45], an ILP (integer linear programming) based approach has been used to schedule data flow graphs for a number of digital filters. Results in Figure 5.55 show the improvements in energy dissipation of multiple-supply designs compared to a design using an optimum supply voltage.

The preceding results permit several observations to be made regarding the effect of latency, circuit resource, and supply voltage constraints on energy savings, area costs, and execution time. Because the primary objective has been to minimize energy dissipation through use of multiple voltages, it will be especially important to compare the multiple-supply-voltage results to minimum single-supply-voltage results. Energy savings ranging from 0 to 50% were observed when comparing multiple- to single-voltage results.

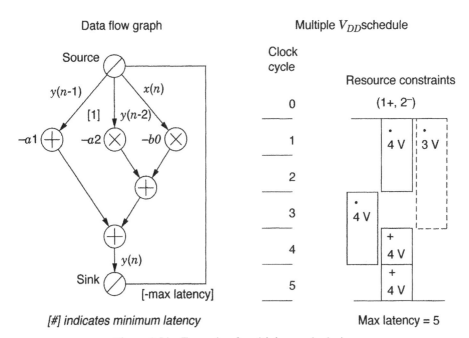

Figure 5.54   Example of multiple-supply design.

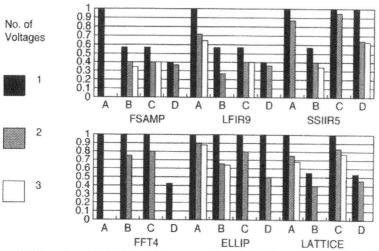

A = Minimum Latency, Unlimited Resources   B = 1.5 × Minimum Latency, Unlimited Resources
C = 1.5 × Minimum Latency, Unlimited Resources   D = Minimum Latency, Unlimited Resources

Figure 5.55   Multiple-supply voltage results.

If one considers the impact of latency constraints alone, effects on area and energy are easier to observe. In most cases, multiple-voltage-area penalties were greatest for the minimum-latency unlimited-resource test cases. One can also observe that increasing latency constraints always led to the same or lower energy for a given number of supply voltages. However, the effect of latency constraints on the single- versus multiple-voltage trade-off varied greatly from one example to another. Results for multiple voltages are most favorable in situations where the single-supply-voltage solution did not benefit from increased latency.

The effects of resource constraints on energy savings are also relatively easy to observe. Not surprisingly, resource constraints tend to produce the lowest area penalties. The only reason for any area penalty at all in the resource-constrained case is that sometimes the minimum single-supply solution does not require all of the resources that were permitted. Energy estimates based on resource-constrained schedules were consistently the same or higher than estimates based on unlimited resource schedules.

There are several design issues that a designer will need to take into consideration when a multiple-voltage design is targeted for fabrication. In particular, the effects of multiple-voltage operation on IC layout and power supply requirements should be considered. Following are some ways that multiple-voltage design may affect IC layout:

1. If the multiple supplies are generated off-chip, additional power and ground pins will be required.
2. It may be necessary to partition the chip into separate regions, where all operations in a region operate at the same supply voltage.
3. Some kind of isolation will be needed between regions operated at different voltages.
4. There may be some limit on the voltage difference that can be tolerated between regions.
5. Protection against latch-up may be needed at the logic interfaces between regions of different voltage.
6. New design rules for routing may be needed to deal with signals at one voltage passing through a region at another voltage.

Isolation requirements between different voltage regions can probably be adequately addressed by increased use of substrate contacts, separate routing of power and ground, increased minimum spacing between routes (for example, between one signal having a 2-V swing and another with a 5-V swing), and slightly increased spacing between wells. While these practices will increase circuit area somewhat, the effect should be small in comparison to increased circuitry (adders, multipliers, registers, etc.) needed to support parallel operations at reduced supply voltages. Area for isolation will be further mitigated by grouping together resources at a particular voltage into a common region. Isolation is then only needed at the periphery of the region.

Some of these layout issues can be incorporated into multiple-voltage scheduling. Perhaps the greatest impact will be related to grouping operations of a particular supply voltage into a common region. Closely intermingled operations at different voltages could lead to complex routing between regions, increased need for level conversions, and increased risk of latch-up. Assigning highly connected operations to the same voltage not only could improve routing, but also should lead to fewer voltage regions on the chip, less space lost to isolation between voltage regions, and fewer signals passing between regions operating at different voltages.

5.9 CONCLUSIONS

Low-voltage circuit design has become more and more important because of the increasing demand for low-power mobile electronic equipment. For low-voltage circuits, there is a trade-off between performance and leakage power dissipation. Proper device optimization along with optimum values of supply voltage and transistor thresholds can achieve low power under performance constraints. The testing of circuits with elevated intrinsic leakage is also tricky. Two-parameter testing based on I_{DDQ} and F_{max} shows that the testing technique can be effectively used to screen defects in high-performance, low-power, low-V_{th} CMOS ICs.

REFERENCES

[1] A. P. Chandrakasan and R. W. Brodersen, "Minimizing Power Consumption in Digital CMOS Circuits," *Proc. IEEE*, vol. 83, no. 4, pp. 498–523, 1993.

[2] A. Bellaouar and M. I. Elmasry, *Low Power Digital VLSI Design*, Kluwer Academic, Boston, 1995.

[3] J. Rabaey, *Digital Integrated Circuits: A Design Perspective*, Prentice-Hall, Englewood Cliffs, NJ, 1996.

[4] K. Eshraghian, *Principles of CMOS VLSI Design*, Addison-Wesley, Reading, MA, 1993.

[5] A. Keshavarzi, K. Roy, and C. Hawkins, "Intrinsic I_{DDQ}: Origins, Reduction, and Applications in Deep Sub-μ Low-Power CMOS IC's, *IEEE International Test Conference*, 1997, pp. 146–155.

[6] R. Pierret, *Semiconductor Device Fundamentals*, Addison-Wesley, Reading, MA, 1996.

[7] A. S. Grove, *Physics and Technology of Semiconductor Devices*, Wiley, New York, 1967.

[8] A. W. Righter, J. M. Soden, and R. W. Beegle, "High Resolution IDDQ Characterization and Testing—Practical Issues," *International Test Conference*, Oct. 1996, pp. 259–268.

[9] W. Lee, U. Ko, and P. Balsara, "A Comparative Study on CMOS Digital Circuit Families for Low-Power Applications," *Proc. International Workshop Low Power Design* pp. 129–131, 1996.

[10] S.-L. Lu and M. Ercegovac, "Evaluation of Two-Summand Adders Implemented in EDCL CMOS Differential Logic," *IEEE J. of Solid State Circuits*, vol. 26, no. 8, pp. 1152–1160, Aug. 1991.

[11] Y. P. Tsividis, *Operation and Modeling of the MOS Transistor*, McGraw-Hill, New York, 1987.

[12] J. R. Brews *High Speed Semiconductor Devices*, Ed., S. M. Sze, Wiley, New York, 1990, Chap. 3.

[13] M. Bohr et al., "A High Performance 0.25 mm Logic Technology Optimized for 1.8 V Operation," *IEDM Tech. Dig.*, pp. 847–851, Dec. 1996.

[14] R. Zimmermann and R. Gupta, "Low-Power Logic Styles: CMOS vs. CPL," *European Solid-State Circuits Conference*, 1996.

[15] S. Thompson, I. Young, J. Greason, and M. Bohr, "Dual Threshold Voltages and Substrate Bias: Keys to High Performance, Low Power, 0.1 μm Logic Designs," *1997 Symposium on VLSI Technology*, 1997, pp. 69–70.

[16] D. Somasekhar and K. Roy, "Differential Current Switch Logic: A Low-Power DCVS Logic Family," *IEEE J. Solid-State Circuits*, pp. 981–991, July 1996.

[17] D. Somasekhar and K. Roy, "LVDCSL: Low Voltage Differential Current Switch Logic, a Robust Low Power DCSL Family," *1997 International Symposium on Low Power Electronics and Design*, pp. 18–23.

[18] Z. Chen, M. Johnson, L. Wei, and K. Roy, "Estimation of Standby Leakage Power in CMOS Circuits," *International Symposium on Low Power Electronics and Design*, Monterrey, CA, Aug. 1998.

[19] C. Hu, "Device and Technology Impact on Low Power Electronics," in *Low Power Design Methodologies*, Kluwer Academic, Boston, pp. 21–36, 1996.

[20] M. C. Johnson, K. Roy, and D. Somasekhar, "A Model for Leakage Control by Transistor Stacking," Technical Report TR-ECE 97-12, Purdue University, Department of ECE.

[21] Y. Ye, S. Borkar, and V. De, "Standby Leakage Reduction in High-Performance Circuits Using Transistor Stack Effects," *1998 Symposium on VLSI Circuits*.

[22] J. Sheu et al., "BSIM: Berkeley Short-Channel IGFET Model for MOS Transistors," *IEEE J. Solid-State Circuits*, vol. SC-22, 1987.

[23] M.-C. Jeng, "Design and Modeling of Deep-Submicrometer MOSFETs," Electronics Research Laboratory, Report No. ERL-M90/90, University of California, Berkeley, 1990.

[24] HSPICE User's Manual, Vol. II, 1996.

[25] J. Halter and F. Najm, "A Gate-Level Leakage Power Reduction Method for Ultra Low-Power CMOS Circuits," *IEEE Custom Integrated Circuits Conference*, 1997, pp. 475–478.

[26] L. Wei, Z. Chen, and K. Roy, "Double Gate Dynamic Threshold Voltage (DGDT) SOI MOSFETs for Low Power High Performance Designs," *IEEE SOI Conference*, 1997, pp. 82–83.

[27] I. Y. Yang, C. Vieri, A. Chandrakasan, and D. A. Antoniads, "Back Gated CMOS on SOIAS for Dynamic Threshold Voltage Control," *International Electron Devices Meeting*, 1995, pp. 877–880.

[28] J. P. Denton and G. W. Neudeck, "Fully Depleted Dual-Gated Thin-Film SOI PMOSFET's Fabricated in SOI Islands with an Isolated Buried Polysilicon Backgate," *IEEE Electron Device Letters*, vol. 17, no. 11, pp. 509–511, Nov. 1996.

[29] P. C. Yeh and J. G. Fossum, "Viable Deep-Submicron FD/SOI CMOS Design for Low-Voltage Applications," *IEEE International SOI Conference*, 1994, pp. 23–24.

[30] L. Wei, Z. Chen, and K. Roy, "Design and Optimization of Low Voltage High Performance Dual Threshold CMOS Circuits," *IEEE/ACM Design Automation Conference*, 1998.

[31] Carlin Vieri et al., "SOIAS: Dynamically Variable Threshold SOI with Active Substrate," *IEEE Symposium on Low Power Electronics*, 1995, p. 86.

[32] T. Kobayashi and T. Sakurai, "Self Adjusting Threshold Voltage Scheme (SATS) for Low Voltage High Speed Operation," *IEEE 1994 Custom Integrated Circuit Conference*, 1994, p. 271.

[33] S. Mutoh et al., "1-V Power Supply High-Speed Digital Circuit Technology with Multithreshold-Voltage CMOS," *IEEE J. of Solid-State Circuits*, vol. 30, no. 8, p. 847, 1995.

[34] T. Andoh et al., "Design Methodology for Low Voltage MOSFETs," *IEDM Tech. Dig.*, p. 79, 1994.

[35] Fariborz Assaderaghi et al., "A Dynamic Threshold Voltage MOSFET(DTMOS) for Ultra-Low Voltage Operation," *IEDM Tech. Dig.*, p. 809, 1994.

[36] M. Horiguchi et al., "Switched-Source-Impedance CMOS Circuit for Low Standby Subthreshold Current Giga-Scale LSI's," *IEEE J. of Solid-State Circuits*, vol. 28, no. 11, p. 1131, 1993.

[37] T. Fuse et al., "0.5V SOI CMOS Pass-Gate Logic," *ISSCC Dig. Tech. Papers*, p. 88, 1996.

[38] T. Douseki et al., "A 0.5V SIMOX-MTCMOS Circuit with 200ps Logic Gate," *ISSCC Dig. Tech. Papers*, p. 84, 1996.

[39] A. Keshavarzi, K. Roy, and C. Hawkins, "Intrinsic I_{DDQ}: Origins, Reduction, and Applications in Deep Sub-μ Low-Power CMOS IC's," *IEEE International Test Conference*, 1997, pp. 146–155.

[40] J. D. Meindl, "Low Power Microelectronics: Retrospect and Prospect," *Proc. IEEE*, vol. 83, no. 4, p. 619, April 1995.

[41] S. M. Sze, *Physics of Semiconductor Devices*, 2nd ed., Wiley-Interscience, New York, 1981.

[42] C. H. Wann, K. Noda, T. Tanaka, M. Yoshida, and C. Hu, "A Comparative Study of Advanced MOSFET Concepts," *IEEE Trans. Electron Devices*, vol. 43, no. 10, pp. 1742–1753, 1996.

[43] R. Yan, A. Ourmazd, and K. F. Lee, "Scaling the Si MOSFET: From Bulk to SOI to Bulk," *IEEE Trans. Electron Devices*, vol. 39, no. 7, pp. 1704–1710, 1992.

[44] Z. Chen et al., "0.18 μm Dual Vt MOSFET Process and Energy-Delay Measurement," *IEDM Dig.*, p. 851, 1996.

[45] M. Johnson and K. Roy, "Datapath Scheduling with Multiple Supply Voltages and Voltage Converters," *ACM Trans. on Design Automation Electron. Syst.*, July 1997.

[46] K. Usami et al., "Automated Low-Power Techniques Exploiting Multiple Supply Voltages Applied to a Media Processor," *IEEE Custom Integrated Circuits Conference*, 1997, pp. 131–134.

[47] C. Gebotys, "An ILP Model for Simultaneous Scheduling and Partitioning for Low Power System Mapping," Technical Report (April), University of Waterloo, Department of Electrical and Computer Engineering, VLSI Group.

[48] J. Chang and M. Pedram, "Energy Minimization Using Multiple Supply Voltages," *Proc. Int. Symp. Low Power Electron. Design*, pp. 157–162, 1996.

[49] S. Raje and M. Sarrafzadeh, "Variable Voltage Scheduling," *International Symposium on Low Power Design*, 1995, pp. 9–14.

[50] K. Usami and M. Horowitz, "Clustered Voltage Scaling Technique for Low-Power Design," *Proc. Int. Symp. Low Power Design*, pp. 3–8, 1995.

CHAPTER 6

LOW-POWER STATIC
RAM ARCHITECTURES

DINESH SOMASHEKAR and KAUSHIK ROY

6.1 INTRODUCTION

Memory structures have become an indivisible part of modern VLSI systems. Semiconductor memory is now present not just as stand-alone memory chips but also as an integral part of complex VLSI systems. While the predominant criterion for memory optimization is often to squeeze in as much memory as possible in a given area, the trend toward portable computing (without impacting performance) has led to power issues in memory coming to the forefront.

The basic storage elements of semiconductor memory have remained fundamentally unchanged for quite some time. This does not imply that other forms of storage cells cannot be conceived of; rather these cells offer the best trade-off between design factors such as layout efficiency, performance, and noise sensitivity. In this chapter, we are concerned with read/write MOS random-access memories (RAMs). Traditional RAMs have been classified into dynamic RAMs (DRAMs) and static RAMs (SRAMs). The former is typically implemented using a single-transistor storage element, where the cell state is stored as charge on a capacitor. The term *dynamic* refers to the need to periodically refresh the charge on nonideal storage capacitors. Static RAMs, in contrast, use a bistable element such as an inverter loop to store the cell state as a voltage differential. These elements can hold their state without the need for refreshing as long as power is applied. The basic SRAM cell is considerably more complex and occupies a larger area as compared to a DRAM cell.

We shall focus primarily on the design of low-power MOS SRAMs. Two modes of operation can be conceived of for SRAMs: normal mode of operation where reads and writes are being performed in the core and a standby mode of operation where the RAM holds previously written values. Low-power design considerations will be separated based upon the mode of operation. We begin by presenting the common forms of storage cells for SRAMs. The architecture of a typical SRAM is described to understand the various components that contribute to overall power. We describe common design and architectural techniques for minimizing power. To illustrate the breadth of design techniques that have been employed for reducing power consumption, some representative approaches reported in published literature are discussed.

6.2 ORGANIZATION OF A STATIC RAM

While there are a vast array of possible RAM configurations, a generic RAM organization that illustrates the basic features of most RAMs is shown in Figure 6.1. We identify the following blocks in a SRAM:

Memory Core The actual storage of information is done in a two-dimensional array of memory cells called the memory core. This structure can access any cell by accessing a particular row and a column. A row is activated by a global line spanning all cells of the row and is termed the word line. This line enables the cells to be written into or read out. Every column can now be accessed individually through vertical bit lines that span individual columns of the memory array. In general, there is no restriction on the number of word lines or bit lines per row or column. The most common configuration uses a single word line to connect cells on to a differential pair of bit lines. Notice that activating a row enables all cells in a row and discharges every pair of bit lines. Such a structure, while not energy efficient, is to be preferred for packing as many cells as possible in a given area and for allowing a simple enough access scheme. A group of columns that may be less than the total number of columns are read out at a time to form a word of data. To summarize, individual cells in the memory core can be accessed by activating a word line and reading data out on a particular set of bit lines.

Word Decoders The set of cells that generate the word line signals form the word decoders. This structure takes a set of *n* address lines and

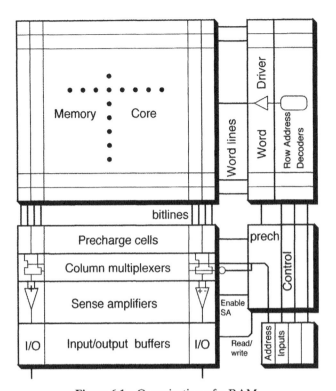

Figure 6.1   Organization of a RAM.

generates 2^n word lines. At most one of the word lines is active at a time.

Column Decoders   Column decoders select particular bit lines for being connected to sense amplifiers. This is accomplished either by sensing every bit line and gating a few of them out or by using pass gates to enable them to a few sense amplifier inputs.

Precharge   Differential read/write schemes are used for memory cells. This requires bit lines to be set up in a well-defined state before an access. Precharge cells accomplish this. Dynamic schemes use a clocked precharge cell to charge bit lines. Static schemes leave the precharge continuously on, in which case they resemble a load cell on the bit lines. Precharge cells also accomplish equalization of differential bit line voltages.

Sense Amplifiers   These circuits accomplish the conversion of bit line differentials to logic levels. Sense amplifiers are analog circuits, often resembling simple differential amplifiers.

We describe some of the above components in greater detail in the following sections.

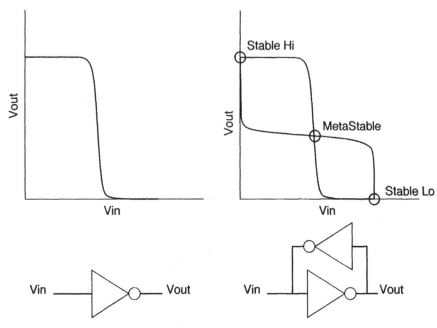

Figure 6.2 Transfer characteristic of an inverter loop.

6.3 MOS STATIC RAM MEMORY CELL

Any static bistable element, say a flip-flop, is a possible candidate for a memory cell. Of interest is the need to place as many of these cells in a given area. We can translate this to the layout issue of minimizing area and the total number of global interconnects. The type of predominant cell is formed of a pair of inverters in a closed loop. Such an element is bistable. We illustrate this in Figure 6.2, which shows the input–output characteristics of a pair of inverters superimposed to show the stable points. While CMOS is the MOS family of choice for implementing logic, the need for minimized area often leads to all NMOS implementation of inverters. Consequently two forms of SRAM cells have become popular and are classified based upon the number of transistors in the cell. These are the four-transistor (4T) SRAM cell, which is all NMOS, and the six-transistor (6T) CMOS SRAM cell.

6.3.1 The 4T SRAM Cell

The circuit topology of the 4T NMOS cell is shown in Figure 6.3. Inverters in this cell use an NMOS transistor with a resistive load. The resistive load is accomplished by generating high-valued resistors using undoped polysilicon. It should be apparent that this often leads to the cell being at a disadvantage in a normal logic process that uses strongly doped polysilicon. The greatest advantage this cell enjoys is the high degree of compactness in its layout.

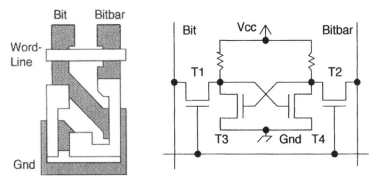

Figure 6.3 Four-transistor SRAM bit cell. A possible layout of the cell in a double-polysilicon process is shown.

A lower transistor count, the absence of n-well areas, and process techniques that allow contacts from active areas of polysilicon all contribute to a lower area.

It may seem that the cell should have a heightened power consumption because of continuous current drawn by the NMOS-only inverters; however, a look at typical values of resistances employed in commercial DRAMs show that the current requirement of such cells is extremely small. While it is clear that a higher resistance reduces current consumption and leads to lower standby power consumption, it seriously impacts the noise sensitivity of the cell. The extreme case of the resistances being absent leads to a 4-T DRAM cell.

The design of the 4T SRAM cell is concerned with getting an appropriate ratio of the pass transistors T1/T2 with respect to the pull-down transistor T3/T4. We shall refer to this as the aspect ratio of the cell. It is clear that the aspect ratio should ensure that at no time internal cell voltages rise to a value large enough to upset the cell state. We should note that the energy dissipation due to discharge of bit line capacitances is considerably larger than the energy required to switch the cell.

6.3.2 The 6T SRAM Cell

The CMOS implementation of the 4T SRAM cell replaces the inverters by fully complementary CMOS inverters. The supply current drawn by this 6T SRAM cell is limited to the leakage current of transistors in the stable state (Figure 6.4). The PMOS devices play the role of the resistors and are often minimum geometry. The inverters usually have a large NMOS width as compared to PMOS width and are not comparable to inverters found in CMOS cell libraries. This often causes the switch threshold of the inverters to be close to the NMOS threshold voltages. Considerations for getting the aspect ratio is the *same* as that for the 4T SRAM cell.

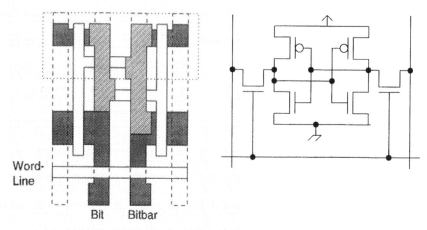

Figure 6.4 Six-transistor RAM cell.

6.3.3 SRAM Cell Operation

An SRAM cell can be in one of three possible states. It can be in the stable state with the cell holding a value or it can be in the process of carrying out a read or a write. The stable state of the cell occurs with the word line connected to T1/T2 being driven low. The cell is effectively disconnected in this mode of operation.

A read operation is initiated by precharging the bit lines high and activating the word line. One of the bit lines discharges through the bit cell, and a differential voltage is set up across the bit line. This voltage is sensed and amplified to logic levels. A write to the cell is carried out by driving the bit lines to the required state and activating the word line.

In standard logic, energy consumption is reduced by reducing one of four factors: total switched capacitance, voltage swing, activity factor, or frequency of operation. In memory cells, it is usually the voltage swing at various points that can easily be manipulated by a circuit designer without impacting other specifications of the memory. It is clear that to achieve the goal for lower power, reduced voltage swings on bit lines and word lines are essential. The lower limit of supply voltages is determined by our inability to resolve small voltage differentials at adequate speed, degraded signal integrity, and increasing soft error rates (SERs).[1]

---

[1]Soft error rates are the observed phenomenon of randomly upset bits in a memory core. These arise because of high-energy particles traversing through silicon. Large numbers of electron–hole pairs created in this process aggregate toward internal nodes of a memory cell. Alpha particles (doubly ionized helium) are the primary source of electron–hole pair generation. It is intuitively apparent that the higher the stored charge at dynamic nodes, the better the alpha particle immunity of the circuit. This factor drives one to chose higher voltages and larger capacitances at bit cell internal nodes.

6.4 BANKED ORGANIZATION OF SRAMs

Banking is an organization technique that targets total switched capacitance to achieve reduced power and improved speed. Our previous discussion of the SRAM memory core shows that an n-bit memory needs a $R \times C$ memory core, where R is the number of rows and C is the number of columns. An unfortunate outcome of such an organization is that any access to the core causes R cells to be enabled and the entire set of bit lines to be toggled. In general, if C_{cell} is the capacitance of the bit line per cell, a total capacitance of $R \times C \times C_{cell}$ is switched.

Clearly splitting the memory into a set of smaller memories is an efficient way to reduce the total switched capacitance. Figure 6.5 shows the organization of a banked SRAM. An additional set of decoders is now necessary to select one of B banks (we have $B = 4$ in Figure 6.5). While an additional delay is incurred in selecting the bank, it is offset by the much lower capacitance that is switched per bank, which is clearly $(R \times C \times C_{cell})/B$. In actual cases it is possible to find an optimum number of banks given the SRAM size and the form of decoding employed. Banking is clearly a technique to be preferred for large RAMs [11, 12], as it reduces the overall power consumption by the bank size. Banking also allows some flexibility in

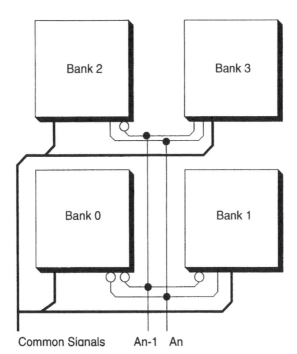

Figure 6.5 Banked organization of SRAMs.

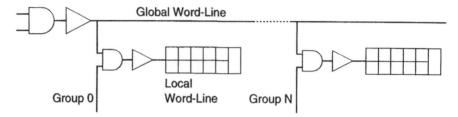

Figure 6.6   Divided word line architecture.

SRAM design since failure of a bank still allows one to package a part with a smaller amount of total memory.

6.4.1   Divided Word Line Architecture

Divided word line architecture (Figure 6.6) applies the above technique to the word lines alone. Word lines are no longer a single line spanning all memory cells in a row of the memory core. We differentiate word lines into global and local word lines. Memory cells along a row are grouped into a smaller number of cells. Each group has local gating and a driver to generate the local word lines from the global word lines. The scheme is actually developed to overcome word line delays along very long word lines.

Both of the above techniques, banking and divided word lines, attempt at reducing power by reducing the total area of memory core that toggles bit lines. Extra overheads incurred in dividing the memory core into smaller chunks limit the extent to which the above approaches can be applied.

6.5   REDUCING VOLTAGE SWINGS ON BIT LINES

For achieving performance, SRAMs use a set of sense amplifiers to detect differential voltages developed across bit lines. A read operation in an SRAM can be curtailed as soon as this detection is accomplished. In other words, it is only necessary that the bit cells develop just the correct differential for operation. This technique of limiting the voltage swing on the bit lines allows saving a fraction of the power expended in the core during reads. In general, if ΔV is the voltage swing across bit lines, V_{core} is the supply voltage to the core, r is the fraction of operations that are reads, and f is the frequency of core operations, then we can model the power expended during reads as $\frac{1}{2}C_{eff}V_{core}\,\Delta V rf$. The naive observation that increasing sense amplifier sensitivity and reducing ΔV, is often not feasible in practice, since it affects noise sensitivity, complexity of sense amplifiers, and the performance of the RAM core. Figure 6.7 illustrates this technique and shows waveforms for a RAM with limited bit cell swings.

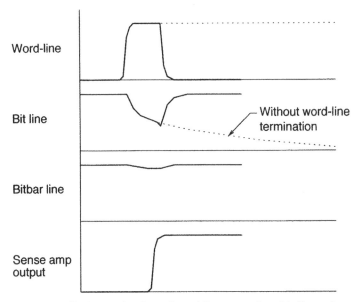

Figure 6.7   Early termination of word lines to reduce bit line swings.

6.5.1 Pulsed Word Lines

To limit the bit line voltage discharge, it is sufficient to enable word lines for precisely the time needed to develop the bit cell voltage discharge. A circuit that accomplishes this using a pulse generator is shown in Figure 6.8. This circuit gates the word line and the sense amplifiers by a pulse generated using delay cells (inverters in Figure 6.8). While the technique is fairly simple, it has the disadvantage that it does not track the actual operation of the sense amplifiers. A designer needs to estimate the actual access time of the RAM and insert a sufficient margin to determine the worst-case pulse width.

6.5.2 Self-Timing the RAM Core

In general, the speed of access to various rows is not identical. Clearly the rows closest to the sense amplifiers should give the fastest access times. Similarly columns closest to the word line drivers are enabled first. To utilize a pulsed word line to its best advantage, we should tailor the width of the pulse according to the access time of the RAM.

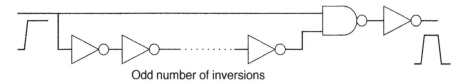

Odd number of inversions

Figure 6.8   Pulse generator.

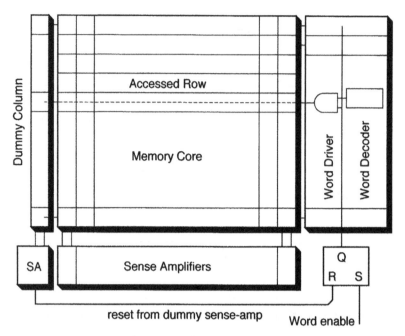

Figure 6.9   Dummy column.

The technique for achieving this uses a "dummy column" in the RAM to time the flow of signals through the core. A dummy column is an additional column of bit cells, sense amplifier, and support circuit placed at the side farthest from the word drivers. Bit cells in the dummy column are forced to a known state by shorting one of the internal nodes to a given voltage. The circuit that accomplishes the pulse generation is shown in Figure 6.9.

The sequence of operations that occurs is as follows: The SR flip-flop is set and the word line is triggered. Cells along the row are enabled with the dummy column being the last cell enabled. By the time the sense amplifier corresponding to the dummy cell generates a high signal, the rest of the columns would have been sensed. The high signal from the sense amplifier resets the SR flip-flop and turns off the word line. This method handles the case of nonuniform access time across rows. The dummy column often adds insignificant overhead to the entire RAM. Consequently it is often a preferred technique for pulsing the word line. This circuitry is also at times termed word line kill circuitry.

6.5.3   Precharge Voltage for Bit Lines

SRAMs employ one of two forms of precharge techniques. The first technique uses a static load formed by a MOS device functioning as a resistive load cell. Enhancement mode NMOS devices operating in their active region, depletion mode NMOS, or standard PMOS devices operating in the saturation region are employed. The second technique clocks the precharge circuitry to ensure that precharge occurs when the core is not being accessed. It should be apparent that the clocked technique avoids static power consumption at the expense of power required to drive the clocked precharge. Which method is more advantageous can only be determined on a case-by-case basis.

Reducing the precharge voltage clearly reduces the power consumption, since the effective voltage swings on bit lines is reduced. We would expect precharge through enhancement mode NMOS devices to be more effective from a power viewpoint, since the bit line voltage rises to at most $V_{CC} - V_{tn}$. While it would be reasonable to assume that setting the precharge voltage to $V_{cc}/2$ would be optimal, since the voltage swings would be minimal, it may not be practically feasible because of the structure of the bit cell. We should remember that a read operation activates the access transistors of the SRAM cell, which leads to the internal nodes of the bit cell being forced toward the bit line voltage. The relatively large NMOS in the cell prevents the internal nodes from rising high. However, a bit line precharged to a low enough voltage (lower than $V_{CC} - V_{tp}$) forces internal nodes of the cell low. This is counteracted by the PMOS in the cell, whose drive is very small (which has to be the case or else it would be impossible to write into the cell). We can then foresee a situation where the read operation destroys the old data held in the cell. It is this factor that limits the reduction of the precharge voltage. A reduced aspect ratio may be used due to the reduced precharge voltage (see Section 6.3.1). This results in cell size reduction. It is possible to get over the above problem of reads corrupting the cell state by employing different voltages on the word line for read and write. In particular, we arrange circuitry to ensure that a lower voltage on the word line is employed for the read as opposed to that for the write.

6.6 REDUCING POWER IN THE WRITE DRIVER CIRCUITS

In this section we shall tackle the problem of reducing power consumption of write driver circuitry, namely, the word line drivers and word line decoders. The reader should recognize that the write driver, while being large, because of the need for driving long, heavily loaded word lines, contributes to the overall power to a small extent, since at most one of the drivers is activated at a time. As opposed to power reduction techniques described previously, which attempt to reduce voltage, write circuits often have supply voltages determined by other factors. In other words, the word line voltages employed is determined by the core, while the decoders work at supply voltages required by surrounding logic. Word line drivers are usually simple buffers designed to fit in the row pitch of a memory cell. Little opportunity for power optimization is present in these circuits.

In general, we would prefer row decoding to be as fast as possible, since it directly affects access time for a RAM. While there exists a huge variety of row decoder structures, we shall restrict our attention to two of the most

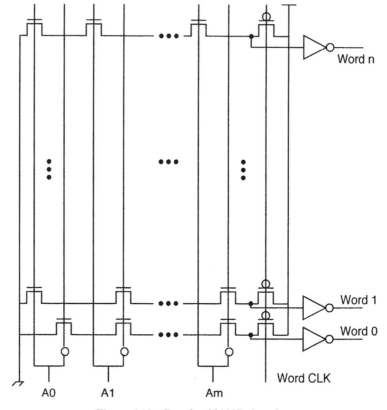

Figure 6.10 Domino NAND decoder.

common forms, the NOR-type decoder and the NAND-type row decoder. A NAND decoder changes the output of the decoder along one row. Its structure is shown in Figure 6.10. In contrast, a NOR decoder activates the outputs of all but one row. Its structure is shown in Figure 6.11. A NOR decoder is consequently faster because of the smaller height of the evaluation NMOS stack. Unfortunately, its operation results in $N - 1$ rows changing their state, as opposed to only one row changing state. Consequently the activity of signals is much higher in a NOR decoder, resulting in a higher power consumption than in NAND decoders.

If we accept the viewpoint that the NAND form decoder is desirable from a power perspective, we should attempt to improve the performance of the NAND decoder. A technique similar to the previously mentioned banked architecture for memories may be employed for decoder structures. Instead of decoding A address lines into one of 2^A word lines, we can split the decode process into a number of steps. Specifically, we first decode $A_1 < A$ address lines and use the 2^{A_1} lines to activate one of a set of second-stage decoders. The second-stage decoders now decode $A - A_1$ lines into $2^{(A-A_1)}$ lines. In all we have 2^{A_1} second-stage decoders, giving us a total $2^{A_1} \times 2^{(A-A_1)} = 2^A$ lines. This procedure can be repeated to give us a tree

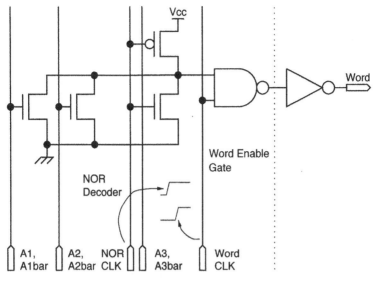

Figure 6.11 NOR decoder.

structure with a number of intermediate decoders and is illustrated in Figure 6.12. At the extreme end when we decode only one address line at a time, we get a tree decoder.

6.7 REDUCING POWER IN SENSE AMPLIFIER CIRCUITS

The function of the sense amplifier is to amplify small differential bit line voltages into logic levels. This operation should be performed as fast as possible.

While numerous sense amplifiers exist, some of the common forms employed for SRAMs are either simple differential amplifiers (Figure 6.13) or charge amplifiers that are similar to bit cells (Figure 6.14). A natural trade-off occurs between speed and power for such sense amplifiers. Larger currents improve sense amplifier speed. The analog nature of sense amplifiers often results in their consuming an appreciable fraction of the total power. Sense amplifiers are enabled by a sense amplifier enable signal.

Schemes employed for reducing the power requirements of sense amplifiers can be differentiated based upon the point at which they are activated. The first form limits sense amplifier currents by precisely timing the activation of the sense amplifier for just the period required. The second scheme employs sense amplifiers that automatically cut off after the sense operation.

We have in fact already detailed the technique required to achieve the former. The self-timed RAM core can be extended for obtaining the sense

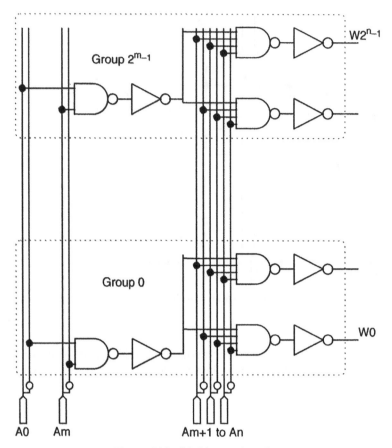

Figure 6.12 Multistage decoder.

amplifier enable signal. We show this in Figure 6.15. The enable sense amplifier signal is used to set up an SR flip-flop in the set state. Once the dummy sense amplifier has finished sensing, it resets the SR flip-flop, which in turn disables the enable for the sense amplifiers. We rely on every sense amplifier having completed the sense action before the dummy sense amplifier.

An alternative method shapes the tail current of differential amplifiers by activating pull-down transistors of different transistors in sequence.

Self-latching sense amplifiers accomplish an automatic limiting of currents after sense. The structure of a latched sense amplifier is shown in Figure 6.16. The structure can be visualized as a cross-coupled amplifying inverter loop, with additional transistors to transfer bit line voltages to the inverter loop.

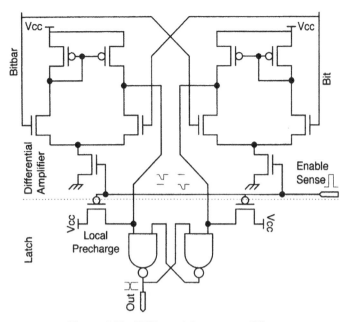

Figure 6.13   Differential sense amplifier.

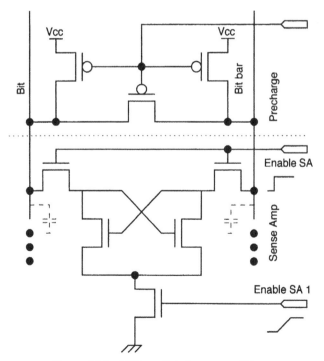

Figure 6.14   Differential charge amplifier.

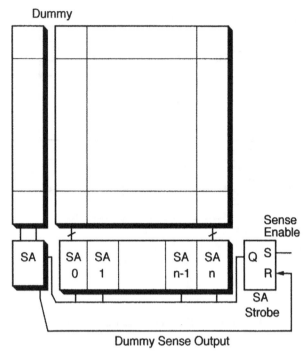

Figure 6.15 Self-timing the enable for sense amplifiers.

6.8 METHOD FOR ACHIEVING LOW CORE VOLTAGES FROM A SINGLE SUPPLY

Many of the techniques mentioned previously exploit reduction in supply voltage as a technique for reducing power consumption. This is especially true for the memory core, where a square law relationship exists for both standby and dynamic power with respect to the core voltage. While it is obvious that a lower core voltage is beneficial, we have side-stepped the issue of generating this lower supply voltage. Commodity RAMs often have a single external supply voltage, which must be stepped down in a power-efficient fashion to get the lower core voltage.

Reference [7] details a technique for achieving half the supply voltage for the core. The method described places two identical portions of the DRAM core in series. If we assume that the average power consumption from the RAM core is fairly constant, placing the RAM cores in series effectively results in a potential divider. The top core sees a voltage of $V_{CC} - V_{CC}/2 = V_{CC}/2$, while the lower core sees a voltage of $V_{CC}/2 - 0 = V_{CC}/2$. Such a structure is, however, complicated by the need for voltage level shifts at some section. This technique may or may not be considered practical, but it does illustrate a general way of using existing circuitry to form potential dividers. The simplicity of the technique is probably its greatest merit. The problem of designing step-down circuits at low DC voltages with a significant current

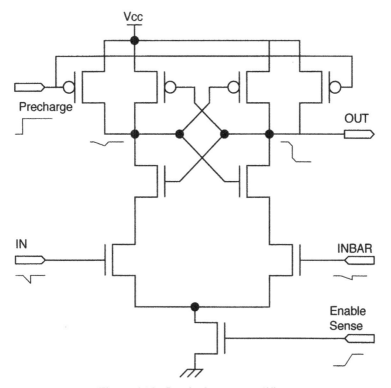

Figure 6.16   Latched sense amplifier.

drain is not trivial. While charge pump circuits have been used to generate static voltages for low-drain, high-impedance supplies, power-efficient circuits for high-drain supplies at very low voltages are difficult to implement. On-chip voltage step-down circuits cannot rely on inductors and pulse width control methods. Typically, they are implemented by charging N capacitors in series from the supply V_{CC}. A parallel connection of the capacitors is obtained by opening and closing a set of switches. This results in the voltage being stepped down to V_{CC}/N. Switching operations when carried out at a sufficiently high rate result in a smooth output waveform. To achieve high efficiencies, these techniques need near-ideal switches and reactive elements (capacitors in this case), both of which are difficult to achieve in monolithic integrated circuits.

6.9   SUMMARY

Commonly used techniques for minimizing power consumption in CMOS SRAMs have been presented in this chapter. These techniques can be subdivided into two classes: those that attempt to reduce the total capacitance that is switched and those that reduce the voltage swing across switched capacitances.

Often times, techniques that reduce the total switched capacitance are presented from the viewpoint of improving performance with the attendant power benefit being considered an added advantage. Banked organization of RAMs and divided word lines fall into this class. It should be remembered that these two techniques are equally applicable to DRAM and regular array designs. Decoder circuits presented in this chapter are also equally applicable in DRAM designs.

Techniques that reduce the voltage swings on lines can be viewed as being similar to methods used to reduce the power consumption of on-chip buses. These include early cutoff of word-lines, reducing bit line swings, and self-timing the RAM core. We mention in passing that reducing leakage currents in SRAMs is often times the primary design constraint for certain types of applications, for example, zero power of battery back-up RAMs. Circuit design techniques of back-biasing the substrate utilizing MOS device stacks are applicable in this area.

REFERENCES

[1] T. Seki, E. Itoh, C. Furukawa, I. Maeno, T. Ozawa, H. Sano, and N. Suzuki, "6-ns-1-Mb CMOS SRAM with Latched Sense Amplifier," *IEEE J. Solid State Circuits*, vol. 28, no. 4, pp. 478–482, 1993.

[2] M. Heshami and B. A. Wooley, "250-MHz Skewed-Clock Pipelined Dual-Port Embedded DRAM," *Proc. Custom Integrated Circuits Conf.*, pp. 143–146, 1995.

[3] T. Kobayashi, K. Nogami, T. Shirotori, and Y. Fujimoto, "Current-Controlled Latch Sense Amplifier and a Static Power-Saving Input Buffer for Low-Power Architecture," *IEEE J. Solid State Circuits*, vol. 28, no. 4, pp. 523–527, 1993.

[4] K. Sasaki, K. Ishibashi, K. Ueda, K. Komiyaji, T. Yamanaka, and N. Hashimoto, "A 7-ns 140-mW 1-Mb CMOS SRAM with Current Sense Amplifier," *IEEE J. Solid State Circuits*, vol. 27, no. 11, pp. 1511–1518, 1992.

[5] K. Sasaki, K. Ueda, K. Takasugi, H. Toyoshima, K. Ishibashi, T. Yamanaka, N. Hashimoto, and N. Ohki, "16-Mb CMOS SRAM with a 2.3-mu m2 Single-Bit-Line Memory Cell," *IEEE J. Solid State Circuits*, vol. 28, no. 11, pp. 1125–1130, 1993.

[6] M. Ukita, S. Murakami, T. Yamagata, H. Kurijama, Y. Nishimura, and K. Anami, "Single-Bit-Line Cross-Point Cell Activation (SCPA) Architecture for Ultra-Low-Power SRAMs," *IEEE J. Solid State Circuits*, vol. 28, no. 11, pp. 1114–1118, 1993.

[7] T. Sakata, "2D Power-Line Selection Scheme for Multi-Gigabit DRAMs," *IEEE J. Solid State Circuits*, vol. 29, no. 8, pp. 887–894, 1994.

[8] H. Hikada et al., "Twisted Bit-Line Architectures for Multi-Megabit DRAM's," *IEEE J. Solid State Circuits*, vol. 24, no. 1, pp. 21–34, 1989.

[9]   R. C. Foss, "Taking DRAM from 4 MBytes/s to 4 GBytes/s," *Proc. ESSIRC (European Solid State Circuits Conference)*, 1996.

[10] T. Tanizaki, T. Fujino, M. Tsukude, F. Morishita, T. Amano, H. Kato, M. Kobayashi, and K. Arimoto, "Practical Low Power Design Architecture for 256 Mb DRAM," *Proc. ESSIRC (European Solid State Circuits Conference)*, 1997.

[11] T. Sugibayashi et al., "A 30ns 256 Mb DRAM with Multi-Divided Array Structure," *ISSCC Dig. Tech. Papers*, pp. 50–51, 1993.

[12] M. Nakamura et al., "A 29ns 64Mb DRAM with Hierarchical Array Architecture," *ISSCC Dig. Tech. Papers*, pp. 246–247, 1995.

[13] T. Sakata, "Subthreshold-Current Reduction Circuits for Multi-Gigabit DRAMs," *IEEE J. Solid State Circuits*, vol. 29, no. 7, pp. 761–769, 1994.

CHAPTER 7

LOW-ENERGY COMPUTING USING ENERGY RECOVERY TECHNIQUES

The popularity of complementary MOS technology can be mainly attributed to inherently lower power dissipation and high levels of integration. However, the current trend toward ultra-low power has made researchers search for techniques to recover/recycle energy from circuits. In the early studies [1–4], researchers largely focused on the possibility of having physical machines that consume almost zero energy while computing and tried to find the lower bound of energy consumption. They found that if the physical processes associated with computing are nondissipative, the natural laws require that the physical entropy is conserved. Entropy conservation means that the processes must be physically reversible. One of the conclusions from the earlier studies is that the abstract logic operations composing the computing tasks must be reversible, that is, the information entropy must be conserved, in order to be performed by physically nondissipative hardware. Nevertheless, logical reversibility is only the necessary condition. Reversible logic operations can be realized by either reversible or nonreversible hardware.

The energy recovery techniques are sometimes referred to as *adiabatic* or *quasi-adiabatic* computing. The word *adiabatic* comes from a Greek word that describes a process that occurs without any loss or gain of heat. In real-life computing, such an ideal process cannot be achieved because of the presence of dissipative elements like resistances in a circuit. However, one can achieve very low energy dissipation by slowing down the speed of operation and only switching transistors under certain conditions. The word adiabatic is used rather loosely in this chapter, and in most of the cases we are really talking about quasi-adiabatic or *energy recovery* techniques. In order to understand

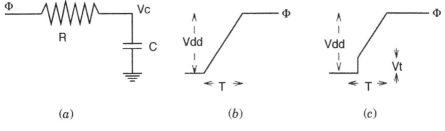

Figure 7.1   An *RC* model.

the concept behind adiabatic logic, let us use a linear *RC* model to analyze the energy dissipation due to the channel resistance of a transistor. Then a nonlinear *RC* model will be used to calculate the dissipation due to threshold voltage.

7.1 ENERGY DISSIPATION IN TRANSISTOR CHANNEL USING AN *RC* MODEL

In this section, we use a simple *RC* model to compute the energy dissipation in a transistor channel while working in the linear region. Let us consider a *p*MOS pass transistor, as shown in Figure 7.1*b*. When the voltage at the power/clock terminal swings from 0 to V_{dd} to charge node capacitance through a transistor channel, there is a voltage drop (and hence energy dissipation) in the channel due to the channel resistance. The *RC* model representing such a phenomenon is shown in Figure 7.2*a*. Let us consider the amount of energy dissipated when charging capacitance C from 0 to V_{dd} in time T with a linear power supply voltage of Figure 7.2*b*. We have

$$RC\left(\frac{dV_c}{dt}\right) + V_c = \Phi \tag{7.1}$$

where

$$\Phi = \begin{cases} 0 & t < 0 \\ \left(\dfrac{V_{dd}}{T}\right)t & 0 \le t < T \\ V_{dd} & t \ge T \end{cases}$$

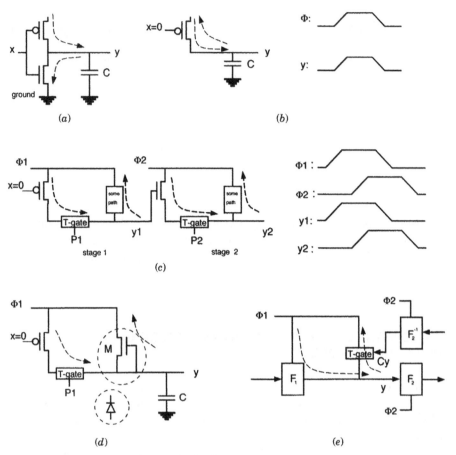

Figure 7.2 Basic recovery processes.

The solution of the above equation is given by

$$
V_c = \begin{cases}
0 & t < 0 \\
\Phi - \left(\dfrac{RC}{T}\right)V_{dd}(1 - e^{-t/RC}), & 0 \le t < T \\
\Phi - \left(\dfrac{RC}{T}\right)V_{dd}(1 - e^{-T/RC})e^{-(t-T)/RC} & t \ge T
\end{cases}
\tag{7.2}
$$

The energy dissipation in the above charging process can be calculated as follows:

$$
E_{\text{linear}} = \int_0^\infty iV_R\, dt = \int_0^T iV_R\, dt + \int_T^\infty iV_R\, dt
\tag{7.3}
$$

The first term of Eq. (7.4) can be written as

$$\int_0^T iV_R \, dt = \int_0^T \frac{(\Phi - V_c)^2}{R} \, dt$$

$$= \int_0^T \frac{[(V_{dd}/T)RC(1 - e^{-t/RC})]^2}{R \, dt}$$

$$= \frac{R^2C^2}{T^2} CV_{dd}^2 \int_0^{T/RC} (1 - e^{-t/RC})^2 \, d\left(\frac{t}{RC}\right)$$

$$= \left(\frac{RC}{T}\right) CV_{dd}^2 \left[1 - \frac{3}{2}\left(\frac{RC}{T}\right) + 2\left(\frac{RC}{T}\right)e^{-T/RC} - \frac{1}{2}\left(\frac{RC}{T}\right)e^{-2T/RC}\right]$$

And the second term can be written as

$$\int_T^\infty iV_R \, dt = \int_T^\infty \frac{(\Phi - V_c)^2}{R} \, dt$$

$$= \frac{RC}{T^2} CV_{dd}^2 (1 - e^{-T/RC})^2 \int_T^\infty e^{-2[(t-T)/RC]} \, dt$$

$$= \left(\frac{RC}{T}\right)^2 CV_{dd}^2 \left[\frac{1}{2}(1 - e^{-T/RC})^2\right]$$

Finally we have

$$E_{\text{linear}} = \left(\frac{RC}{T}\right) CV_{dd}^2 \left(1 - \frac{RC}{T} + \frac{RC}{T}e^{-T/RC}\right) \tag{7.4}$$

Let us consider the two extreme cases. When $T \gg RC$,

$$E_{\text{linear}} = \left(\frac{RC}{T}\right) CV_{dd}^2 \tag{7.5}$$

and when $T \ll RC$, as in normal CMOS,

$$E_{\text{linear}} = \left(\frac{RC}{T}\right) CV_{dd}^2 \left\{1 - \frac{RC}{T} + \frac{RC}{T}\left[1 - \frac{T}{RC} + \frac{1}{2}\left(\frac{T}{RC}\right)^2\right]\right\}$$

$$= \frac{1}{2} CV_{dd}^2 \tag{7.6}$$

It is clear from Eq. (7.5) that the energy dissipation through the dissipative medium can be made arbitrarily small by making the transition time T

arbitrarily large. This observation also points to the fact that for low-power dissipation a MOS device (or switch) should not be turned on unless the potential across it is zero or a switch should not be disabled if current is flowing through it. Such requirements will be considered in the next section while designing energy recovery logic circuits.

The above analysis ignores the threshold voltage drop of a transistor. Let us consider Figure 7.1a. When the voltage Φ at the power terminal swings from 0 to V_{dd} (as shown in the figure) to charge node capacitance, the pMOS transistor does not turn on until Φ exceeds the threshold voltage V_{th}. There is voltage drop $V_{ds} \approx V_{th}$ between the drain and source ends when the transistor jumps from the cutoff region to the linear region, which results in energy dissipation. Since an amount of CV_{th} charge is required to build the voltage to the V_{th} level, the energy loss due to the threshold voltage can be approximated by

$$E_{th} \approx \tfrac{1}{2}CV_{th}^2 \tag{7.7}$$

Due to channel resistance, there is still a small voltage drop (and hence energy dissipation) in the channel when the transistor works in the linear region. We use E_{linear} to represent this amount of energy loss.

Let us use the model shown in Figure 7.1a to calculate the energy dissipation. Let us consider charging C from 0 to V_{dd} in time T with the linear power supply voltage of Figure 7.1c (note that the power supply voltage shown in the figure considers the effect of transistor threshold voltage drop). We have

$$RC\left(\frac{dV_c}{dt}\right) + V_c = \Phi \tag{7.8}$$

where Φ is shown in Figure 7.1b. The solution of the above equation is given by

$$V_c = \begin{cases} 0 & t < t_0 \\ \Phi - \left(\dfrac{RC}{T}\right)V_{dd}(1 - e^{(t-t_0)/RC}) + V_{th}e^{-(t-t_0)/RC} & 0 \le t_0 < t < T \\ \Phi - \left(\dfrac{RC}{T}\right)V_{dd}(1 - e^{-(T-t_0)/RC})e^{-(t-T)/RC} - V_{th}e^{-(t-t_0)/RC} & t \ge T \end{cases} \tag{7.9}$$

where $t_0 = (V_{th}/V_{dd})T$.

The energy dissipation in the above charging process can be calculated as follows:

$$E_{dissipated} = \int_0^\infty iV_R\, dt = \int_0^T iV_R\, dt + \int_T^\infty iV_R\, dt \tag{7.10}$$

The above equation results in

$$E_{\text{dissipated}} = \tfrac{1}{2}CV_{\text{th}}^2 + \left(\frac{RC}{T}\right)CV_{dd}^2\left(1 - \frac{RC}{\beta T} + \frac{RC}{\beta T}e^{-\beta T/RC}\right)$$

$$+ \left(\frac{RC}{T}\right)CV_{\text{th}}V_{dd}\left(\frac{RC}{\beta T} - e^{-\beta T/RC} - \frac{RC}{\beta T}e^{-\beta T/RC}\right)$$

$$= \left(\frac{RC}{T}\right)CV_{dd}^2 + \tfrac{1}{2}CV_{th}^2 + O\left(\left(\frac{RC}{T}\right)^2\right)$$

$$\approx E_{\text{linear}} + E_{\text{th}} \tag{7.11}$$

where $\beta = 1 - V_{\text{th}}/V_{dd}$, $E_{\text{linear}} = (RC/T)CV_{dd}^2$, and $O((RC/T)^2)$ represents all other terms of the order of $(RC/T)^2$, which are very small for the energy recovery circuits.

Assume $V_{\text{th}} = 1.2$ V and $V_{dd} = 5$ V; then we have $(\tfrac{1}{2}CV_{\text{th}}^2)/(\tfrac{1}{2}CV_{dd}^2) = 0.0576$. Discharging consumes the same amount of energy, and hence, 11.5% of energy is consumed because of the threshold voltage. Since this threshold dissipation is independent of the transition time, it dominates the power consumption when the operating frequency is low, while linear dissipation is more significant in the higher frequency region.

7.2 ENERGY RECOVERY CIRCUIT DESIGN

In standard CMOS circuits charges are fed from the power supply, steered through MOSFET devices, and then dumped into the ground terminal. To change a node's voltage with associated capacitance C, as shown in Figure 7.2a, $V_{dd}Q$ ($= CV_{dd}^2$) of energy is extracted from the V_{dd} terminal. Half of the energy ($\tfrac{1}{2}CV_{dd}^2$) is stored in the capacitance temporarily, and the other half is dissipated in the path. Although energy is dissipated in the channel resistance and wire resistance, the amount of dissipated energy ($\tfrac{1}{2}V_{dd}Q$) only depends on the voltage and the amount of charge that flows through and is independent of the resistance. Later, when this node is connected to the ground, the stored energy is again dissipated. In a cycle, all $V_{dd}Q$ of energy is converted into heat.

Whenever there is a conducting path, energy is dissipated if there is a potential difference between the endpoints of the path (e.g., power supply terminal and the internal node in the circuit). The amount of energy dissipated is $\Delta V_{\text{avg}}Q$, where ΔV_{avg} is the average potential difference between the endpoints and Q is the amount of charge that flows through the path. This implies that the circuit should switch adiabatically to avoid the energy dissipation. Let us consider Figure 7.2b, where the power supply terminal swings gradually from 0 to V_{dd}, stays at V_{dd} for a while, and then swings back to zero. If $x = 0$, and the initial charge on capacitance C is zero, then node y follows the power supply to V_{dd} during its upward swing. Similarly, while discharging, the potential of node y follows the power supply terminal, swinging gradually from V_{dd} to zero, so that there is little potential difference across the path (from the power supply terminal to node y) in the whole transition process. Hence only a small amount of energy is dissipated. The question is: Can an internal node's voltage level follow the change in the power supply terminal? Let Δt be the transition time of the power supply terminal from zero to V_{dd} (or from V_{dd} to zero), and the time constant in a conducting path is RC, where R is the effective resistance in the path and C is the effective load capacitance. If the transition at the power supply terminal is sufficiently slow, that is, $\Delta T \gg RC$, then the voltage drop between the power supply terminal Φ and node y is small at any time instant during the transition, and hence, there is very low power dissipation. A simple model to estimate the power dissipation in this case is (from Section 7.1)

$$E_{\text{dissipation}} = \left(\frac{RC}{\Delta t}\right) CV_{dd}^2 \qquad (7.12)$$

Since $RC \sim 1$ ns for a moderate fanout and $\Delta t \sim 1/f$, $E_{\text{dissipation}}$ is small for operating frequency $f < 10$ MHz.

So far, the load is simply viewed as a dummy capacitance. In a circuit, an output of a gate drives other gate(s). Let us consider Figure 7.2c. The power supply/clock waveform Φ_1 can be divided into four phases: Φ_1 is in the *idle phase* when $\Phi_1 = 0$ and is in the *evaluation phase* when Φ_1 goes up from zero to V_{dd}. The time interval in which Φ_1 remains high, that is, $\Phi_1 = V_{dd}$, is defined as the *hold phase*. Finally, the phase when Φ_1 goes from V_{dd} down to zero is defined as the *restoration phase*. After evaluation, the output y_1 of the first stage must hold for a while in order to be sampled by the second stage. There must also exist some device, say, a transmission gate, denoted by a "T-gate" in Figure 7.2c, to isolate the input x and the output y_1. Otherwise x cannot change its value as long as y_1 needs to hold its value, and y_1 cannot change its value as long as y_2 needs to hold its value, which implies that an input has to be held constant until the signal propagates all the way to the last stage of the logic level. A consequence of the isolation is that the charge may not flow back to the power/clock terminal along the original charging path. Thus, another path has to be created to let the potential of y_1 follow Φ_1 going down to zero. However, unlike the charging path controlled by the inputs, the turning on of the discharging path depends on the logic value of y_1. When $y_1 = 1$, a charge can flow back to the power terminal through the path shown in Figure 7.2c during the restoration phase. If $y_1 = 0$, this path must be open to prevent current leaking from Φ_1 to y_1, which not only will consume energy but might set a wrong logic value to node y_1 as well.

Then the question is: How can the voltage level at the output of a gate be detected and the recovery path be controlled? It is basically the control schemes that differ in the various designs currently available. The output voltage to control the restoration path is simple and is shown in Figure 7.2d. Such control schemes are referred to as self-control schemes.

Self-control is not perfect because the voltage controlling the channel of M is also varying along with the voltage levels being controlled. The gate, the source, and the drain ends cannot have the same potential at the same time when the channel is conducting. The voltage drop across the channel of the control transistor, $V_{ds} \approx V_{th}$, causes as much as $CV_{th}V_{dd}$ of energy dissipation in each restoration. Note that the control transistor M is equivalent to a diode since its gate terminal and drain terminal are connected together. The perfect controlling signal should hold a correct and constant voltage during the restoration process. Let us consider Figure 7.2e. Such a controlling signal C_y can be inversely generated from the next stage signals, that is, from the inverse logic function F_2^{-1}, if the logic function F_2 itself is reversible. Since C_y is isolated from node y and is in a phase later than that of y, it holds constant (in the hold phase) during the restoration phase of node y.

The power supply/clock waveforms can be generated using some simple schemes [5, 6] that consist of two stages, as shown in Figure 7.3. The first stage is the DC power supply, and the second stage generates alternating current/clock wave forms, which are controlled by external clock signal(s) to maintain the constant frequency. Multiple phases of supply are required such that logic gates can be cascaded in the circuits. If the circuit viewed by the power/clock generator is modeled as a simple and constant capacitive load, the entire system is effectively an RLC resonator. The dissipated energy is replenished by the first-stage DC power supply by restoring the peak voltage of the second-stage circuit to the V_{dd} level in each cycle. However, there might be a problem in some existing designs since the number of gates charging and discharging varies from cycle to cycle during computation, which violates the constant-capacitance condition. When the amount of energy feeding into the circuit and recovering from it behaves like a random

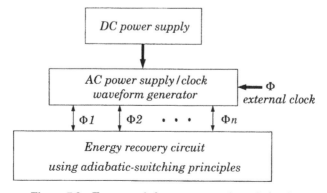

Figure 7.3 Framework for power supply and circuit.

process, voltage waveforms and the frequency in the second-stage power supply will be unstable, which in turn forces the first-stage DC power supply to stabilize the waveform. Restoration of the voltage waveform is accompanied by power dissipation.

7.3   DESIGNS WITH PARTIALLY REVERSIBLE LOGIC

We discussed in Section 7.2 that the adiabatic operation can be implemented if ideal charging and discharging paths are set up correctly. If every logic function primitive in the circuit is reversible, the controlling signal for the discharging path can be produced by an inverse function with correct timing. Theoretically, only reversible logic operations can be performed without accompanying energy dissipation [1–3]. Unfortunately, many common logic primitives, like NAND, NOR, and XOR are irreversible. Irreversibility does not mean that all energies involved have to be wasted. The theoretical lower bound for an isolated irreversible operation is kT. However, in CMOS technology, the lower bound is related to the threshold voltage V_{th} and the amount of charges to build the voltage level to V_{th}.

We present three designs and their modified versions, which are not based on reversible logic primitives. However, the designs recover most of the energy involved in the operations. The obvious advantage of these designs includes the simplicity as well as practicality for implementation. These designs are essentially based on the self-control scheme, where the recovery path is controlled by a signal generated from output. The goal is to achieve lowest possible dissipation without requiring inverse logic functions.

Let us first consider a diode-based logic family that dissipates about $CV_{dd}V_{th}$ of energy, where C is the output capacitance, V_{dd} is the supply voltage, and V_{th} is the transistor threshold voltage. The logic family is referred to as 2N-2N2D by Denker et al. [7] and considers a constant-capacitance condition. Let us explain the operations of a 2N-2N2D inverter/buffer using Figure 7.4 The logic uses differential signaling, and hence, each signal is represented by itself and its complement. A logical 0 is represented by a downward pulse on c while a logical 1 is represented by a downward pulse on d. Signals can be inverted by exchanging the signal wires.

The four stages of the clock cycle required for this design is shown in Figure 7.2. At the beginning of the cycle, the clock is high (V_{dd}) and the diodes ensure that the outputs are high. The evaluate phase begins when the clock ramps down from V_{dd} to 0 V. During this phase, the inputs have to be valid. Let us assume that input $a = 0$ and $b = 1$. Since $a = 0$, no current flows in that branch and the output c remains high. However, all transistors in the right evaluation branch are on because $b = c = 1$. Hence, signal d follows the clock down. The evaluation phase is followed by the hold phase when the output is a valid logic signal and can be sampled by other logic gates. Also, the input does not have to be valid at this time since both the

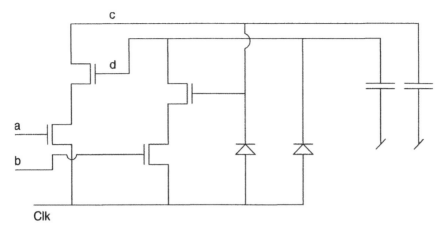

Figure 7.4   A 2N-2N2D inverter/buffer.

clock and signal d are low at this time, and turning the right evaluation branch on or off does not affect the output. Similarly, just because signal d is low, the left evaluation branch is off, and the output d remains at logical 1 irrespective of the inputs. The last phase is the recharge phase when the clock ramps up from 0 V to V_{dd}. Because of the diodes, whichever output was low follows the clock to a logical high. Note that for cascading such logic gates, the second gate has to be operated from a different clock because the output is valid only during the hold phase while the inputs are required to be valid during the evaluation phase.

A basic logic gate (NAND/AND) is shown in Figure 7.5(a) [8]. It uses differential signaling: Both input signals and their complements are required and an output signal and its complement are generated. The circuit consists of two branches. In each cycle, one branch is charging and the other is not. Thus, the constant-load-capacitance condition is satisfied. Without loss of generality, let us assume $x_1 = 1$ and $x_2 = 1$. At the beginning of a cycle, Φ is at zero and the gate is in the idle phase. Both outputs y and $\bar{y}$ are not valid and are equal to zero at this phase. During the evaluation phase, Φ swings to V_{dd}. Node y follows Φ to high as IS_1 turns on while $\bar{y}$ stays at zero. The output is now valid and can be sampled by other gates in the subsequent stages it is driving. Importantly, the inputs and the outputs are in effect isolated at this phase because the isolation transistor IS_2 is turned off and IS_1 remains on and does not affect the output y in its branch. Thus, the input needs do not remain valid and can restore to 0's while the output of the gate still holds its logic values. Following the hold phase, let us examine how the recovery paths are set up in the restoration phase. The transmission gates in recovery paths are turned on in this phase. The controlling signal C_1 is from $\bar{y}$ and is equal to zero. It turns on the discharging path and node y follows Φ, going downward to zero. The other controlling signal, C_2, is from

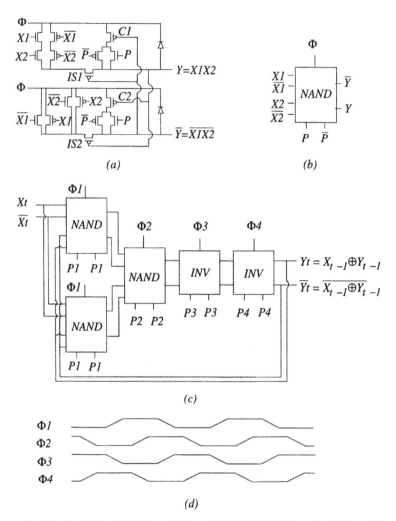

Figure 7.5   Design with adiabatic switchings: (*a*) NAND; (*b*) symbol; (*c*) serial adder (mod 2); (*d*) timing diagram.

y. Since $C_2(=y)$ is changing together with Φ, the *p*MOS control transistor, C_2, will not turn on in the whole discharging phase. The energy dissipation comes from the threshold voltage of the isolation transistor IS_1 and control transistor C_1, which are in the cutoff region until $V_{gs} \geq V_{th}$. The dissipation due to the threshold voltage is modeled and computed in Section 7.1. A diode is put in each branch to ensure that nodes y and $\bar{y}$ restore to a low voltage in the idle phase.

Extension to complex gates is straightforward. A logic function F can be constructed by two branches of a T-gate tree, which are in dual form.

Figure 7.5*c* shows the four-phase power supply required in this design. As an example, to show how a circuit can be built using the logic primitives, let us consider a serial adder (mod 2) in Figure 7.5b. Three NAND gates form the logic function. By using differential signaling, all NAND, NOR, AND, and OR gates have the same structure with different connections at the inputs and the outputs. Two additional buffers are used to fit the four-phase power/clock line. Note that the transmission gate control signals P_1 and $\bar{P}_1$ only turn on in the restoration phase and therefore can be overlapped with Φ_2 and Φ_4, that is, $P_1 = \Phi_2$, $\bar{P}_2 = \Phi_4$ and so on. Clock phases will be further discussed in the next section.

One can modify the design to further reduce power dissipation by replacing the isolation transistor by a transmission gate. Dissipation due to the threshold voltage of an isolation transistor is avoided and the channel resistance is reduced. However, due to additional control signals, six-phase power/clock lines are now required in this modified version.

Measurement of energy savings for adiabatic circuits can be complicated. The energy recovered from the circuits is dependent on the efficiency of the power supply oscillator. We will consider a power supply design in more detail later in this chapter. In the following analysis, we will assume that the power supply is 100% efficient and estimate the energy recovery of the circuit itself. To measure the power savings in the circuits, one can compute the "net energy" $E(t)$, which is defined as follows:

$$E(t) = \int_0^t i(t)\Phi(t)\, dt \tag{7.13}$$

Here $E(t)$ is the net energy flowing into the circuit from the power supply line. The voltage and the current waveforms can be obtained from SPICE simulation.

As discussed earlier in this chapter, energy transfers between the controlling signals and the controlled signals. Therefore, the net energy can be computed by summing up the energy in power supply lines. Within a cycle (charging and discharging), energy flows into the circuit and is recovered back from it. The level difference of $E(t)$ in two consecutive cycles reflects the energy loss in a full cycle. The effect of energy transfers between the controlling and the controlled signals lead to the energy transferring from one phase of power supply to another. However, if the amount of energy transferred is steady in the design, not varying from cycle to cycle, then energy can be efficiently recycled.

The circuits described earlier in this section were simulated using SPICE to obtain the voltage and current waveforms, from which the net energy function $E(t)$ was calculated. The transistor model is from 1.2-μm CMOS processing technology from MOSIS. The difference of the net energy between two consecutive cycles is the amount of energy dissipated in the circuit in such a cycle. With a few exceptions mentioned later, transistors used in this analysis are of the same size, with $W = 3.6$ μm and $L = 1.2$ μm. The "energy dissipation percentage" is defined as the ratio of the amount of

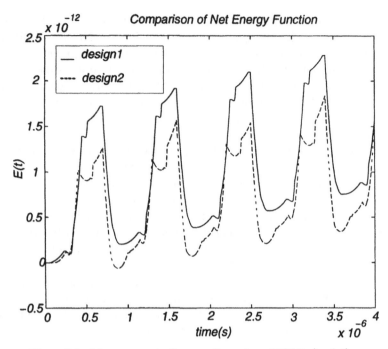

Figure 7.6 Measurement of energy flows from SPICE simulation.

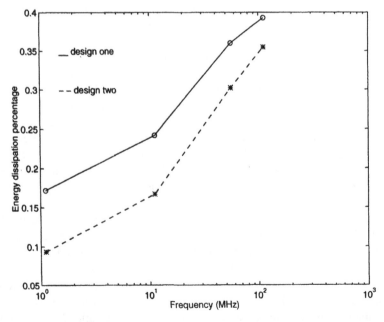

Figure 7.7 Energy dissipation of a gate (buffer) in different operating frequencies.

dissipated energy and the amount of energy entering the circuit from the power supply in a complete cycle.

The design shown in Figure 7.5 along with a modified version of the design (the isolation transistor replaced by transmission gate), referred to as "design 2," was simulated. For an inverter (buffer), the net energy function $E(t)$ from these two designs is shown in Figure 7.6. The load in each branch is 0.04 pF, which is fairly large considering the fact that the gate capacitance $C_g < 0.01$ pF for the transistors used in the design. In Figure 7.7, the "energy dissipation percentages" for these two designs have been compared. Results show that this percentage is around 10% at 1 MHz and increases when the frequency goes higher. The revised version has smaller power dissipation but needs six phases of power supply/clock, while the design 1 needs only four.

The serial adder (mod 2) with partially reversible logic (shown in Figure 7.5) was also simulated. At 1.1 MHz, the serial adder can recover about 90% of energy, the reversible buffer chain recovers 93% of energy. At 111 MHz, the circuit can still recover more than 40% of energy entering the circuit.

7.3.1 Designs with Reversible Logic

Reversible logic is desired since nondissipative operations can only be realized in this type of logic. As discussed in the previous sections, we need the inverse logic function to control the discharging path of the original gate.

We begin with the naturally reversible logic, the buffer (inverter) chain. Figure 7.8a shows the design of such a buffer (inverter) [8]. One can observe that it is constructed with transmission gates, which reduces the channel resistance and has no threshold voltage drop when it is "on." The functionality of the four transmission gates in each branch is as follows: the input x_i and $\bar{x}_i$ control the charging path to set up the logic value of the gate output. The isolation gate controlled by IS_i and $\overline{IS}_i$ essentially ensures that the charging path is "on" only during certain time instants. The gate in the discharging path (controlled by signals C and $\bar{C}$) functions similarly, turning on during the discharging phase. The discharging path is set up by another transmission gate, controlled by the next gate outputs x_{i+2} and $\overline{x_{i+2}}$, instead of the local signal as in the previous designs. Thus, a buffer chain can be constructed as in Figure 7.8c.

Six phases of power/clock are required for this design (Figure 7.8). Here IS_i has the phase earlier than that of x_i in order to perform the isolation, and x_i has the phase earlier than Φ_1. Hence, IS_i is two phases earlier than Φ_i. Moreover, IS_i is still in the hold phase when Φ_i is in the evaluation phase. One can immediately observe that six phases are required in this design. Although it is possible to reduce the number of phases by signal encoding [5] for a buffer chain, the six-phase clocks are generally required for reversible logic. Also note that there is no need to introduce additional control signals for the isolation gate and control gate by observing that $IS_i = \Phi_{i-2}$, $\overline{IS}_i = \Phi_{i+1}$, $C_i = \Phi_{i+2}$, $\bar{C}_i = \Phi_{i-1}$, and so on.

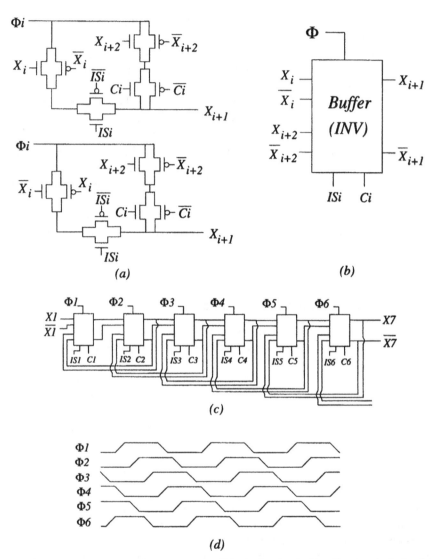

Figure 7.8 A design example with reversible logic, the buffer (inverter) chain. (*a*) buffer (inverter); (*b*) symbol; (*c*) buffer (inverter) chain; (*d*) timing diagram.

The generalization to other gates is straightforward; however, the signal to set up the discharging path is not available unless reversible logic is used.

There are two directions to design reversible logic circuits. One is to introduce reversible logic primitives, as the Fredkin gate in [2]. The Fredkin gate has some good properties since it is not only logically reversible but also conservative. However, besides the difficulty to implement the gate in electronic circuit, reversible logic primitives may have more outputs than one may need; that is, produce redundant (garbage) outputs.

Although conventional logic primitives are mainly irreversible, it is easy to modify them to be reversible; for example, by simply buffering the input signals and augmenting them to the outputs. Hence, one can obtain a reversible logic circuit from synthesis of conventional logic gates. But again, the reversible logic circuit based on some conventional gate will usually produce redundant output signals, and the energy involved in the redundant output signals is wasted. However, in theory, one can always find a reversible logic design in which the number of garbage signals is, at most, the number of primary input bits for the whole circuitry [2, 4].

The reversible buffer chain of Figure 7.8 was also implemented and simulated with SPICE. The technology was the same as the one described earlier. The buffer chain can recover 93% of energy at 1.1 MHz (assuming the power supply to be 100% efficient).

7.3.2 Simple Charge Recovery Logic Modified from Static CMOS Circuits

Although we focused on the reversible and partially reversible logic earlier, it is attractive to have a quasi-adiabatic circuit with simplicity comparable to the static CMOS circuit [9]. It should be observed that the adiabatic circuits described in this chapter have constant energy dissipation irrespective of the switching activity of the inputs. In the extreme case when the switching activities of the inputs are zero, the energy dissipation in standard static CMOS circuits will only be due to leakage current, while the adiabatic circuits will still dissipate constant energy. Hence, power management techniques are required to switch between standard static CMOS and adiabatic circuits. Figure 7.9 shows such an adder directly modified from the static CMOS circuit [10]. Two supply lines with complementary phases Φ and $\overline{\Phi}$ suffice for this circuit. The supply waveform consists of two phases, evaluation and hold, as shown in Figure 7.9. The circuit is designed to work in two modes, the *adiabatic mode* or *CMOS mode*, which is selected by the external signal SEL.

Let us consider the adiabatic (or quasi-adiabatic) mode. When Φ and $\overline{\Phi}$ are in the evaluate phase, there is a conducting path(s) in either pMOS devices or nMOS devices. Node $\overline{\text{CARRY}}$ may evaluate from low to high or from high to low or remain unchanged, which resembles the static CMOS circuit. Thus, there is no need to restore the node voltage to zero (or V_{dd}) every cycle. When Φ and $\overline{\Phi}$ are in the hold phase, node $\overline{\text{CARRY}}$ holds its value in spite of the fact that Φ and $\overline{\Phi}$ are changing their values. The reader can find that such is the case by observing the function of diodes and the fact that the inputs of a gate have a different phase with the output (A, B, and C are in the evaluation phase while $\overline{\text{CARRY}}$ is in the hold phase).

Finding the actual energy savings of these designs over static CMOS is nontrivial. The area increases by more than twofold, and hence, the capacitance increases. Signal switching activities are important for power dissipa-

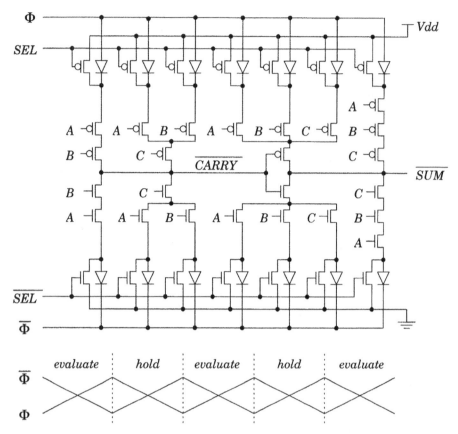

Figure 7.9   Adiabatic adder modified from static CMOS circuit.

tion and are virtually 1 in the adiabatic circuits with both partially and completely reversible logic. In CMOS circuits, signal activities are lower and depend on the applications under consideration. However, in adiabatic designs, the signals are generated with their complements, and also are self-buffered. Hence, some inverters and registers can be eliminated. Let us consider the serial adder (mod 2) of Figure 7.5. The CMOS circuit would use two NAND gates, one NOR gate, two inverters, and one register. Thus, two inverters and one register can be eliminated in this design in principle. Also, area overhead could be significantly reduced if efficient circuit topologies are used.

Direct comparison with CMOS circuits is possible for the adiabatic circuits of Figure 7.9, which can work in both the adiabatic mode and the static CMOS mode, as discussed in this section. A 2 × 2 multiplier using this type of logic was measured in both modes. Figure 7.10 shows the four output signal waveforms (inverted) in the adiabatic mode. The ratio of energy

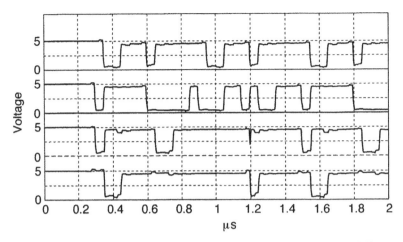

Figure 7.10 Four-output signal waveforms in 2×2 adiabatic multiplier.

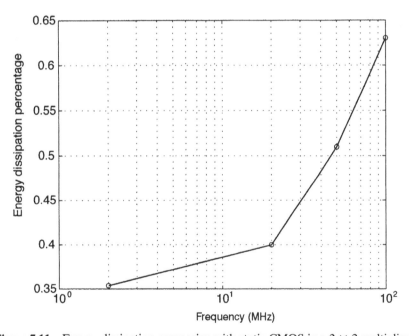

Figure 7.11 Energy dissipation comparing with static CMOS in a 2×2 multiplier.

dissipation between the adiabatic mode and the static CMOS mode is shown in Figure 7.11. Sixty percent of energy can be saved over static CMOS at 20 MHz, and there is still 35% less energy consumption at 100 MHz. For high frequencies (> 50 MHz), the transistors in the CARRY generator should be appropriately sized for proper operations since the CARRY is driving 10 transistors. Alternatively, a buffer can be inserted between two cascaded CARRY circuits to avoid sizing up the transistors. The fairness in the comparison is still not easy to justify in the high-frequency region, since the additional efforts required for adiabatic circuits are not required for static CMOS circuits.

7.3.3   Adiabatic Dynamic Logic

The adiabatic dynamic logic (ADL) circuits proposed by Dickinson and Denker [11] combines adiabatic theory with conventional dynamic CMOS. A conventional dynamic CMOS inverter is shown in Figure 7.12. When the clock Clk is low, the pMOS transistor is on and the output is precharged to logical 1. During the evaluation phase, the clock is high and the output capacitor C conditionally discharges if the input to the inverter is logical 1. The charging and the discharging are abrupt and hence nonadiabatic in nature. There is also some energy dissipation due to current flowing from supply to ground due to the n- and p-transistors being simultaneously on for a short period of time. The corresponding adiabatic inverter is very simple and is shown in Figure 7.13.

Let us now consider the operation of the ADL inverter. The clock or the supply voltage swings are shown in Figure 7.13 The *precharge* phase is defined by the clock swing from zero to V_{dd} when the diode is turned on and the output voltage V_{out} follows the clock swing to $V_{dd} - V_D$, where V_D is the voltage drop across the diode. In the evaluate phase, the clock voltage ramps

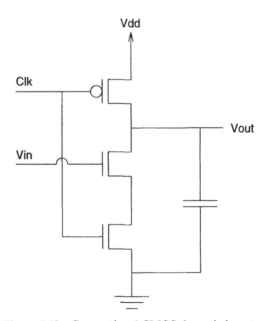

Figure 7.12   Conventional CMOS dynamic inverter.

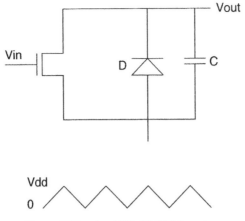

Figure 7.13   An ADL CMOS inverter.

down from V_{dd} to 0. Notice that the diode is in the reverse-bias condition and the output will follow the clock down to zero if V_{in} is high. Else, if V_{in} is low, the output capacitance retains the charge stored on it. Also note that an inversion occurs between the input and the output.

As in the case of other adiabatic circuits described in this chapter, several other considerations need to be addressed for cascading ADL gates. Note that the clock signal has to be modified to include a stage of constant voltage of V_{dd} between the precharge and the evaluate phase so that the input voltage can safely make a transition without making a nonadiabatic transition within the circuit. Similarly, a constant voltage stage of 0 V is added between evaluate and precharge to ensure that the output is latched for a finite time and can be sampled by the next stage. The clock waveform along with four interconnected ADL inverters is shown in Figure 7.14 [11]. Let us now consider how the set of cascaded gates work. To connect two logic gates, it is necessary to synchronize their respective clock supplies such that when the output of the first stage is latched, the second stage starts evaluating. Another consideration is that when the first stage is evaluating, the second stage should not undergo any nonadiabatic transition. Hence, one can use the clocking scheme shown in Figure 7.14. Note that four clock phases are necessary for cascading such gates. Because there are four stages in a clock cycle, it is necessary to place four gates in series before the last gate can feed the first one.

Another important consideration during cascading is due to clock voltage. When precharging the output of the nMOS transistor to $V_{dd} - V_D$, the next stage (precharge low pMOS inverter) is in the hold phase, and hence, the pMOS transistor should not turn on during this phase. However, if $V_{dd} - V_D$ is less than the turn-on point of the pMOS transistor ($V_{dd} - V_{th}$), it will turn on the pMOS transistor, that is, $V_D > V_{th}$. This problem can be solved either

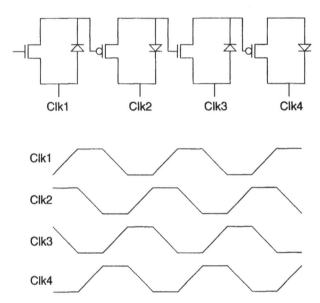

Figure 7.14 Cascaded ADL gates.

by ensuring that $V_D < V_{th}$ or by off-setting the clock voltages. For example, the first and the third gates of the inverters of Figure 7.14 should have a positive off-set compared to that of the second and the fourth gate for correct operations.

Modification of the ADL inverter to any complex logic gate is straightforward. Figure 7.15 shows a two-input NAND gate. The reader can verify that the circuit really behaves as an ADL NAND gate.

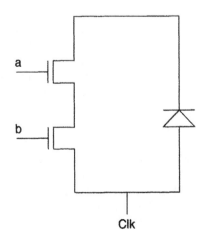

Figure 7.15 Two-input ADL NAND gate.

7.3.4 Energy Recovery SRAM Core

Today's VLSI systems integrate both random logic and assorted memories. Hence, it is natural to apply the adiabatic switching principle to memories to achieve similar large savings as in random logic. However, the application of such methods should not cause drastic increases in either size or circuit complexity. Let us consider the design of a *static RAM* (SRAM, which is capable of recovering on the order of 75% of energy for both *read* and *write* operations. This can be achieved without increasing the complexity of the memory cell and with a low area overhead over conventional SRAM.

In this section, we first briefly describe the organization of SRAM and then describe the topology of the memory cell and SRAM core in details [12]. Figure 7.16 shows the energy recovery SRAM organization. Compared to the standard CMOS SRAM, a row driver is inserted that generates the appropriate voltage signals to drive the memory core. Also note that sense amplifiers are replaced by the voltage level shifters.

Figure 7.17 shows the memory cell, which is identical in topology to the six-transistor RAM cell used in the standard SRAM. The cell consists of a cross-coupled inverter pair and a pair of read/write access transistors. The pair of access transistors is enabled by the *word* line. A block of SRAM core is composed of a multiplicity of these cells arrayed horizontally and vertically, and the memory core is again composed of multiple blocks. Within a block, the *word* lines of adjacent cells are connected along the horizontal axis, while the *bit* and $\overline{bit}$ lines are connected for all cells in a column.

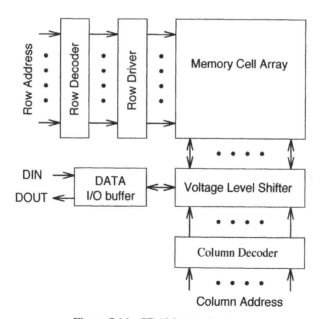

Figure 7.16 SRAM organization.

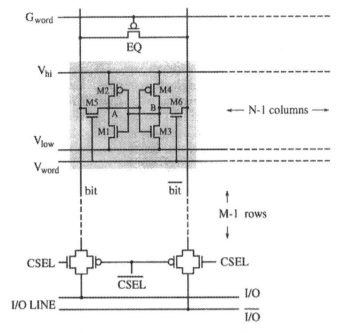

Figure 7.17 Memory core organization.

With reference to Figure 7.17, a conventional SRAM ties V_{hi} and V_{low} to the supply V_{dd} and ground, respectively. For a discussion on the operation of a six-transistor SRAM, we refer readers to Chapter 6 on low-power SRAMs. The design of the cell ratios the size of $M3$ and $M5$ ($M4$ and $M6$) so that the value stored in the cell is not upset. The layout of the memory core emphasizes compactness and is usually flipped for two adjacent rows to share the power supply. The dominant component of energy dissipation arises from switching the large capacitance on the *bit* lines and the *word* lines. Other significant components of energy consumption are row and column address decoders and the sense amplifier. We will discuss energy reduction techniques in each component by applying the energy recovery principles.

7.3.4.1 *Operations of Adiabatic SRAM Core*

In the energy recovery SRAM core of Figure 7.17, V_{hi} and V_{low} are no longer static. The parameters V_{hi}, V_{low}, and V_{word} are generated by the row driver circuitry, which is shown in Figure 7.18. The row driver will be discussed in detail in Section 7.3.9. For the time being, it is sufficient for readers to realize the following: Row selection signals W_0, $W_1, \ldots, W_{M-1}$, which are generated from the row address decoder, enable the drivers for a particular row. Now, V_{hi}, V_{low}, and V_{word} of the enabled row may be controlled independently by global supply lines G_{hi}, G_{low}, and G_{word}, respectively. For

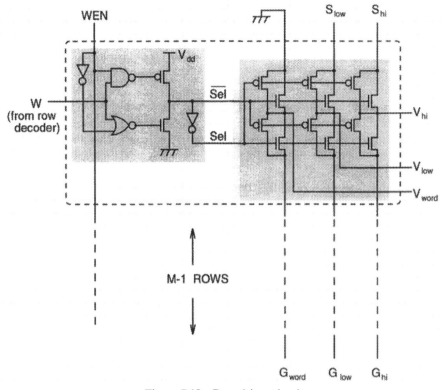

Figure 7.18   Row driver circuit.

the unselected rows, V_{hi}, V_{low}, and V_{word} are connected to the static power supply lines S_{hi}, S_{low}, and ground, respectively. Figure 7.17 also shows the bit line equalization transistor. Bit line precharge circuits of the standard CMOS SRAM are not required in adiabatic SRAM.

We now show that by the proper application of stimulus at G_{hi}, G_{low}, and G_{word}, it is possible to operate the memory core in an adiabatic fashion. For the purpose of discussion, we shall assume the DC supply voltage of $S_{hi} = V_{dd} = 5$ V, $S_{low} = 2$ V, and a transistor threshold $|V_t|$ of 1 V. The SRAM core starts out in the rest state with all rows disabled. The row driver circuit ensures that V_{hi} is at $S_{hi} = 5$ V, V_{low} is at $S_{low} = 2$ V, and V_{word} is pulled to ground. The bit-line is assumed precharged midway to 2 V. A read operation starts with the row selection being applied and the V_{word} being smoothly ramped up to 3 V by G_{word}. Now V_{hi} and V_{low} are ramped down to 3 and 0 V, respectively. (The waveforms are shown later in Figure 7.28. The reader can also refer to the circuit model of Figure 7.26b for the *read* operation in Section 7.3.10.) If we assume that internal node A was low $= (V_{low})$ and B was high $(= V_{hi})$, both M1 and M5 are on and, hence, *bit* follows V_{low}, smoothly ramping down to zero. On the other hand, $\overline{bit}$ remains at 2 V since node B is at V_{hi} ($\geq$ 3 V) and V_{word} is at 3 V, which prevents access transistor $M6$ from turning on. The bit line differential is amplified by an adiabatic level shifter to generate the logic output. Unlike the conventional SRAM, where precharge circuitry is required to precharge the bit-lines after each operation, bit-lines are charged back to 2 V by the same cells being read. Indeed, *bit* reverts back to the rest state through $M5$ and $M1$ when V_{hi} and V_{low} ramp up to their rest state. This process replenishes the charge on *bit*, and hence, the bit line peripheral circuitry can be eliminated in adiabatic SRAM. Subsequently, G_{word} is pulled low, turning off the row and equalizing the bit lines. The stimulus applied on V_{hi} and V_{low} ensures a constant cell voltage all the time.

The write operation requires that the bit information stored in a cell be overwritten by signals carried in the bit lines. This is accomplished by applying the row selection, pulling up V_{word} to enable the word line, and pulling down V_{hi} to 3 V. Thus, the voltage difference across the cell is $V_{hi} - V_{low} = 1 \text{ V} = V_t$, which is high enough to ensure that the cell state is held for columns that are not being written and low enough to easily flip the cell state for the selected columns. Now V_{low} and $\overline{bit}$ are simultaneously and smoothly ramped down from 2 V to 0 V. Then B is pulled down by $\overline{bit}$ and the cell state flips (as we have assumed in the *read* operation, A was low and B was high). Thus, a 0 is written into B, as illustrated later in Figure 7.28. Returning to the rest state is accomplished in a fashion similar to the read operation when V_{low} and *bit* revert back to 2 V and *word* is disabled.

7.3.5 Another Core Organization: Column-Activated Memory Core

One phenomenon in the operation of SRAM is that all cells in the selected row within a memory block are activated. Although only the selected columns in a block are multiplexed to sense amplifiers and I/O buffers, all columns are active, discharging the bit lines and charging them back to *high* subsequently. For instance, considering a 64 × 64 block of memory core in an SRAM chip with eight I/O pins, a *read* only reads out 8 bit-lines while the rest of the 56 bit lines also perform the read operations. The energy involved in charging these 56 bit lines is actually wasted. If we only activate the selected columns, less energy will be dissipated in each operation. Another memory core organization is presented that only enables the selected columns and consumes significantly less energy under certain memory organizations.

Figure 7.19 shows this memory core organization. The unique modification from the core configuration of Figure 7.17 is that V_{hi} and V_{low} now run vertically and are generated by column driver circuitry, which can be implemented analog to the row driver circuitry. The selection signals generated from the column address decoder enable the driver for the selected columns. For these columns V_{hi} and V_{low} are then controlled by G_{hi} and G_{low}, respectively. A row driver is still required because V_{word} runs horizontally.

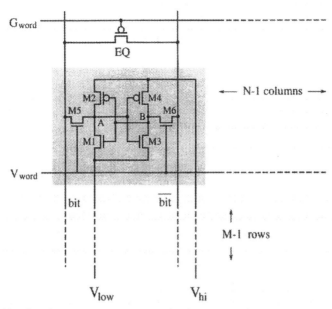

Figure 7.19 Another memory core organization; V_{hi} and V_{low} run vertically and are generated by column driver circuitry. In this core organization, only selected columns are activated.

The cell structure remains the same. The memory core starts out in the rest state with all rows disabled. The column driver circuitry ensures that V_{hi} is at $S_{hi} = 5$ V, V_{low} is at $S_{low} = 2$ V, and V_{word} is pulled down to ground by the row driver circuitry. A read operation starts with V_{word} in the selected row being ramped up to 2.5 V by G_{word}. In the *selected columns* V_{hi} and V_{low} are now ramped down to 3 and 0 V, respectively. The selected columns operate identically as described in Section 7.3.4. Let us consider the unselected columns. The gate terminals of access transistors $M5$ and $M6$ are tied to V_{word} and at 2.5 V during the read operation. However, V_{hi} and V_{low} of the cell are not ramping and stay at 5 and 2 V, respectively. Assume that internal node A is at low = V_{low} and B at high = V_{hi} (refer to the cell shown in Figure 7.19). Since V_{word} ramps up to 2.5 V only, neither of the access transistors $M5$ and $M6$ is turned on. Thus, the pair of bit lines of the cell stays at rest state 2 V. We have obtained a memory core in which only the selected columns are active. A *write* operation is performed in a similar way such that only the writing columns are active.

   In this core organization, each active column is driven by its own driver, and hence only small-size transmission gates are required to drive V_{hi} and V_{low} for each column in the column driver circuitry. As a comparison, the previous core organization needs substantially larger transmission gates in the row driver circuitry because V_{hi} and V_{low} in one selected row are driving all the bit lines (remember that every cell in the selected row is discharging and charging its bit line). Cell size in this new core organization is slightly larger than the previous case. The actual power advantage depends on the overall SRAM memory organization, which will be further discussed in Section 7.3.7.

7.3.6 Energy Dissipation in Memory Core

The choice of voltages described earlier is governed by having a safe differential V_{keep} (selected as 1 V) across the cell to hold its state. Assuming that V_{dd} is the supply voltage, V_{hi} swings between V_{dd} and $(V_{dd} + V_{keep})/2$ and V_{low} swings between $(V_{dd} - V_{keep})/2$ and zero. Very high energy recovery can be achieved if the read process is gradual enough. This is because the charging and discharging paths for read operation are identical, and no cell flips its state. Hence, the energy dissipation is dominated by resistive loss, which is modeled by Eq. (7.1). Threshold loss is negligible due to the fact that $V_{word} - V_{low} \approx V_{th}$ at the beginning of a read operation.

The write operation is not truly reversible. The bit of information stored in the cell is destroyed by switching the inverter loop through a voltage V_{keep}. Thus, there is a certain amount of energy dissipated in the cell being written, which is inevitable because of the irreversible nature of erasing information. However, the capacitances of the long bit lines are much more significant than the capacitance of a single cell. Hence, the major portion of charging and discharging is performed adiabatically. The above analysis concludes that the energy consumption of the memory core for both read and write operations can be greatly minimized if the processes are sufficiently slow. Resistive loss is the dominant factor of energy dissipation, which is inversely proportional to the signal transition time T.

7.3.7 Comparison of Two Memory Core Organizations

Let us compare energy dissipation of the two memory core organizations, which might have a major impact on total power dissipation and chip architecture. If the signal transition time T and the supply voltage V_{dd} are fixed and are the same for both core organizations, the energy dissipation depends on the capacitances to be charged/discharged, which are modeled and compared below. The main difference between the two core organizations is that the second one activates selected columns only, which usually make up a small portion of the total number of columns in a memory core block. However, all cells on the selected columns participate in the voltage swings, while in the first core organization, only those cells in a single selected row are enabled. As a matter of fact, the total capacitance associated with cells in one column is more significant than the peripheral capacitance of the bit line, which implies that the actual capacitance being charged

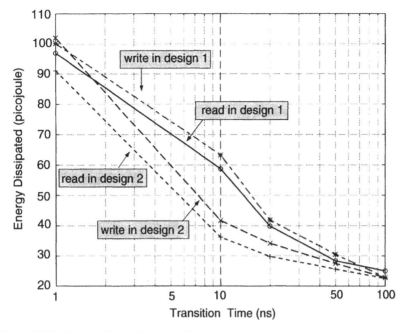

Figure 7.20   Comparison of energy dissipation in two memory core organizations.

during an operation depends on the geometry of the core block. Assume that both designs have the same recovery rate, and one column of the cells is associated with capacitance C_{cell} and the bit line has the peripheral capacitance C_{bit}. Further assume that a block of memory core consists of $M \times N$ bits, with n bits read or written in each operation. Then the ratio of the effective capacitances of the two core organizations is given by

$$\frac{C_{1st\ organization}}{C_{2nd\ organization}} = \frac{NC_{bit} + (N/M)C_{cell}}{n(C_{bit} + C_{cell})} \approx \left(\frac{N}{n}\right)\frac{C_{bit}}{C_{bit} + C_{cell}} \quad (7.14)$$

Figure 7.20 compares the SPICE simulation results of the two core organizations for a 64×64 block (with $n = 8$ I/O pins). The column-activated approach consumes slightly less energy than the regular row-activated organization. However, as one can observe from Eq. (7.14), when the ratio N/n is large, the second approach will certainly be superior to the first one. For example, in a 256×256 block, the row-activated approach would consume approximately 4 times the energy of the column-activated counterpart. Thus, the designs can provide a potential space for architectural level optimization of the SRAM.

7.3.8 Design of Peripheral Circuits

7.3.8.1 Energy Recovery Address Decoder

Another major component of power dissipation in SRAMs is due to the address decoder. In order to reduce the total power consumption, it is desirable to design the address decoder, which can be operated adiabatically as well. Figure 7.21 shows two implementations of 2-to-4 adiabatic address decoders—NAND and NOR implementation. The only difference in the configuration of the adiabatic decoder from the conventional dynamic decoder is that the precharge pMOS transistors are replaced by nMOS transistors, which no longer function as precharge transistors. Let us first show how the NOR decoder can operate in the adiabatic fashion. The decoder starts out in the rest state with V_{low} at V_{dd} and all row lines W_0, W_1, W_2, W_3 at $V_{dd} - V_{th}$ (guaranteed by the transistors at the leftmost column). After the address signals settle down, V_{low} gradually swings down from V_{dd} to zero, and all the row lines follow V_{low} down to zero except the selected row staying at $V_{dd} - V_{th}$. Note that the transistors in the leftmost column are disabled in this discharging process since their gate terminals are also tied to V_{low}. During the period when V_{low} stays at 0 V, the row selection signals W_0, W_1, W_2, W_3 are valid and are sampled by the row driver, which then enables the selected row in the memory core. Reverting back to the rest state is accomplished by ramping up V_{low} to V_{dd}, which pulls up all row lines to their rest state at $V_{dd} - V_{th}$. Subsequently, all address signals return to zero. The waveforms of the adiabatic NOR decoder are shown in Figure 7.22.

In a similar fashion, the adiabatic decoder can be implemented in a NAND array, which is shown in Figure 7.21b. Again, the precharge transistor in each row of a dynamic NAND decoder is replaced by an nMOS transistor, which ensures that every row line returns to high at $V_{dd} - V_{th}$ after a

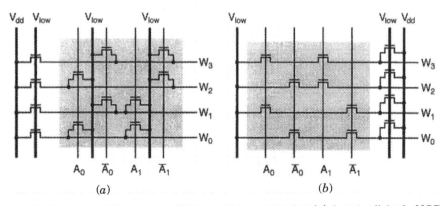

Figure 7.21 Configurations of adiabatic address decoder: (a) 2-to-4 adiabatic NOR decoder; (b) 2-to-4 adiabatic NAND decoder.

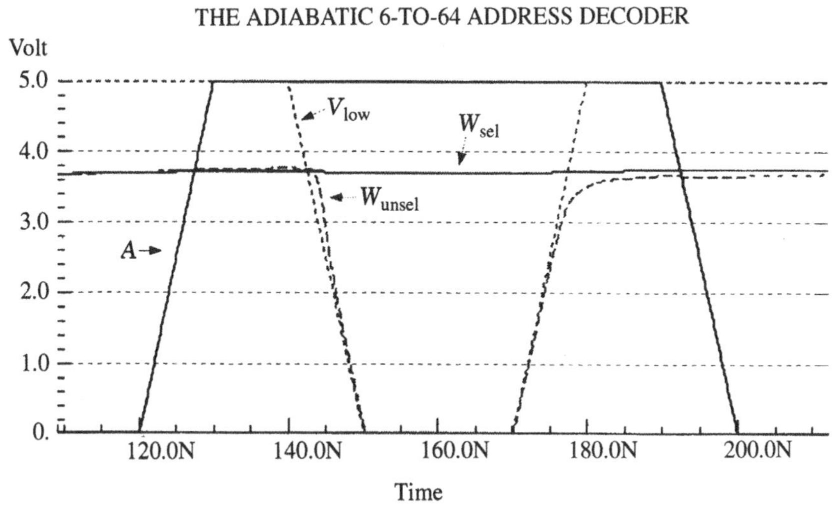

THE ADIABATIC 6-TO-64 ADDRESS DECODER

Figure 7.22 Waveforms of adiabatic address decoder.

decoding operation. The operation process is similar to the NOR decoder. In the NAND decoder, however, only one row is selected each time. The selected row line follows V_{low} ramping down to zero while all the other row lines stay at the rest state voltage $V_{dd} - V_{th}$. Because of this, the adiabatic NAND decoder consumes much less energy than the NOR decoder. In a NAND structure decoder, transistors in each row are serially connected, which makes it inappropriate to be used in large decoders. Hence, the choice between the NOR and NAND structure depends on the size of the decoder and the speed required.

Adiabatic decoders can also be realized in two stages with *predecoding*, in which less transistors are used. However, two phases of decoding have to be introduced as well, and hence, the total time required for address decoding is 2 times the transition time of V_{low}. By using the predecoding stage, the total effective capacitance of the decoder is substantially smaller since less transistors are required. Moreover, the *RC* time constant in each stage is reduced, which results in higher percentage of energy recovery. Two-level implementation is a better choice for adiabatic decoders of large size.

Unlike the standard CMOS SRAM, where row decoders and column decoders are usually implemented in different styles, the same decoding scheme is adopted for both row and column decoders in the adiabatic SRAM design. It is also interesting to note that the geometric structure of the address decoder is identical to the read-only memories (ROMs), differing only in the data pattern. Thus, the same designs for address decoders can be applied to adiabatic ROMs.

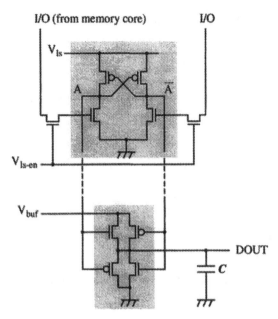

Figure 7.23   Adiabatic level shifter and I/O buffer.

7.3.8.2 *Energy Recovery Level Shifter and I/O Buffer*

The pair of bit lines from the memory core has a voltage difference of 2 V in the designs shown, which need to be amplified and buffered in order to drive the I/O bus lines. In standard CMOS SRAMs, *sense amplifiers* (SAs) are used to amplify the small voltage difference into full scale. The sense amplifier is usually clocked by an *enable* signal generated by the address transistion detection (ATD) circuitry to reduce the energy loss due to *short-circuit current*. A slightly different approach is used in order to minimize the short-circuit current and avoid the use of ATD circuitry, which consumes considerable amount of power [13].

The approach used in [12] is shown in Figure 7.23, which consists of two stages, a *voltage level shifter* and a *buffer*. The level shifter is essentially a pMOS cross-coupled sense amplifier with two access transistors. It functions as follows: Initially V_{ls} is at 0 V and the shifter is disabled. The two internal nodes A and $\bar{A}$ are also equalized. After the access transistors are turned on and *bit* and $\overline{bit}$ arrive, the voltage difference between two sides of the cross-coupled sense amplifier is built. Then V_{ls} gradually ramps from 0 V to V_{dd}. Since the voltage difference has been built before the sense amplifier is enabled, short-circuit current is not significant due to the positive-feedback effect. Figure 7.24 shows that node A is pulled up very rapidly and the stable state is achieved immediately; hence the short-circuit current is negligible. Subsequently, the two access transistors are turned off to isolate the level

THE ADIABATIC 64×64 SRAM CORE WITH LEVEL-SHIFTER AND I/O BUFFER

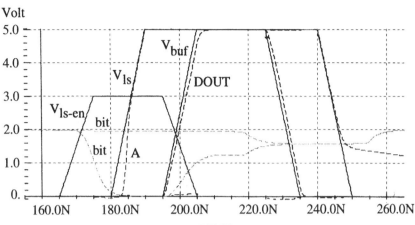

Figure 7.24   Waveforms for the energy recovery level shifter.

shifter from the bit-lines such that A and $\overline{A}$ hold their states while the bit lines return to the rest state. The buffer is now ready to drive the I/O bus line. The transmission gate constructs of the buffer ensures that the charging and discharging processes are performed adiabatically, and simulation results indicate that more than 90% of energy recovery can be achieved. Due to the threshold voltage of the transistors, A is not able to revert back to 0 V (stops at V_{th} instead) when V_{ls} returns to 0 V. Equalization is then applied to A and $\overline{A}$. However, the energy loss in the level shifter is not significant since it only drives a buffer. The major portion of charging and discharging is performed adiabatically.

7.3.8.3 Row Driver Circuitry

Let us consider another interface circuitry, the row driver, which uses the signals from the row address decoder to enable a particular row of the memory core. Each row of the row driver consists of two stages, a standard CMOS *tristate* buffer and adiabatic transmission gate drivers. Let us consider the tristate buffer first and assume that the NAND structure decoding scheme is used. The decoded signal W for the selected row is low at 0 V when it is in the valid period. Then the *enable signal* WEN switches from zero to V_{dd}, which enables the tristate buffer and results in SEL = V_{dd} and $\overline{SEL}$ = 0 V. Unlike other signals that switch gradually during a transition, WEN switches in a step function to avoid the short-circuit current in both NAND and NOR gates. When W is still valid, WEN switches back to zero, which simultaneously turns off the pMOS and the nMOS transistors. Thus, SEL and $\overline{SEL}$ are disconnected from their inputs and kept at V_{dd} and zero,

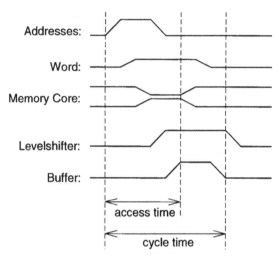

Figure 7.25   Overall supply schemes for the adiabatic SRAM.

respectively. The reason for using the standard CMOS tristate buffer is as follows: only one of the M rows is selected at each time; hence two rows have transition at most, that is, the selected row in the last cycle is unselected, and a new row is selected. Therefore, the rest of $M - 2$ rows do not have transition by using the tristate buffer, and hence, do not consume energy (ignoring the leakage current).

The second stage of the row driver is three transmission gate drivers for V_{hi}, V_{low}, and V_{word}, respectively. For the particular row selected, SEL $= V_{dd}$ and $\overline{\text{SEL}} = 0$. Hence, V_{hi}, V_{low}, and V_{word} are controlled by G_{hi}, G_{low}, and G_{word}, respectively. For all other $M - 1$ rows, V_{hi}, V_{low}, and V_{word} are connected to the DC supplies S_{hi}, S_{low}, and ground, respectively. The overall clocking scheme for the SRAM is shown in Figure 7.25.

7.3.9   Optimal Voltage Selection

In conventional CMOS digital systems, energy consumption decreases in proportion to the square of the supply voltage V_{dd}. This is *not* the case for adiabatic circuits in general. Instead, there might exist an optimal voltage swing that leads to the minimal energy dissipation in adiabatic circuits [14]. In this section, let us examine the optimal voltage swing problem and derive a method to find the optimal voltage for the adiabatic SRAM as well as for general adiabatic circuits.

Let us consider charging a load capacitance C_L through a MOSFET to deliver a charge $C_L V_{dd}$ over a time period T. The energy dissipation through the channel of the MOSFET is given by

$$E_{diss} = \int_0^T \Delta V i \, dt = \int_0^T (\Delta V)^2 g \, dt = T \langle (\Delta V)^2 g \rangle \tag{7.15}$$

where g is the MOSFET channel conductance, ΔV is the voltage drop across the channel, and $\langle \cdot \rangle$ denotes the average over time period T. Now,

$$C_L V_{dd} = \int_0^T i\, dt = \int_0^T \Delta V g\, dt = \langle \Delta V g \rangle T \qquad (7.16)$$

Hence,

$$\langle \Delta V g \rangle = C_L V_{dd}/T \qquad (7.17)$$

We use the following approximation to simplify E_{diss}:

$$\langle (\Delta V)^2 g \rangle \langle g \rangle \approx \langle \Delta V g \rangle^2 \qquad (7.18)$$

Substituting Eq. (7.17) and (7.18) into Eq. (7.15), we get

$$E_{diss} = T \left(\frac{C_L V_{dd}}{T} \right)^2 \frac{\langle (\Delta V)^2 g \rangle}{\langle \Delta V g \rangle^2} = \left(\frac{C_L}{\langle g \rangle T} \right) C_L V_{dd}^2 \qquad (7.19)$$

The approximation of Eq. (7.18) turns out to be exact if ΔV is a constant throughout the charging process. More detailed analysis and simulations suggest only a small variance of ΔV, which promises the accuracy of the approximation.

We now consider a transmission gate shown in Figure 7.26a, which is a fundamental circuit construct for many adiabatic approaches, to drive a

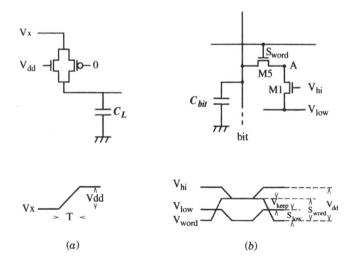

(a)   (b)

Figure 7.26 Circuit models of adiabatic charging/discharging.

capacitive load C_L. The conductance of the nMOS channel is

$$
g_n = \begin{cases} k_n(V_{dd} - V_x - V_t) & V_x < V_{dd} - V_t \\ 0 & V_x \geq V_{dd} - V_t \end{cases} \tag{7.20}
$$

where k_n is a constant given by $(\mu_n \epsilon / t_{ox})(W/L)$, where W and L are the width and length of the transistor, μ_n is the mobility of electrons, ϵ is the permittivity of silicon dioxide, and t_{ox} is the gate oxide thickness. The average conductance is then given by

$$
\begin{aligned}
\langle g_n \rangle &= \frac{1}{T} \int_0^T g_n \, dt \\
&= \frac{1}{V_{dd}} \int_0^{V_{dd} - V_{th}} k_n(V_{dd} - V_x - V_{th}) \, dV_x \\
&= \frac{k_n}{2V_{dd}} (V_{dd} - V_{th})^2
\end{aligned} \tag{7.21}
$$

The waveform has been assumed to be switching linearly so that the average over time can be obtained by integrating over voltage. Similarly,

$$
\langle g_p \rangle = \frac{k_p}{2V_{dd}} (V_{dd} - V_{th})^2 \tag{7.22}
$$

Thus, we have

$$
\langle g \rangle = \langle g_n \rangle + \langle g_p \rangle = \frac{k_n + k_p}{2V_{dd}} (V_{dd} - V_{th})^2 \tag{7.23}
$$

and hence, the energy dissipation is

$$
E_{\text{T-gate}} = \frac{2C_L^2}{(k_n + k_p)T} \frac{V_{dd}^3}{(V_{dd} - V_{th})^2} \tag{7.24}
$$

The energy dissipation $E_{\text{T-gate}}$ has a minimum at $V_{dd} = 3V_{th}$. When the difference in threshold voltage between nMOS and pMOS is accounted for, the optimal V_{dd} lies between $3V_{tn}$ and $3V_{tp}$. Although the second-order effects (e.g., the body effect) have not been taken into account, SPICE simulations verify that $3V_{th}$ is close to the minimum energy dissipation. In SRAM designs considered in this chapter, transmission gate drivers have been used in I/O buffers, the supply drivers for V_{hi} and V_{low} of the memory core, the word line drivers, and the address signal drivers.

Let us now consider another major component of energy dissipation—the bit lines in the memory core. A circuit model of discharging a bit line capacitance is shown in Figure 7.26b. The conductance of transistors $M1$ and $M5$ are

$$g_{M1} = k_n(V_{hi} - V_{low} - V_{th}) = k_n(S_{word} - V_{th})$$

$$g_{M5} = k_n(S_{word} - V_A - V_{th})$$

where $V_A \approx {}_{low}$. Set $V_{keep} = V_{th}$. We have $S_{word} - V_{keep} = S_{word} - V_{th} = S_{low}$. The serial connection of $M1$ and $M5$ gives

$$\frac{1}{g} = \frac{1}{g_{M1}} + \frac{1}{g_{M5}} \tag{7.25}$$

Hence,

$$
\begin{aligned}
g &= \frac{g_{M1} g_{M5}}{g_{M1} + g_{M5}} \\
&= \frac{k_n(S_{word} - V_{th})(S_{low} - V_{low})}{S_{word} + S_{low} - V_{th} - V_{low}} \\
&= k_n\left[(S_{word} - V_{th}) - \frac{(S_{word} - V_{th})^2}{S_{word} + S_{low} - V_{th} - V_{low}}\right] \tag{7.26}
\end{aligned}
$$

$$
\begin{aligned}
\langle g \rangle &= \frac{1}{S_{low}}\int_0^{S_{low}} g \, dV_{low} \\
&= k_n(S_{word} - V_{th}) - \frac{k}{S_{low}}\int_0^{S_{low}} \frac{(S_{word} - V_{th})^2}{S_{word} + S_{low} - V_{th} - V_{low}} \, dV_{low} \\
&= k_n(S_{word} - V_{th}) - k_n\frac{(S_{word} - V_{th})^2}{S_{low}} \ln\left(\frac{S_{word} + S_{low} - V_{th}}{S_{word} - V_{th}}\right) \tag{7.27}
\end{aligned}
$$

On the other hand, we have $S_{word} = (V_{dd} + V_{keep})/2 = (V_{dd} + V_{th})/2$ and $S_{low} = (V_{dd} - V_{keep})/2 = (V_{dd} - V_{th})/2$. Substituting S_{low} and S_{word} into the above equation gives

$$
\begin{aligned}
\langle g \rangle &= k_n\left(\frac{V_{dd} - V_{th}}{2}\right) - k_n\left(\frac{V_{dd} - V_{th}}{2}\right)\ln 2 \\
&= \frac{k_n(1 - \ln 2)}{2}(V_{dd} - V_{th}) \tag{7.28}
\end{aligned}
$$

Thus, we obtain the energy dissipation on a bit line to be

$$
\begin{aligned}
E_{\text{bit}} &= \left(\frac{C_{\text{bit}}}{\langle g \rangle T} \right) C_{\text{bit}} S_{\text{low}}^2 \\
&= \frac{2C_{\text{bit}}}{k_n(1 - \ln 2)T} \frac{C_{\text{bit}}\left[(V_{dd} - V_{\text{th}})/2 \right]^2}{V_{dd} - V_{\text{th}}} \\
&= \frac{C_{\text{bit}}}{2k_n(1 - \ln 2)T} C_{\text{bit}}(V_{dd} - V_{\text{th}})
\end{aligned}
\tag{7.29}
$$

which is proportional to $V_{dd} - V_{\text{th}}$. Hence, the smaller the voltage swing, the smaller is the energy dissipation. However, there are other constraints on the voltage swing. For the level shifter considered in this chapter, $S_{\text{low}} > V_{\text{th}}$ must hold to make it work properly, which results in $V_{dd} > 3V_{\text{th}}$. Therefore, for the design of the SRAM, the optimal voltage swing lies between $3V_{\text{th}}$ and $4V_{\text{th}}$. The more accurate value can be determined by simulations. It also depends on the particular design of the level shifter (remember that there are plenty of choices for the level shifter).

In [2], the circuits were implemented using a MOSIS 1.2-μm CMOS process. The layout of the cell of the two core organizations (referred to as design 1 and design 2) is shown in Figure 7.27. The cell size of the two designs is $27\lambda \times 40\lambda$ and $32\lambda \times 38\lambda$, respectively, where $\lambda = 0.6\ \mu$m. Unlike standard RAM cells, a memory cell working in an adiabatic fashion does not require the high aspect ratio between the pass and the pull-down

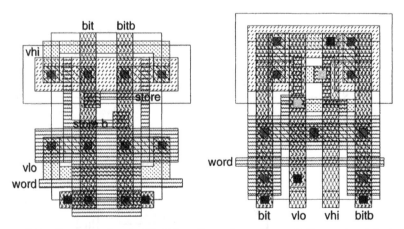

Figure 7.27 Layout of two SRAM cell configurations. Cell on the left is for row-activated memory core organization, and cell on the right is for column-activated memory core organization, in which only selected columns are active. Supply lines V_{hi} and V_{low} run horizontally in the left cell and vertically in the right cell.

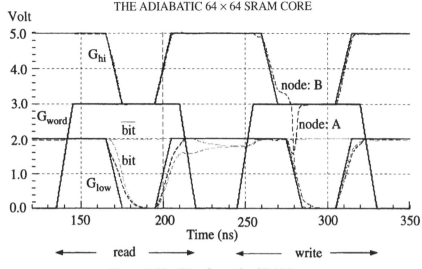

Figure 7.28   Waveforms for SRAM core.

transistor. The ratio of transistors $M3$ and $M5$ ($M4$ and $M6$) is unity in the designs. Although the topology of the cell in design 1 is identical to a standard six-transistor RAM cell, the fact that separate V_{hi} and V_{low} lines are needed for each row results in additional area overhead. For comparison, a conventional cell takes area $27\lambda \times 36.6\lambda$. The cell of design 2 takes the largest area due to the vertically laying of V_{hi} and V_{low} lines.

Results of SPICE simulations show that for a block of SRAM core of 64 rows by 64 show large savings in power dissipation. Waveforms from simulations for both core organizations are similar and results for the first core organization are shown in Figure 7.28. The first operation was a *read* followed by a *write*. Figure 7.29 shows the fraction of energy recovered at various speeds of operation. For the stimulus with transition time of 10 ns, energy recovery was around 50% for both read and write operations.

The circuits of 6-to-64 NOR and NAND decoders were also simulated in the same technology, the results of which are shown in Figure 7.30. Results indicate that approximately 90% of energy can be recovered for both organizations when the stimulus has a transition time of 10 ns. However, as Figure 7.31 suggests, the energy dissipation of the NAND array decoder is far less than the NOR array decoder. The reason is due to the fact that only a single row is pulled down in the NAND decoder while $N - 1$ rows are pulled down in the NOR decoder.

Although a level shifter with high energy recovery is difficult to design, it should be noted that it only drives a buffer. The energy loss at the voltage level-shift stage is insignificant. The I/O buffer, which has heavy capacitive load, can perform in an adiabatic fashion and can recover most of the energy

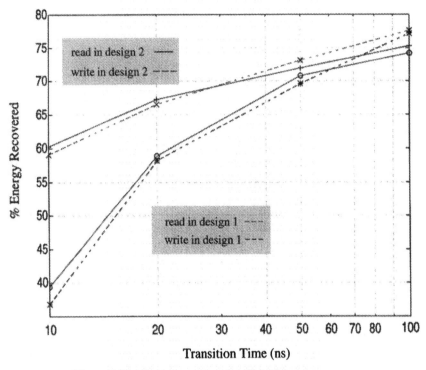

Figure 7.29 Energy recovery in adiabatic memory core.

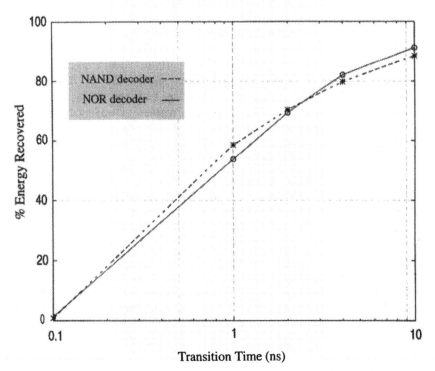

Figure 7.30 Energy recovery in 6-to-64 adiabatic NAND and NOR decoders.

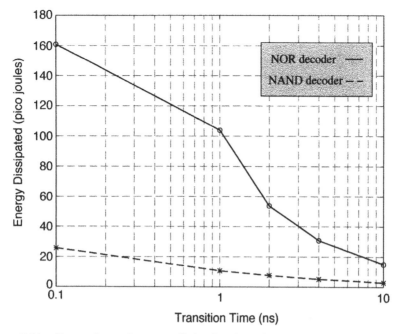

Figure 7.31 Comparison of energy dissipation in adiabatic NAND and NOR decoder.

by using the transmission gate construct. The simulation results indicate that more than 90% of energy recovery can be achieved with 1 pF capacitive load at 10 ns stimulus transition time.

7.4 SUPPLY CLOCK GENERATION

Research on the design of highly efficient resonant drivers for generating adiabatic supply clocks has started in earnest [6, 11, 15, 16]. The clock generator should be able to operate with high efficiency at high clock frequencies. In addition, the clock generator should work at a frequency determined by an external reference frequency.

The underlying idea of adiabatic clock generation circuits is to use a resonant driver. Let us first consider a generic resonant driver, shown in Figure 7.32. Ideally, the circuit oscillates between zero and $2V_{ref}$. The circuit starts to oscillate when S_0 is turned on and ceases oscillating when S_0 is turned off. There is a *pull-up* path and a *pull-down* path that can replenish the energy dissipated by the resistances in the load. The pull-up pMOS transistor S_p is turned on and the pull-down nMOS transistor S_n is turned off when voltage at node y is higher than V_{ref}. Conversely, the pull-down path is on and the pull-up path is off when voltage at y is below V_{ref}. Thus, the control signals at S_p and S_n are 180° out of phase. The size of replenish

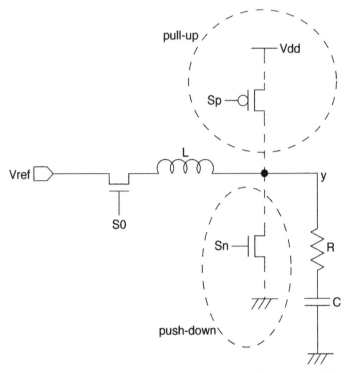

Figure 7.32   Generic resonant scheme for adiabatic clock (power supply) generation circuits.

transistors S_p and S_n should be determined to maintain the desired amplitude of oscillation.

The generic circuitry exhibits the following problems:

- The serial-connected control transistor S_0 has finite resistance, which decreases the energy efficiency substantially.
- The control signals for S_p in the pull-up path and S_n in the pull-down path have to be generated by extra circuitry.
- The circuitry requires an additional reference voltage source V_{ref} other than the supply V_{dd}.
- All energy used for charging the gate capacitances of S_0, S_p, and S_n is dissipated. Since these transistors have large size, the energy dissipation can be significant.
- The circuitry generates a single-phase clock, which is not enough for general adiabatic circuits. Hence, more than one clock generation circuitry is required.

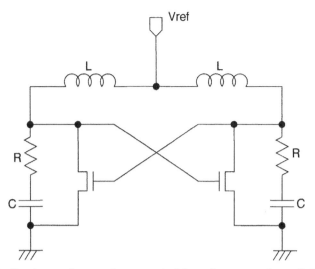

Figure 7.33 Previous scheme of resonant driver for generating adiabatic power supply/clocks.

A clock generation scheme is proposed in [15], which is shown in Figure 7.33. It consists of two branches that are 180° out of phase. Two almost nonoverlapping clock signals are generated. Only a pull-down path is used in each branch for replenishing energy and maintaining undamped oscillation. Nonsinusoidal "blip" waveforms are produced for lack of pull-up path. For resonant circuits, a sinusoidal waveform has the highest energy recycling percentage. Any other waveform contains a base sinusoidal component and higher order harmonics. The component of base frequency f_0 can be efficiently recycled (determined by the Q value of the resonant circuit), while all other components in higher frequencies $(2f_0, 3f_0, 4f_0, \ldots)$ are almost completely dissipated, as illustrated in Figure 7.34. Therefore, the energy efficiency of this scheme can still be substantially improved. In addition, two inductors and an extra reference voltage source is used in this scheme.

Two possible variations are illustrated in Figure 7.35 and 7.36, respectively. Note that the reference voltage source can be removed and the two inductors can be further replaced by one inductor in both schemes. The two branches in the scheme of Figure 7.35 are not symmetric. A pull-up path is used in one branch while a pull-down branch is used in the other branch. In the scheme of Figure 7.36, both pull-up and pull-down paths are used in each branch; thus the generated waveforms are closer to the sinusoidal curve. While the sinusoidal waveform is important for higher energy efficiency, this scheme has a severe shortcoming. Within a certain period of time in every cycle, both nMOS and pMOS transistors are on when the voltage V_1 or V_2 is in the vicinity of $V_{dd}/2$. Hence, short-circuit current dissipates a significant amount of energy.

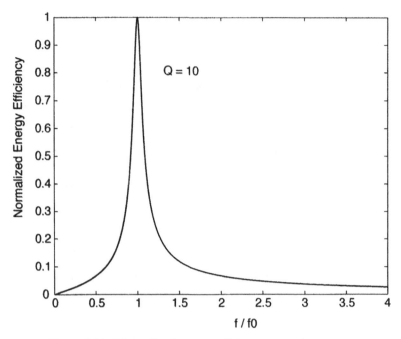

Figure 7.34 Normalized energy efficiency versus frequency.

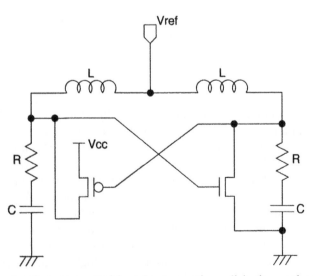

Figure 7.35 Another resonant driver for generating adiabatic supply clocks. Two branches are not symmetric. Pull-up path is used in one branch while pull-down branch is used in the other branch.

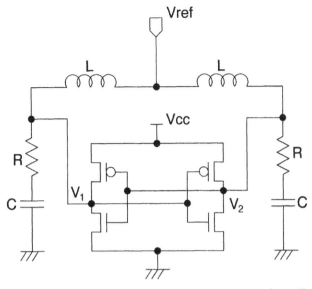

Figure 7.36 Another scheme of a resonant driver for generating adiabatic supply clocks. Both pull-up and pull-down paths are used in each branch.

Another scheme for clock generation considers some of the shortcomings of the previous scheme to generate better sinusoidal waveforms while achieving lower short-circuit current. The clock generation circuitry is shown in Figure 7.37. The oscillator generates two complementary phases of nearly sinusoidal waveforms. Two pMOS transistors (P_1 and P_2) and two nMOS transistors (N_1 and N_2) are used for energy replenishment and frequency phase lock-up. The circuitry starts to oscillate when the control signal *enable* = 1 and ceases to oscillate when *enable* = 0 (because the pull-down transistors N_1 and N_2 are turned on and the pull-up transistors P_1 and P_2 are turned off). The size of these four transistors are determined by the operating frequency and the capacitive loads the circuitry is driving. The higher the frequency and/or the larger the capacitive load, the larger the transistors should be, such that the peak oscillating voltage V_{dd} is achieved. The optimal size can be found by simulations. Unlike the previous scheme, no reference voltage source is needed. There is no serial-connected transistor in the driver circuitry, and hence the energy efficiency is mainly determined by the load, that is, the resistance R in the clock distribution lines and the capacitive load C. One inductor is sufficient. The value of the inductor is determined by the resonant condition at the given frequency. Simulation results indicated that L should be slightly larger than the theoretically calculated value for optimal efficiency. The PLL (phase-locked loop) samples the clock signal(s) at the load and produces two control signals C_1 and C_2 at the frequency of the reference clock, which in turn forces the circuitry to

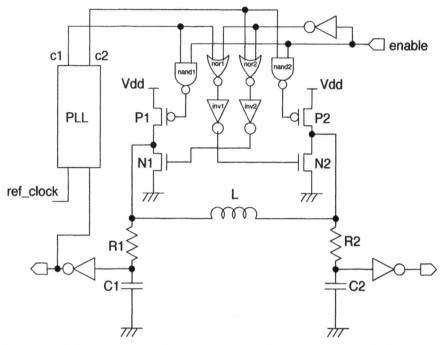

Figure 7.37 New scheme of resonant driver for generating adiabatic power supply/clocks.

oscillate at the frequency of the reference clock. The two control signals C_1 and C_2 are 180° out of phase, each of which is a pulse signal with 25% duty cycle only. The transistors of the inverters (INV1, INV2), NAND gates (NAND1, NAND2) and NOR gates (NOR1, NOR2), which are controlling the replenishing transistors, should be much smaller than the replenishing transistors (e.g., approximately 1/20 of the replenishing transistors at 100 MHz). Thus, the voltage at the gate of a replenishing transistor has finite rise and fall times. Approximately a triangular waveform is produced at the gate of each replenishing transistor. The optimal rise and fall times should be close to 25% of a clock cycle such that each replenish transistor is on for approximately 50% of time. The advantages of the above scheme are as follows:

- The replenishing transistors turn on and off gradually so that only small interferences are imposed on resonant circuitry, which ensures that the waveforms at both sides of the inductor are nearly sinusoidal. Sinusoidal waveforms have the highest energy efficiency. Energy for the higher order harmonic components are almost completely dissipated in the resistances of clock distribution lines and resistive loads.
- The pMOS and nMOS replenishing transistors in one side (e.g., P_1 and N_1) are never turned on simultaneously to prevent the short-circuit current.

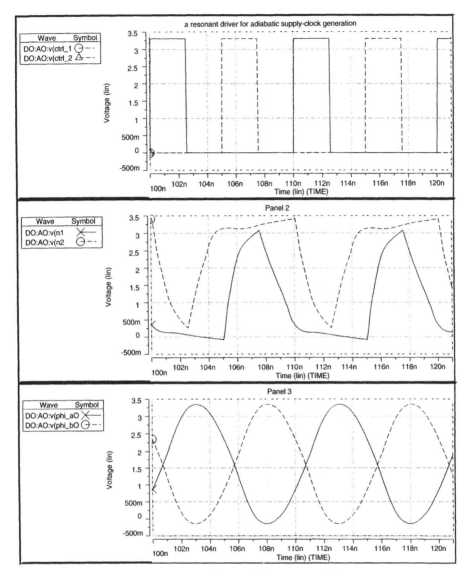

Figure 7.38   Waveforms of the resonant driver.

- All transistors in the control gates (INV1, INV2, NAND1, etc.) are small. The energy consumed in these control gates are negligible.

Figure 7.38 shows the waveforms of a control signal C_1 from the PLL (top of Figure 7.38), voltage at the gate of replenishing transistors P_1 and N_1 (middle of Figure 7.38), and the clock waveforms produced at the output (bottom of Figure 7.38).

Let us define the *energy efficiency* of the supply clock generation as the ratio of energy delivered from DC (equals to the energy dissipated) and CV^2,

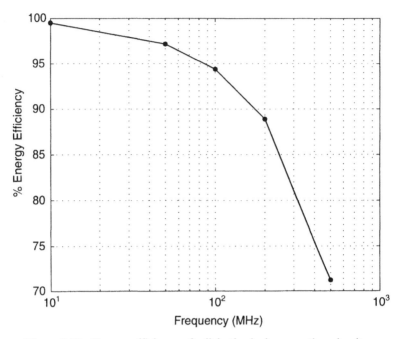

Figure 7.39   Energy efficiency of adiabatic clock generation circuitry.

where V is the peak voltage. The replenishing transistors are sized to satisfy $V_{peak} \approx V_{dd}$.

Figure 7.39 shows the energy efficiency of the scheme, which is obtained from simulations for $R = R_1 = R_2 = 0.5\ \Omega$ and $C = C_1 + C_2 = 100$ pF. The energy efficiency is approximately 95% at 100 MHz for the chosen R and C values. It is essential to minimize R in the supply clock distribution network to obtain high energy efficiency. When the load capacitances are excessively large, multiple clock generation circuits connected in parallel are needed to have high energy efficiency.

There are other possible variations of this clock generation circuitry. Figure 7.40 shows a slightly different version of the scheme. The main difference is that the pair of pMOS transistors are self-controlled in a cross-coupled fashion. The advantage of this change is that energy for charging the pMOS transistors can be recycled while the oscillating frequency is still locked to the external reference frequency.

Simulation also indicates that the proposed clock generation circuitry offers large tolerance on the process and load variation. There is only a slight decrease in energy efficiency for 10% mismatch between C_1 and C_2. When C_1 and C_2 have 50% mismatch, the clock frequency is still locked to the reference clock, although it severely deviates from a sinusoidal waveform. The impacts from other process variations are less significant. It should be noted that the power supply generation circuitry described in this chapter

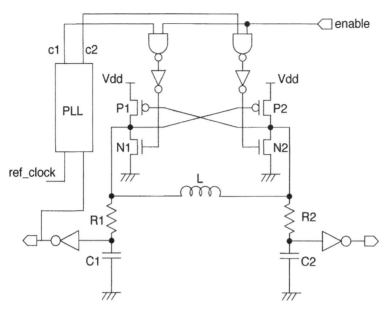

Figure 7.40 Alternative scheme of resonant driver for generating adiabatic power supply/clocks, in which two *p*MOS replenishing transistors are self-controlled in a cross-coupled fashion.

is ideally suited for the quasi-static energy recovery techniques described earlier.

7.5 SUMMARY AND CONCLUSIONS

For low-energy computing, recovered energy circuitry can produce a large improvement in energy dissipation at the cost of area and slowing down the clock speed. However, there are several practical aspects that need to be considered to harness the maximum possible energy recovery from these circuits. First, the switching power supply has to be very efficient with a very high Q factor. Second, the use of totally reversible logic can produce a number of redundant signals. The redundant signals can dissipate energy. Hence, methodologies for logic synthesis of reversible logic is required such that the number of redundant signals is optimized. Hence, the use of quasi-static CMOS circuits that resemble fully static circuits is more practical. Some changes in the standard CMOS processing technology would also be helpful for implementing adiabatic circuits. The standard CMOS processes do not have diodes, though parasitic diodes are produced as a by-product of the CMOS process. The current methodology for simulating diodes in CMOS connects the gate and source of a transistor. However, the voltage drop and the performance of such diodes can be intolerable for adiabatic circuits.

REFERENCES

[1] C. H. Bennett, "The Thermodynamics of Computation—a Review," *Int. J. Theoret. Phys.*, vol. 21, no. 12, pp. 905–940, 1982.

[2] E. Fredkin and T. Toffoli, "Conservative Logic," *Int. J. Theoret. Phys.* vol. 21, no. 3/4, pp. 219–253, 1982.

[3] R. Landauer, "Dissipation and Noise Immunity in Computation and Communication," *Nature*, vol. 335, pp. 779–784, 1988.

[4] R. P. Feynman, "Quantum Mechanical Computers," *Found. Phys.* vol. 16, no. 6, 1986.

[5] W. C. Athas, L. "J". Svensson, J. G. Koller, N. Tzartzanis, and E. Chou, "A Framework for Practical Low-Power Digital CMOS Systems Using Adiabatic-Switching Principles," *International Workshop on Low Power Design*, Napa Valley, CA, 1994, pp. 189–194.

[6] S. G. Younis and T. F. Knight, "Asymptotically Zero Energy Split-Level Charge Recovery Logic," *International Workshop on Low Power Design*, Napa Valley, CA, 1994, pp. 177–182.

[7] J. S. Denker, S. C. Avery, A. G. Dickinson, A. Kramer, and T. R. Wik, "Adiabatic Computing with the 2N-2N2D logic Family," *International Workshop on Low Power Design*, Napa Valley, CA, 1994, pp. 183–187.

[8] Y. Ye and K. Roy, "Ultra Low Power Circuit Design using Adiabatic Switching Principle," *IEEE Midwest Symposium on Circuits and Systems*, 1995.

[9] Y. Ye and K. Roy, "QSERL: Quasi-Static Energy Recovery Logic," *IEEE J. of Solid-State Circuits*, accepted for publication.

[10] Y. Ye and K. Roy, "Energy Recovery Circuits Using Reversible and Partially Reversible Logic," *IEEE Trans. on Circuits and Syst. I*, pp. 769–778. Sept. 1996.

[11] A. G. Dickinson and J. S. Denker, "Adiabatic Dynamic Logic," *IEEE J. Solid-State Circuits*, vol. 30, no. 3, pp. 311–315, 1995.

[12] D. Domasekhar, Y. Ye, and K. Roy, "An Energy Recovery Static RAM Memory Core," *IEEE Symposium on Low Power Electronics*, San Jose, Sept. 1995.

[13] S. Flannagan et al., "8-ns CMOS 64K × 4 and 256K × 1 SRAM's," *IEEE J. Solid-State Circuits*, vol. 25, no. 5, pp. 1049–1056, 1995.

[14] W. C. Athas, L. "J". Svensson, J. G. Koller, N. Tzartzanis, and E. Chou, "Low-Power Digital Systems Based on Adiabatic-Switching Principles," *IEEE Trans. VLSI Systems*, vol. 2, no. 4, pp. 398–407, 1994.

[15] W. C. Athas, L. "J". Svensson and N. Tzartzanis, "A Resonant Signal Driver for Two-Phase, Almost-Non-Overlapping Clocks," *ISCAS-96*.

[16] Y. Ye, K. Roy, and G. Stamoulis, "Quasi-Static Energy Recovery Logic and Supply Clock Generation Circuits," *IEEE/ACM International Symposium on Low Power Electronics and Design*, Monterey, CA, Aug. 1997.

[17] V. De and J. D. Meindl, "Complementary Adiabatic MOS Logic Families for Gigascale Integration," *Dig. Tech. Papers, ISSCC*, pp. 298–299, 1996.

[18] M. C. Knapp, P. J. Kindlmann and M. C. Papaefthymiou, "Implementing and Evaluating Adiabatic Arithmetic Units," *IEEE Custom Integrated Circuits Conference*, San Diego, CA, 1996, pp. 115–118.

[19] I. M. Gottlieb, *Power Supplies, Switching Regulators, Inverters, and Converters*, McGraw-Hill, New York, 1994.

CHAPTER 8

SOFTWARE DESIGN
FOR LOW POWER

MARK C. JOHNSON and KAUSHIK ROY

8.1 INTRODUCTION

Most efforts in controlling power dissipation of digital systems have been and continue to be focused on hardware design. There is good reason for this since hardware is the physical means by which power is converted into useful computation. However, it would be unwise to ignore the influence of software on power dissipation. In systems based on digital processors or controllers, it is software that directs much of the activity of the hardware. Consequently, the manner in which software uses the hardware can have a substantial impact on the power dissipation of a system. An analogy drawn from automobiles can help explain this further. The manner in which one drives can have a significant effect on total fuel consumption.

Until recently, there were no efficient and accurate tools to estimate the overall effect of a software design on power dissipation. Without a power estimator there was no way to reliably optimize software to minimize power. Some software design techniques are already known to reduce certain components of power dissipation, but the global effect is more difficult to quantify. For example, it is often advantageous to optimize software to minimize memory accesses, but there may be energy-consuming side effects, such as an increase in number of instructions. Not all low-power software design problems have been solved, but progress has been made. In this chapter, we present a variety of software design techniques for low power. Some will address specific sources of power dissipation, but there are others that take a more global approach.

We will start by exploring the sources of power dissipation that should be most influenced by software. The next section will look at how power dissipation can be modeled to facilitate power estimation of a particular application program. The core of this chapter will be a presentation of a variety of software power optimization techniques found in recent literature. The techniques will encompass these general categories: algorithm optimizations, minimizing memory accesses, optimal selection and sequencing of machine instructions, and power management. We find that low-power software design, much like low-power hardware design, requires optimizations at several levels of abstraction. In the case of software, these levels range from algorithm design all the way down to machine code or even microcode. Ultimately, one cannot fully optimize a design without considering trade-offs between hardware and software. The chapter will close with a summary of the most important techniques to consider when designing software for a low-power system.

8.2   SOURCES OF SOFTWARE POWER DISSIPATION

There are several contributors to CPU power dissipation that can be influenced significantly by software. The memory system takes a substantial fraction of the power budget (on the order of one-tenth to one-fourth) for portable computers [1], and it can be the dominant source of power dissipation in some memory-intensive DSP applications such as video processing [2]. System buses (address, instruction, and data) are a high-capacitance component [3] for which switching activity is largely determined by the software. Data paths in integer arithmetic logic units (ALUs) and floating-point units (FPUs) are demanding of power, especially if all operational units (e.g., adder, multiplier) or pipeline stages are activated at all times. Power overhead for control logic and clock distribution are relevant since their contribution to a program's energy dissipation is proportional to the number of execution cycles of the program [4]. Power management [5] should further increase the influence of software on power in at least two ways. As the power of idle components are reduced, there should be an increase in the proportion of power given to carrying out the work actually ordered by software. In addition, some processors allow for direct control by software of some power savings modes.

Memory accesses are expensive for several reasons. Reading or writing a memory location involves switching on highly capacitive data and address lines going to the memory, row and column decode logic, and word and data lines within the memory that have a high fanout. The mapping of data structures into multiple memory banks can influence the degree to which parallel loading of multiple words are possible [6]. Parallel loads not only improve performance but are more energy efficient [7]. The memory access patterns of a program can greatly affect the cache performance of a system. Unfavorable access patterns (or a cache that is too small) will lead to cache misses and a lot of costly memory accesses. In multidimensional signal processing algorithms, the order and nesting of loops can alter memory size and bandwidth requirements by orders of magnitude [8]. Compact machine code decreases memory activity by reducing the number of instructions to be fetched and reducing the probability of cache misses [9]. Cache accesses are more energy efficient than main memory accesses. The cache is closer to the CPU than is main memory, resulting in shorter and less capacitive address and data lines. The cache is also much smaller than main memory, leading to smaller internal capacitance on word and data lines.

Buses in an instruction processing system typically have high load capacitances due to the number of modules connected to each bus and the length of the bus routes. The switching activity on these buses is determined to a large degree by software [3]. Switching on an instruction bus is determined by the sequence of instruction op-codes to be executed. Similarly, switching on an address bus is determined by the sequence of data and instruction accesses. Both effects can often be accounted for at compile time. Data-related switching is much harder to predict at compile time since most input data to a program are not provided until execution time.

Data paths for ALUs and FPUs make up a large portion of the logic power dissipation in a CPU. Even if the exact data input sequences to the ALU and FPU are hard to predict, the sequence of operations and data dependencies are determined during software design and compilation. The energy to evaluate an arithmetic expression might vary considerably with respect to the choice of instructions. A simple example is the common compiler optimization technique of reduction in strength where an integer multiplication by 2 could be replaced by a cheaper shift-left operation. Poor scheduling of operations might result in unnecessary energy-wasting pipeline stalls. There are many other software design decisions that can affect data path power, but these will be discussed in Section 8.4.

Some sources of power dissipation such as clock distribution and control logic overhead might not seem to have any bearing on software design decisions. However, each execution cycle of a program incurs an energy cost from such overhead. The longer a program requires to execute, the greater will be this energy cost. In fact, Tiwari et al. [4] found that the shortest code sequence was invariably the lowest energy code sequence for a variety of microprocessor and DSP devices. In no case was the lower average power dissipation of a slightly longer code sequence enough to overcome the overhead energy costs associated with the extra execution cycles. However, this situation may change as power management techniques are more commonly used. In particular, power management commonly takes the form of removing the clock from idle components. This reduces the clock load as well as prevents unwanted computations.

8.3   SOFTWARE POWER ESTIMATION

The first step toward optimizing software for low power is to be able to estimate the power dissipation of a piece of code. This has been accomplished at two basic levels of abstraction. The lower level is to use existing gate level simulation and power estimation tools on a gate level description of an instruction processing system. A higher level approach is to estimate power based on the frequency of execution of each type of instruction or instruction sequence (i.e., the execution profile). The execution profile can be used a variety of ways. Architectural power estimation [10] determines which major components of a processor will be active during each execution cycle of a program. Power estimates for each active component are then taken from a look-up table and added into the power estimate for the program. Another approach is based on the premise that the switching activity on buses (address, instruction, and data) is representative of switching activity (and power dissipation) in the entire processor. Bus switching activity can be estimated based on the sequence of instruction op-codes and memory addresses [3]. The final approach we will consider is referred to as *instruction level power analysis* [11]. This approach requires that power costs associated with individual instructions and certain instruction sequences be characterized empirically for the target processor. These costs can be applied to the instruction execution sequence of a program to obtain an overall power estimate.

8.3.1   Gate Level Power Estimation

Gate level power estimation [12, 13] of a processor running a program is the most accurate method available, assuming that a detailed gate level description including layout parasitics is available for the target processor. Such a detailed processor description is most likely not available to a software developer, especially if the processor is not an in-house design. Even if the details are available, this approach will be too slow for low-power optimization of a program. Gate level power estimates are important in evaluating the power dissipation behavior of a processor design and in characterizing the processor for the more efficient instruction level power estimation approaches.

8.3.2   Architecture Level Power Estimation

Architecture level power estimation is less precise but much faster than gate level estimation. This approach requires a model of the processor at the major component level (ALU, register file, etc.) along with power dissipation estimates for each component. It also requires a model of the specific components which will be active as a function of the instructions being executed. The architecture level approach is implemented in a power estimation simulator called ESP [10] (Early design Stage Power and performance simulator). ESP simulates the execution of a program, determining which system components are active in each execution cycle and adding the power contribution of each component.

8.3.3 Bus Switching Activity

Bus switching activity is another indicator of software power based on a simplified model of processor power dissipation. In this simplified model, bus activity is assumed to be representative of the overall switching activity in a processor. Modeling bus switching activity requires knowledge of the bus architecture of a processor, op-codes for the instruction set, a representative set (or statistics) of input data to a program, and the manner in which code and data are mapped into the address space. By simulating the execution of a program, one can determine the sequence of op-codes, addresses, the data values appearing on the various buses. Switching statistics can be computed directly from these sequences of binary values. Su, Tsui, and Despain implement bus switching activity in their *cold scheduling* algorithm for instruction scheduling [3]. In this algorithm, only instruction- and address-related switching are considered due to the unpredictability of data values. However, for signal processing applications it is often possible to even predict switching activities due to correlations in sampled signal data [14].

8.3.4 Instruction Level Power Analysis

Instruction level power analysis (ILPA) [4] defines an empirical method for characterizing the power dissipation of short instruction sequences and for using these results to estimate the power (or energy) dissipation of a program. A detailed description of the internal design of the target processor is not required. However, some understanding of the processor architecture is important in order to appropriately choose the types of instruction sequences to be characterized. This understanding becomes even more important when looking at ways to optimize a program based on the power estimate.

The first requirement for ILPA is to characterize the average power dissipation associated with each instruction or instruction sequence of interest. Three approaches have been documented for accomplishing this characterization. The most straightforward method, if the hardware configuration permits, is to directly measure the current drawn by the processor in the target system as it executes various instruction sequences. If this is not practical, another method is to first use a hardware description language model (such as Verilog or VHDL) of the processor to simulate execution of the instruction sequences. An actual processor can then be placed in an automated IC tester and exercised using test vectors obtained from the simulation. An ammeter is used to measure the current draw of the processor in the test system. A third approach is to use gate level power simulation of the processor to obtain instruction level estimates.

The choice of instruction sequences for characterization is critical to the success of this method. As a minimum, it is necessary to determine the *base cost* of individual instructions. Base cost refers to the portion of the power dissipation of an instruction that is independent of the prior state of the processor. Base cost excludes the effect of such things as pipeline stalls, cache misses, and bus switching due to the difference in op-codes for consecutive instructions. The base cost of an instruction can be determined by putting several instances of that instruction into an infinite loop. Average power supply current is measured while the loop executes. The loop should be made as long as possible so as to minimize estimation error due to loop overhead (the jump statement at the end of the loop). However, the loop must not be made so large as to cause cache misses. Power and energy estimates for the instruction are calculated from the average current draw, the supply voltage, and the number of execution cycles per instruction.

Base costs for each instruction are not always adequate for a precise software power estimate. Additional instruction sequences are needed in order to take into account the effect of prior processor state on the power dissipation of each instruction. Pipeline stalls, buffer stalls, and cache misses are obvious energy-consuming events whose occurrence depends on the prior state of the processor. Instruction sequences can be created that induce each of these events so that current measurements can be made. However, stalls and cache misses are effects that require a large scale analysis or simulation of program execution in order to appropriately factor them into a program's energy estimate. There are other energy costs that can be directly attributed to localized processor state changes resulting from the execution of a pair of instructions. These costs are referred to as *circuit state effects*. Consider the following code sequence as an example:

$$ADD \quad A \leftarrow B, C$$
$$MULT \quad C \leftarrow A, B$$

The foremost circuit state effect is probably the energy cost associated with the change in the state of the instruction bus as the op-code switches from that of an addition operation to that of a multiplication operation. Other circuit state effects for this example could include switching of control lines to disable addition and enable multiplication, mode changes within the ALU, and switching of data lines to reroute signals between the ALU and register file. Although this example examines a pair of adjacent instructions, it is also possible for circuit state effects to span an arbitrary number of execution cycles. This could happen if consecutive activations of a processor component are triggered by widely separated instructions. In such cases, the state of a component may have been determined by the last instruction to use that component.

The circuit state cost associated with each possible pair of consecutive instructions is characterized for instruction level power analysis by measuring power supply current while executing alternating sequences of the two instructions in an infinite loop. Unfortunately, it is not possible to separate the cost of an $A \rightarrow B$ transition from a $B \rightarrow A$ transition, since the current measurement is an average over many execution cycles.

Equation (8.1) from [4] concisely describes how the instruction level power measurements can be used to estimate the energy dissipation of a program:

$$E_P = \sum_i (B_i \times N_i) + \sum_{i,j} (O_{i,j} \times N_{i,j}) + \sum_k E_k \qquad (8.1)$$

where E_P is the overall energy cost of a program, decomposed into base costs, circuit state overhead, and stalls and cache misses. The first summation represents base costs, where B_i is the base cost of an instruction of type i and N_i is the number of type i instructions in the execution profile of a program. The second summation represents circuit state effects where $O_{i,j}$ is the cost incurred when an instruction of type i is followed by an instruction of type j. Because of the way $O_{i,j}$ is measured, we have $O_{i,j} = O_{j,i}$. Here, $N_{i,j}$ is the number of occurrences where instruction type i is immediately followed by instruction type j. The last sum accounts for other effects, such as stalls and cache misses. Each E_k represents the cost of one such effect found in the program execution profile.

A simple example will illustrate the estimation method. Let there be a hypothetical DSP called MyDSP whose instruction set includes those listed in Table 8.1. MyDSP has registers named A, B, C, and D. Each instruction on MyDSP executes in one cycle, which is 25 ns. Assume that someone has measured the currents for each instruction and instruction pair in the manner described earlier. The currents were then used to calculate the base energy cost for each instruction and circuit state effect energy for each

TABLE 8.1 Instruction Set, Base Costs, and Circuit State Effects for MyDSP

Instruction Name	Base Cost (pJ)	Circuit State Effects (pJ)					
		LOAD	DLOAD	ADD	MULT	LOAD ADD	LOAD MULT
LOAD	1.98	0.13	0.15	1.19	0.92	1.25	1.06
DLOAD	2.37		0.17	1.19	0.92	1.32	1.06
ADD	0.99			0.26	0.53	0.86	0.99
MULT	1.19				0.66	0.79	0.96
LOAD; ADD	2.10					0.40	0.53
LOAD; MULT	2.25						0.79

possible instruction pair. These costs are shown in Table 8.1. In reality, these numbers were contrived to resemble a Fujitsu DSP documented in [4]. The numbers will be used later to illustrate the effect of some code optimizations. Let us analyze the energy and average power associated with the following code fragment running on MyDSP that evaluates the expression $(x + y) + z$, where x, y, and z are memory variables. We will assume that the code fragment does not cause any stalls or cache misses:

```
DLOAD A ← x, B ← y          # prior instruction was ADD
LOAD C ← z; MULT D ← A, B    # LOAD & MULT packed together
ADD A ← C, D                 # A now contains (x × y) + z
```

The energy and power estimate can be tabulated as follows:

Instruction Name	Base Cost (pJ)	Circuit State (pJ)	
DLOAD	2.37	1.19	
LOAD;MULT	2.25	1.06	
ADD	0.99	0.99	
Totals	5.61	3.24	Total energy = 8.85 pJ
			Average power = 8.85 pJ/75 ns = 118 μW

The base cost for each instruction is taken from the second column of Table 8.1. We have to look at adjacent pairs of instructions in order to determine the circuit state costs. We will assume that the DLOAD instruction was preceded by an ADD. The circuit state cost for DLOAD can then be found at the intersection of the row labeled DLOAD and the column labeled ADD. Similarly, the intersection of row DLOAD and column LOAD;MULT gives the circuit state cost for LOAD;MULT. The base and circuit state costs are all added together to obtain the energy cost of the code sequence. Energy is divided by execution time to obtain average power.

The previous example demonstrates the concept of ILPA for a trivial code fragment. The procedure is automated and integrated into a process for analyzing the power dissipation of real-world applications. Basic blocks are first extracted from the assembly or object code for a program. Instruction level base costs and circuit state effects are totaled for each block. The frequency and cost of stalls for each basic block are estimated and added into the basic block cost. A program profiler determines how many times each basic block is executed in order to accumulate the total basic block costs for the program. Finally, a cache simulator is used to estimate the frequency of cache misses so that cache miss costs can be added to the total program energy.

8.4 SOFTWARE POWER OPTIMIZATIONS

Software optimizations for minimum power or energy tend to fall into one or more of the following categories: selection of the least expensive instructions or instruction sequences, minimizing the frequency or cost of memory accesses, and exploiting power minimization features of hardware. The following sections will discuss each of these categories, present several optimization techniques that have been documented, and relate the techniques back to the sources of software power dissipation identified in Section 8.2.

A prerequisite to optimizing a program for low power must always be to design an algorithm that maps well to available hardware and is efficient for the problem at hand in terms of both time and storage complexity. Given such an algorithm, the next requirement is to write code that maximizes performance, exploiting the performance features and parallel processing capacity of the hardware as much as possible. In doing so, we will have gone a long way to minimize the energy consumption of the program [4]. Maximizing performance also gives increased latitude to apply other power optimization techniques such as reduced supply voltages, reduced clock rates, and shutting off idle components.

It should be clarified that the energy optimization objectives differ with respect to the intended application. In battery-powered systems, the total energy dissipation of a processing task is what determines how quickly the battery is spent. In systems where the power constraint is determined by heat dissipation and reliability considerations, instantaneous or average power dissipation will be an important optimization objective or constraint. An instantaneous power dissipation constraint could lead to situations where one would need to deliberately degrade system performance through software or hardware techniques. A hardware technique is preferable since hardware performance degradation can be achieved through energy-saving techniques (lower voltage, lower clock rate). Software performance degradation is likely to increase energy consumption by increasing the number of execution cycles of a program.

8.4.1 Algorithm Transformations to Match Computational Resources

The general problem of efficient algorithm design is a large area of study that goes well beyond the scope of this chapter. However, many algorithm design approaches have significant implications for software-related power dissipation. One impact of algorithm design is on memory requirements, but we will defer that problem to the next section. Another impact is on the efficient use of computational resources. This problem has been addressed extensively in parallel processor applications and in lower-power DSP synthesis.

In parallel processor applications, a typical problem is to structure software in a way that maximizes the available parallelism. Parallel computing resources can then be used to speed up program execution. In low-power DSP synthesis, a typical problem is to design an algorithm to allow a circuit implementation that minimizes power dissipation given throughput and area constraints. Often a low-power DSP design will also exploit parallelism in an algorithm, but the objective is to shorten critical paths so that supply voltages can be lowered while maintaining overall performance [15]. A similar trade-off is possible in multiple-processor or superscalar systems. If a piece of software makes sufficient use of parallel resources, it may become possible to lower the supply voltage and clock rate of the processor to achieve power savings. This is already an option if one is designing an application-specific processor. Researchers are working to make such a trade-off possible in general-purpose processors through dynamic scaling of supply voltage and clock rate under control of the operation system [16].

Reduction operations are an example of a common class of operations that lend themselves to a trade-off between resource usage and execution time. A simple example is the summation of a sequence. Figures 8.1–8.3 illustrate what can be done as the number of adder units is changed.

If only one adder is available, then Figure 8.1 is a sensible approach. Parallelizing the summation would only force us to use additional registers to store intermediate sums. If two adders are available, then the algorithm illustrated in Figure 8.2 makes sense because it permits two additions to be performed simultaneously. Similarly, Figure 8.3 fits a system with four adders by virtue of making four different independent operations available at each time step.

In the general case, one cannot manipulate the parallelism of an algorithm quite so conveniently. However, the principle is still applicable. The basic principle is to try to match the degree of parallelism in an algorithm to the number of parallel resources available.

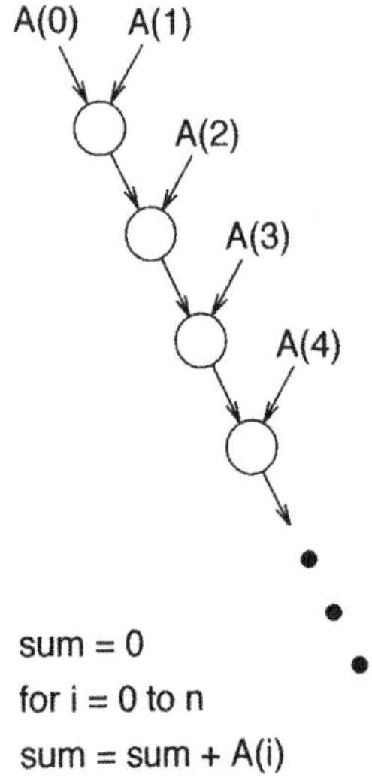

```
sum = 0
for i = 0 to n
    sum = sum + A(i)
```

Figure 8.1 Summation with single adder.

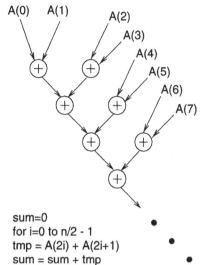

```
sum=0
for i=0 to n/2 - 1
tmp = A(2i) + A(2i+1)
sum = sum + tmp
```

Figure 8.2   Summation with two adders.

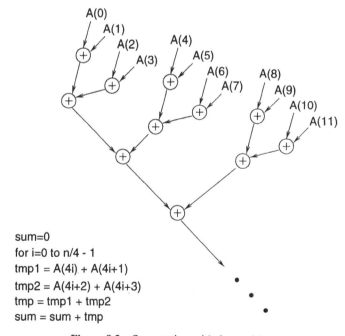

```
sum=0
for i=0 to n/4 - 1
tmp1 = A(4i) + A(4i+1)
tmp2 = A(4i+2) + A(4i+3)
tmp = tmp1 + tmp2
sum = sum + tmp
```

Figure 8.3   Summation with four adders.

8.4.2 Minimizing Memory Access Costs

Since memory often represents a large fraction of a system's power budget, there is a clear motivation to minimize the memory power cost. Happily, software design techniques that have been used to improve memory system efficiency can also be helpful in reducing the power or energy dissipation of a program. This is largely due to memory being both a power and a performance bottleneck. If memory accesses are both slow and power hungry, then software techniques that minimize memory accesses tend to have both a performance and a power benefit. Lower memory requirements also help by allowing total RAM to be smaller (leading to lower capacitance) and improving the percentage of operations that only need to access registers or cache. Power minimization techniques related to memory concentrate on one or more of the following objectives:

Minimize the number of memory accesses required by an algorithm.

Minimize the total memory required by an algorithm.

Put memory accesses as close as possible to the processor. Choose registers first, cache next, and external RAM last.

Make the most efficient use of the available memory bandwidth; for example, use multiple-word parallel loads instead of single-word loads as much as possible.

For multidimensional signal processing applications, the nesting of loops controlling array operations and the order of operations can substantially influence the number of memory transfers and the total storage requirements. Catthoor et al. [8] presented three simple examples, representative of some typical signal processing loop structures, that illustrate the impact of loop nesting and operation ordering. These examples are reproduced below along with brief discussions. In each case, M and N are large integer values.

Example 8.1

Before	After
FOR i := 1 TO N DO	FOR i := 1 TO N DO BEGIN
B[i] := f(A[i]);	B[i] := f(A[i]);
FOR i := 1 TO N DO	C[i] := g(B[i]);
C[i] = g(B[i]);	END;

In Example 8.1, the B array is too large to store in registers, so memory transfers are required to store and retrieve the intermediate values stored in B. Rearranging the loops allows the intermediate values to be kept in a register. The $2N$ memory transfers are replaced by cheaper register accesses and N memory locations are no longer needed.

Example 8.2

Before	After

FOR i := 1 TO N DO
  B[i] := f(A[i]);
FOR j := 1 TO N DO
  D := g(C[j], D);
	FOR j := 1 TO N DO
	  D := g(C[j], D);
	FOR i := 1 to N DO
	  B[i] := f(A[i]);

Example 8.3

Before	After

FOR j := 1 to M DO
  FOR i := 1 TO N DO
    A[i][j] := g(A[i][j − 1], D);
FOR i := 1 to N DO
  OUT[i] := A[i][M];
	FOR i := 1 TO N DO BEGIN
	  FOR j := 1 TO M DO
	    A[i][j] := g(A[i][j − 1], D);
	  OUT[i] := A[i][M];
	END;

The transformations shown in Examples 8.2 and 8.3 do not improve on the number of memory transfers, but space requirements are reduced. In Example 8.2, if C is not needed afterward, operations are reorganized so that B can be mapped into the same storage as C, eliminating N storage locations. In Example 8.3, storage for intermediate values $A[i][M]$ can be reduced from N locations to 1 by restructuring and combining loops.

An example where careful mapping of an algorithm to hardware was beneficial in terms of memory performance, size requirements, and power was accomplished by DeGreef [2]. DeGreef evaluated the mapping of a video motion estimation algorithm onto a variety of digital signal processors. The algorithm requires enormous memory bandwidth since 5.3×10^9 pixels must be processed per second to support real-time processing of a standard television signal. Use of a cache was essential, but cache protocols such as a least recently used replacement policy can lead to unacceptable variations in execution time for such a time-critical application. Many signal processing algorithms such as motion estimation can have very predictable control flow and memory access patterns. Consequently, it may be possible to derive an optimal cache policy. DeGreef used the on-chip memory of a DSP as a data cache for which storage and replacement decisions could be made at compile time. With a cache as small as 23 k bytes, it was possible to reduce accesses to external memory by a factor of more than 100 as compared to no cache at all. Another technique applied by DeGreef was to minimize loop control overhead and insert dummy code to synchronize the DSP software to the video I/O data rate. This allowed I/O, current, and previous video frame buffers to be merged into one frame buffer. Synchronization was already known as a useful tool to reduce buffers, but there should also be a power benefit resulting from a smaller memory and simpler I/O management.

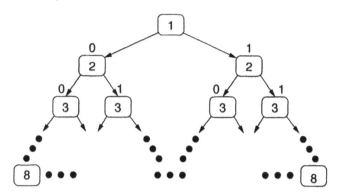

Figure 8.4   Binary codebook search.

Another recent example of a beneficial mapping of an algorithm to an architecture was presented by Lidsky [17]. The application was vector quantization of video data. Each 4×4 pixel region of an image had to be quantized to the nearest vector in a 256-element codebook of precomputed vectors. Decoding a quantized image requires a simple table lookup. However, encoding a 4×4 pixel region requires a search of the entire codebook. Lidsky organized the codebook into a binary search tree, as illustrated in Figure 8.4. The number shown inside each node indicates a level in the tree. Starting at the top node, a vector to be quantized is compared to two codebook entries in the next level down. The closer node determines the branch that is followed in the search tree. The tree below the farther node is eliminated from the search. The process is repeated until a leaf of the tree is reached. This kind of search requires $O(\log_N)$ operations to search the tree. A full linear search would be required $O(N)$ operations.

The binary search algorithm is not only much more efficient than linear search but also permits a fast pipelined implementation. A separate memory and processing element can be used for each level of the search tree. The encoding process can then proceed in a pipelined manner, handling eight vectors simultaneously. The clock rate can be reduced by a factor of 8 from a serial implementation. This architecture increases memory area more than twofold, but short critical paths and reduced clock permit a supply voltage reduction from 3 to 1.1 V. Overall, memory accesses were reduced by a factor of 30 and power was scaled down by a factor of 17 compared to a system with a single serial access memory.

In general-purpose systems it may not be practical to craft the software and hardware so finely together, but there are techniques that are more readily applicable to a wide variety of problems. One of these techniques is to maximize parallel loads of multiple words from memory. Dual loads are beneficial for energy reduction but not necessarily for power reduction because the instantaneous power dissipation of a dual load is somewhat higher than for two single loads. Tiwari et al. [4] demonstrated energy savings

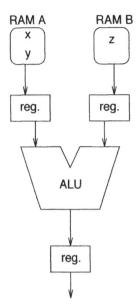

Figure 8.5   Data path example for dual memory load.

of as much as 47% by maximizing dual loads. Achieving this savings required two phases. The first phase optimized the memory allocation of a program to maximize opportunities for using dual loads. Memory accesses were then combined as much as possible. Suppose that the data path depicted in Figure 8.5 is required to evaluate the expression $(x \times y) + z$, as in the example from Section 8.3.4. Suppose also that there are two memory banks, all operations must be performed on register values, and the processor supports dual loads and packing together of ALU and memory operations. Depending on the memory allocation and objective chosen, the optimal instruction sequence choice can vary. Table 8.2 uses the hypothetical characterization of MyDSP (Table 8.1) to illustrate what could happen. In case (a)

TABLE 8.2   Effect of Memory Mapping on Energy and Power of MyDSP

	Case (a)	Case (b)	Case (c)
RAM A	x, y, z	x, y	x, z
RAM B		z	y
	LOAD $B \leftarrow x$	DLOAD $B \leftarrow x, A \leftarrow z$	DLOAD $B \leftarrow x, C \leftarrow y$
	LOAD $C \leftarrow y$	LOAD $C \leftarrow y$	LOAD $A \leftarrow z$;
			MULT $B \leftarrow B, C$
	LOAD $A \leftarrow z$;	MULT $B \leftarrow B, C$	ADD $A \leftarrow A, B$
	MULT $B \leftarrow B, C$		
	ADD $A \leftarrow A, B$	ADD $A \leftarrow A, B$	
Energy (pJ)	10.57	9.32	8.85
Power (μW)	105.7	93.2	118.0

all variables are in the same memory block, so no dual loads were possible. It was possible to overlap a load and a multiply. In (b), x and z could be loaded in parallel but no other operations could be overlapped. In (c), a dual load of x and y and parallel execution of a load and multiply were both possible. The lowest energy solution occurred when the memory allocation permitted maximization parallelism of operations. Lower average power solutions were obtained at the expense of execution cycles and total energy.

The previous example illustrates the energy benefit of favorable memory allocation, but algorithms are needed to obtain this benefit for entire programs, not just code fragments. Researchers have dealt previously with the problem of automated memory allocation for optimal performance; examples include [18] and [6]. Although performance was their target, reduced program execution times translate directly to reduced program energy consumption and can also provide opportunities for voltage and clock frequency reduction. Lee and Tiwari [7] have presented a memory allocation technique targeted specifically for energy reduction on systems with partitioned memory. Davidson [18] listed several practices that minimize memory bandwidth requirements:

Register allocation to minimize external memory references.

Cache blocking, that is, transformations on array computations so that blocks of array elements only have to be read into cache once.

Register blocking, similar to cache blocking, except that redundant register loading of array elements is eliminated.

Recurrence detection and optimization, that is, use of registers for values that are carried over from one level of recursion to the next.

Compact multiple memory references into a single reference.

All of the researchers just cited focus on combining memory references, but they take substantially different approaches. Combining of memory references is of special interest in part because it is an objective that was not well addressed by earlier compiler technology. We will briefly review the methods applied by Davidson, Sudarsanam, and Lee.

Davidson's approach [18] evaluates the benefit of *loop unrolling* on overall performance and on the degree to which *memory reference compaction* (also called *memory access coalescing*) opportunities are uncovered. Loop unrolling transforms several iterations of a loop into a block of straight-line code. The straight-line code is then iterated fewer times to accomplish the same work as the original loop. The purpose of the transformation is to make it easier to analyze data dependencies when determining where to combine memory references. Davidson found that the performance benefit of loop unrolling and memory access coalescing varied a great deal depending on the processor. Speedups as much as 40% for a DEC alpha and 25% for a Motorola 88100 were observed. On the other hand, performance was reduced on a Motorola 68030.

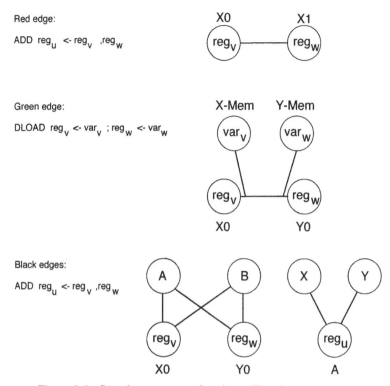

Figure 8.6 Sample memory and register allocation constraints.

Sudarsanam [6] combines register and memory bank allocation into a unified optimization phase after instruction selection. First, all straight-line blocks of code in a program are compacted, combining two memory operations or a memory and an ALU operation as much as possible without changing the order of operations. A *constraint graph* is then generated that represents register or memory allocation constraints that are implied by the compacted code. In the constraint graph, each vertex represents a symbolic register or variable. Each edge represents an allocation constraint on the pair of vertices joined by the edge.

Several types of edges are defined and color coded to represent different types of constraints. Figure 8.6 provides examples of some of these edge types. The label inside the circle for each vertex indicates a variable (e.g., var_v), a symbolic register name (e.g., reg_v), or a physical register name (e.g., X). The label just outside of each circle indicates the assignment of a physical register name to a symbolic register or a memory bank name to a variable. A red edge indicates that two register values are active at the same time and should be assigned to different physical registers. A green edge identifies a parallel memory to register transfer that should be preserved. Brown, blue, and yellow edges all indicate other types of parallel data transfers such as a memory access in parallel with a register–register transfer.

The brown, blue, and yellow edges have a similar structure to the green edge. The vertices representing two destination registers or variables are connected by an edge. At each end of the edge, a pointer is attached to the source register or variable. Black edges indicate limitations on operand usage imposed by the target processor (perhaps only certain registers can be used as input to ALU operations). A black edge is connected at one end to a symbolic register name. The other end is connected to a physical register name. Each black edge identifies an impossible register assignment.

Every edge has a weight (not shown in Figure 8.6) that indicates the penalty of violating the constraint. The register and memory bank allocation is then accomplished by using a simulated annealing algorithm to label each vertex.

Lee and Tiwari [7] present an algorithm for memory bank assignment that is formulated as a graph-partitioning problem and solved by simulated annealing. Each vertex represents a variable. Each edge indicates that there exists an ALU operation that requires a particular pair of variables as arguments. Consider the following code fragment:

$$
\begin{aligned}
&\text{ADD } R_1 \leftarrow a, e \\
&\text{ADD } R_2 \leftarrow e, b \\
&\text{ADD } R_3 \leftarrow a, b \\
&\text{ADD } R_4 \leftarrow b, d \\
&\text{ADD } R_5 \leftarrow a, c \\
&\text{ADD } R_6 \leftarrow c, d
\end{aligned}
$$

The access graph for this code could be drawn as shown in Figure 8.7.

A two-way partition of this graph directly corresponds to a memory bank assignment. All the variables (vertices) in one partition are assigned to the same memory bank. The cost of a partition is evaluated in terms of the number of execution cycles required to accomplish all of the memory transfers indicated by the graph. If an edge lies entirely in one side of the partition, two memory transfers will be required for the indicated variables. If an edge crosses the partition, then the memory accesses for those two variables can be compacted into a dual-load operation at a cost of one execution cycle. The partition shown in Figure 8.8 maximizes the number of edges crossing the partition. The cost of this partition is 7:2 for the a, e edge and 1 for all other edges. Clearly this cost is an upper bound on the optimum

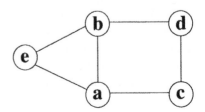

Figure 8.7 Access graph for code fragment.

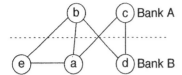

Figure 8.8 Partitioned access graph.

number of loads since appropriate allocation of registers and sequencing of instructions could reduce the cost to two parallel loads and one single word load.

Cache locking, if the processor supports it, provides another way to minimize memory accesses. Locking prevents any memory reads and writes from going to main memory. The real benefit comes from cache write hits not being written through to main memory. For read hits there would be no reference to main memory even if the cache were unlocked. The cost of memory writes have been measured for both locked and unlocked caches [19]. For a Fujitsu SPARClite MB86934 32-bit RISC processor, writing a value of 0 drew 341 mA from the power supply when the cache was unlocked but just 194 mA with the cache locked. Similar results have been reported for a variety of write operations.

8.4.3 Instruction Selection and Ordering

In Section 8.3, we looked at various ways that the power and energy of a program can be estimated from assembly or machine code. Since there are usually many possible code sequences that accomplish the same task, it should be possible to select a code sequence that minimizes power or energy. In fact, many code optimization techniques (for speed or code size) should be equally applicable to power or energy optimization; the main difference is the objective function used in code generation decisions. We will not attempt to give a comprehensive presentation of general code optimization techniques. Such techniques are well documented in compiler literature. Instead, we will attempt to provide a basis for judging how different kinds of coding techniques are likely to affect energy usage. We will also discuss some specific techniques that researchers have evaluated for their energy savings: *instruction packing*, *minimizing circuit state effects*, and *operand swapping*. These techniques are described in greater detail in [11].

Given the benefits of high-performance code with respect to energy savings, optimizations that improve speed should be especially helpful. Code size minimization may be necessary, especially in embedded systems, but it is not quite as directly linked to energy savings. Regarding performance and energy minimization, cache performance is a greater concern. Large code and a small cache can lead to frequent cache misses with a high power penalty. Code size and cache performance are concerns that motivate the effort to maximize code density for embedded processors [9].

Instruction packing is a feature of a Fujitsu DSP [4] and a number of other processors. Instruction packing permits an ALU operation and a memory data transfer to be packed into a single instruction. Much of the cost of each individual instruction is overhead that is not duplicated when operations are executed in parallel. Consequently, there can be nearly a 50% reduction in energy associated with the packing of two instructions. Similar benefits can be achieved by parallel loading of multiple words from different banks of memory. Optimization of memory accesses is a broad topic that is covered in Section 8.4, including an example that illustrates the benefit of parallel loads and instruction packing. Concurrent execution of integer and floating-point operations represents still another opportunity to minimize the total energy of a group of operations. All of these techniques are really special cases of the types of coding that can be done in VLIW (very long instruction word) and superscalar architectures where opportunities for instruction packing will be much more extensive.

Instruction ordering for low power attempts to minimize the energy associated with the circuit state effect. Circuit state effect, if you recall, is the energy dissipated as a result of the processor switching from execution of one type of instruction to another. On some types of processors, especially DSP, the magnitude of the circuit state effect can vary substantially depending on the pair of instructions involved. In Section 8.4, we described techniques for measuring or estimating the circuit state effect for different pairs of consecutive instructions. Given a table of circuit state costs, one can tabulate the total circuit state cost for a sequence of instructions. Instruction ordering then involves reordering instructions to minimize the total circuit state cost without altering program behavior. Researchers have found that in no case do circuit state effects outweight the benefit of minimizing program execution cycles [4]. Circuit state effects are found to be much more significant for DSPs than for general-purpose architectures.

Accumulator spilling and mode switching are two DSP behaviors that are sensitive to instruction ordering [20] and are likely to have a power impact. In some DSP architectures, the accumulator is the only general-purpose register. With a single accumulator, any time an operation writes to a new program variable, the previous accumulator value will need to be spilled to memory, incurring the energy cost of a memory write. In some DSPs, operations are modal. Depending on the mode setting (such as the sign extension mode), data path operations will be handled differently. There has been work on techniques to minimize mode switches [21], but we have not seen mode switching evaluated with respect to the effect on energy costs.

Operand swapping attempts to swap operands to ALU or floating-point operations in a way that minimizes the switching activity associated with the operation. One way operand swapping could help is to minimize switching related to the sequence of input values. If latches are located at the inputs of a functional unit to keep the inputs from switching during idle periods, then there could be an advantage to swapping operands to minimize switching at the inputs to the functional unit. For example, if two consecutive additions have one operand that stays the same (say $x + 7$ and $y + 7$), then switching would be minimized if the common operand was applied to the same adder input both times.

Ordering of operands can have an even greater effect if the operands to a commutative operation are not treated symmetrically. For example, the number of additions and subtractions performed in a Booth multiplier depends on the bit pattern of the second operand. The number of additions and subtractions required is referred to as the *recoding weight* of the second operand. Given a pair of values to be multiplied, power should be reduced if we put the value with the lower recoding weight into the second input of the Booth multiplier. We rarely know ahead of time the value of both operands, but it is not unusual to know the value (and recoding weight) of one operand. Following is a simple and effective heuristic to apply this knowledge. If an operand is known to have a high recoding weight, put it in the first operand position. If an operand is known to have a low recoding weight, put it in the second position. Lee [11] did this for several sets of operand values and demonstrated that in a majority of cases a substantial energy savings per operation on the order of 10–30% was realized. For the few cases when swapping was not beneficial, the penalty was modest, less than 6%.

8.4.4 Power Management

Most of the low-power software optimizations that we have discussed already exploit specific hardware features to achieve maximum benefit. Many of these features are not unique to low-power designs. However, considerable hardware design effort has been applied to designing processors that avoid dissipation of power in circuitry that is idle.

The degree to which software has control over power management varies from one processor to another. It is possible for power management to be entirely controlled by hardware, but typically software has the ability to at least initiate a power-down mode. For event-driven applications such as user interfaces, system activity typically comes in bursts. If the time a system is idle exceeds some threshold, it is likely that the system will continue to be idle for an extended time. Based on this phenomenon, a common approach is to initiate a shutdown when the system is idle for a certain length of time. Unfortunately, the system continues to draw power while waiting for the threshold to be exceeded. Predictive techniques [23] take into account the recent computation history to estimate the expected idle time. This permits the power down to be initiated at the beginning of the period of idle time.

Several processors support varying levels of software control over power management. Some examples include the Fujitsu SPARClite MB86934 [19], Hitachi SH3 [9], Intel486SL [22], and the PowerPC 603 and 604 [23]. The SPARClite has a *power-down* register that masks or enables clock distribution to the SDRAM interface, DMA module, FPU, and floating-point queues.

The Hitachi SH3 provides two power reduction modes, *standby* and *sleep*, along with control over the CPU and peripheral clock frequencies. In the sleep mode, all operations except the real-time clock stop. In the standby mode, the CPU core is stopped, but the peripheral controller, but controller, and memory refresh operations are maintained. The Intel486SL provides a System Management Mode (SMM), a distinct operating mode that is transparent to the operating system and user applications. The SMM is entered by means of an asynchronous interrupt. Software for the SMM is able to enable, disable, and switch between fast and slow clocks for the CPU and ISA bus. The PowerPC 603 and 604 offer both dynamic and static power management modes. Dynamic power management removes clocking from execution units if it is not needed to support currently executing instructions. This allows for significant savings even while the processor is fully up and running. Savings on the order of 8–16% have been observed for some common benchmarks [24]. No software intervention is needed, except to put the processor into the power management mode. The PowerPC has three static power management modes: "doze," "nap," and "sleep". They are listed here in order of increasing power savings and increasing delay to restart the processor. The doze mode shuts off most functional units but keeps bus snooping enabled in order to maintain data cache coherency. The nap mode shuts off bus snooping and can set a timer to automatically wake up the processor. It keeps a phase-locked loop running to permit quick restart of clocking. The sleep mode shuts off the phase-locked loop. Nap and sleep modes can each be initiated by software, but a hardware handshake is required to protect data cache coherency.

A trade-off must be made regarding the merits of software control versus hardware control of power management. Software-based power management has the advantage of additional information upon which to base power management decisions. Purely hardware-based power management has to make its decisions by monitoring processor activity. An incorrect decision to power down can hurt performance and power dissipation due to the cost of restoring power and system state [25]. Processors with integrated power management provide ways to efficiently save and restore the processor state in the event of a power-down. Program execution cycles committed to power management represent a down side to software-based power management.

8.5 AUTOMATED LOW-POWER CODE GENERATION

Section 8.4 described a variety of optimizations and software design approaches that can minimize energy consumption. To be used extensively, those techniques need to be incorporated into automated software development tools. There are two levels of tools: tools that generate code from an algorithm specification and compiler level optimizations.

We have not located any comprehensive high-level tools intended for low-power code development, but several technologies appear to be available to build such tools, especially for DSP applications. Graphical [26] and textual languages [27] have been available for some time that enable one to specify a DSP algorithm in a way that does not obscure the natural parallelism and data flow. HYPER-LP [15] is a DSP data path synthesis system that incorporates several algorithm level transformations to uncover parallelism and minimize critical paths so that the data path supply voltage can be minimized. Even though HYPER-LP is targeted for data path circuit synthesis, the same types of transformations should be useful in adapting an algorithm to exploit parallel resources in a DSP processor or multiprocessor system (in fact HYPER has been used to evaluate algorithm choices prior to choice of implementation platform [28]). MASAI [8] is a tool that reorganizes loops in order to minimize memory transfers and size. Finally, to complete the tool set, there exist compilers [3, 11, 29] that either already optimize DSP code for low power or could be enhanced to apply the techniques that have been discussed.

Compiler technology for low power appears to be further along than the high-level tools, in part because well-understood performance optimizations are adaptable to energy minimization. This is true more for general-purpose processors than for DSP processors. Digital signal processing compiler technology faces difficulties in dealing with a small register set (possibly just an accumuiator), irregular data paths, and making full use of parallel resources. The rest of this section describes two examples of power reduction techniques that have been incorporated into compilers.

Su [3] presented a *cold scheduling algorithm* that is an instruction scheduling algorithm that reduces bus switching activity related to the change in state when execution moves from one instruction type to another. The algorithm is a list scheduler that prioritizes the selection of each instruction based on the power cost (bus switching activity) of placing that instruction next into the schedule. Following is the ordering of compilation phases that Su proposed in order to incorporate cold scheduling:

1. Allocate registers.
2. Pre-assemble: Calculate target addresses, index symbol table, and transform instructions to binary.
3. Schedule instructions using cold scheduling algorithm.
4. Post-assemble: Complete the assembly process.

The assembly process was split into two phases to accommodate the cold scheduler. Cold scheduling could not be performed prior to preassembly because the binary codes for instructions are needed to determine the effect of an instruction on switching activity. Scheduling could not be the last phase since completion of the assembly process requires an ordering of instructions.

Su pointed out that one limitation of this approach is that it becomes difficult to schedule instructions across basic block boundaries because of the early determination of target addresses. Cold scheduling was found to obtain a 20–30% switching activity reduction with a performance loss of 2–4%. It is easy to imagine that the instruction level power model (section 8.3.4) could also be used to prioritize instruction selection during cold scheduling.

Lee et al. proposed a code generation and optimization methodology that encompasses several of the techniques discussed in Section 8.4. Following is the sequence of phases that they proposed:

1. Allocate registers and select instructions.
2. Build a data flow graph (DFG) for each basic block.
3. Optimize memory bank assignments by simulated annealing.
4. Perform *as soon as possible* (ASAP) packing of instructions.
5. Perform list scheduling of instructions (similar to cold scheduling).
6. Swap instruction operands where beneficial.

The optimization phases appear to be sequenced in order from greatest to least incremental benefit. Overall energy savings on four benchmarks ranged from 26 to 73% compared to results that did not incorporate instruction packing or optimize memory bank assignments.

8.6   CODESIGN FOR LOW POWER

The emphasis of this chapter has been on software design techniques for low power. However, it is probably obvious by now that many of these techniques are dependent in some way upon the hardware in order for a power savings to be realized. Hardware/software codesign for low power is a more formal term for this problem of crafting a mixture of hardware and software that provides required functionality, minimizes power consumption, and satisfies objectives that could include latency, throughput, and area. Instruction set design and implementation seems to be one of the most well defined of the codesign problems. In this section, we will first look at instruction set design techniques and their relationship to low power design. Another variation of the codesign problem is to use a processor for which some portion of the interconnect and logic is reconfigurable. Reconfigurable processors allow much of the codesign problem to be delayed until the software design and compilation phase. For a nonreconfigurable processor, there is much less opportunity to optimize the hardware once the processor design is fixed. Finally, we will survey the hardware/software trade-offs that have come up in our consideration of power optimization of software.

8.6.1 Instruction Set Design

Some of the most recent efforts at instruction set optimization were presented by Sato et al. [30], Binh et al. [31], and Huang and Despain [32]. The work of both Sato and Binh is related to the PEAS-I system (Practical Environment for ASIP development-version I). PEAS-I is a system that synthesizes an HDL (hardware description language) description of a CPU core along with a C compiler, assembler, and simulator for that CPU. Inputs to PEAS-I are a set of design constraints (including chip area and power consumption), a hardware module database, and a sample application program and program dataset. PEAS-I optimizes the instruction set and instruction implementations for the application program and dataset that was given. PEAS-I defines an extended set of instructions by starting with a core instruction set that would be needed to implement any C program. The core set is augmented with instructions corresponding to any C operators not already represented by a single instruction. The set could be further expanded to include any predefined or user-defined functions. For each possible instruction, alternative implementations are defined, including hardware, microprogram, and software options. Optimal selection of implementation methods for each instruction is defined as an integer program and solved by a branch-and-bound algorithm. The power and area contribution of each instruction implementation is estimated and added together to include in power and area constraints. The number of execution cycles of each instruction is multiplied by the frequency of occurrence in the sample application and added together to form the objective function for the optimization. Binh's contribution was to extend the instruction selection formulation to account for pipeline hazards.

Huang and Despain's [32] approach is similar in one respect to PEAS-I, in that it optimizes the instruction set for sample applications. The most significant difference is that Huang's approach groups *micro-operations* (MOPS) together to form higher level instructions. Each benchmark is expressed as a set of MOPS with dependencies to be satisfied. MOPS are merged together as a byproduct of the scheduling process. MOPS that are scheduled to the same clock cycle are combined. For any candidate schedule and instruction set, instruction width (bits), instruction set size, and hardware resource requirements are constrained. The scheduling problem is solved by simulated annealing with an objective of minimizing execution cycles and instruction set size.

It would seem that any of the power estimation techniques described in Section 8.3 could fit nicely into PEAS-I or Huang's approach. Either approach relies on instruction level application profiles to prioritize decisions regarding the instruction set. The challenging task would be determining appropriate instruction level power estimates and circuit state effects for an unsynthesized CPU. Op-code selection is another design decision that could be optimized for low power, considering the impact of the op-codes on circuit state effects.

8.6.2 Reconfigurable Computing

"Reconfigurable computing" is a growing area of research in processor design that could motivate radical new requirements for software design tools. It also presents new opportunities and challenges for low-power systems design [33]. In reconfigurable processors, some portion of the interconnect and logic can be modified at run time. This should allow the processor to be closely optimized to each of a wide variety of applications. Reconfiguration can be implemented at any of several levels of granularity, ranging from the gate level up to architectural level changes [33]. The reconfigurable components are commonly implemented using field programmable gate arrays. The portion of the processor given to be reconfigurable can also vary widely. The reconfigurable portion can be limited to a coprocessor or it can encompass a large portion of the processor architecture.

Software and compiler design for reconfigurable architectures is an open area of research. Software design methodologies will vary depending on the archictecture. Nevertheless, many of the hardware/software trade-offs we have discussed should still be relevant. The biggest change is that the software designer will have a much greater opportunity to tailor the software and processor to fit each other even after the hardware design is fixed.

8.6.3 Architecture and Circuit Level Decisions

A survey of the preceding sections of this chapter should make it evident that software optimizations for power minimization are heavily dependent on the available hardware. Up to this point we have emphasized software techniques that exploit particular hardware features to minimize power. However, this information should also provide guidance to a system architect. In Tables 8.3–8.5 we summarize this chapter from the system architect's perspective, highlighting the impact of processor design decisions on opportunities for power minimization in software. These design decisions are organized into three categories: memory system design, processor architecture, and power management. Each table addresses one category.

8.7 SUMMARY

In this chapter, we have dealt with several questions pertinent to the design of software for low-power instruction processing systems. Why does software matter in a low-power system? How does software contribute to system power? How can one estimate the impact of software design decisions on system power? How can one minimize the software contribution to power dissipation? How can the hardware and software design fit together for a low-power system? Several answers to each of these questions have been extracted from recent literature.

TABLE 8.3   Memory System Considerations for Low-Power Software

Memory System Feature	Low-Power Software Impact
Total memory size	Minimizing total memory requirements through code compaction and algorithm transformations allows for a smaller memory system with lower capacitances and improved performance. However, this benefit might have to be traded off against a faster algorithm that requires more memory.
Partitioning into banks; how many, how big	Needed if parallel loads are to be used. The size of application, data structures, and access patterns should determine this.
Wide data bus	Needed for parallel loads.
Proximity to CPU	Memory close to CPU reduces capacitances; makes memory use less expensive.
Cache size	Should be sized to minimize cache misses for the intended application. Software should apply techniques such as cache blocking to maintain high degree of spatial and temporal locality.
Cache protocol	If an application's memory access patterns are well defined, the protocol can be optimized.
Cache locking	If significant portions of an application can run entirely from cache, this can be used to prevent any external memory accesses while those portions of code are executing.

TABLE 8.4   Architectural Considerations for Low Power Software

Architectural Feature	Low-Power Software Impact
Class of processor DSP/RISC/CISC	Application dependent. How well does the instruction set fit the application? Is there hardware support for operations common to the application? Does the hardware implement functions not needed by the application?
Parallel processing VLIW, Superscalar SIMD, MIMD. How much parallelism?	Does the level and type of parallelism in an algorithm fit one of these architectures? If so, the parallel architecture can greatly improve performance and allow reduced voltages and clock rates to be used.
Bus architecture	Separate buses (e.g., address, data, instruction, I/O) can make it easier to optimize instruction sequences, addressing patterns, and data correlations to minimize bus switching.
Register file size	Increased register count eases register allocation and reduces memory accesses, but too large of a register file can make register accesses as expensive as a cache access.

TABLE 8.5   Power Management Considerations for Low-Power Software

Power Management Feature	Low-Power Software Impact
Software vs. hardware control	Software control allows power management to be tailored to the application at the cost of added execution cycles.
Clock/voltage removal: At what level of granularity?	Some kind of shutdown modes are needed in order to realize a benefit from minimized execution times.
Clock/voltage reduction	If there is a schedule slack associated with a software task, reduce power by slowing down the clock and reducing the supply voltage.
Guarded evaluation	Latch the inputs of functional units to prevent meaningless calculations when outputs of units are not being used. This increases the power reduction from reduction in strength and operand swapping.

The primary objective of this activity has been to help in making software design decisions consistent with the objective of power minimization. Key objectives to be considered are the following. Choose the best algorithm for the problem at hand and make sure it fits well with the computational hardware. Failure to do this can lead to costs far exceeding the benefit of more localized power optimizations. Minimize memory size and expensive memory accesses through algorithm transformations, efficient mapping of data into memory, and optimal use of memory bandwidth, registers, and cache. Optimize the performance of the application, making maximum use of available parallelism. Take advantage of hardware support for power management. Finally, select instructions, sequence them, and order operations in a way that minimizes switching in the CPU and data path.

REFERENCES

[1] E. P. Harris, S. W. Depp, W. E. Pence, S. Kirkpatrick, M. Sri-Jayantha, and R. R. Troutman, "Technology Directions for Portable Computers," *Proc. IEEE*, vol. 83, no. 4, pp. 636–657, 1995.

[2] E. DeGreef, F. Catthoor, and H. DeMan, "Memory Organization for Video Algorithms on Programmable Signal Processors," In *Proceedings of the IEEE International Conference on Computer Design: VLSI in Computers and Processors*, pp. 552–557, 1995.

[3] C.-L. Su, C.-Y. Tsui, and A. M. Despain, "Low Power Architecture Design and Compilation Techniques for High-Performance Processors," in *Proceedings of IEEE COMPCON*, pp. 489–498, 1994.

[4] V. Tiwari, S. Malik, A. Wolfe, and M. T.-C. Lee, "Instruction Level Power Analysis and Optimization of Software," *J. VLSI Signal Process*. vol. 13, nos. 2/3, pp. 223–238, 1996.

[5] V. Tiwari, R. Donnelly, S. Malik, and R. Gonzalez, "Dynamic Power Management for Microprocessors: A Case Study," in *Proceedings, Tenth International Conference on VLSI Design*, pp. 185–192, 1997.

[6] A. Sudarsanam, and S. Malik, "Memory Bank and Register Allocation in Software Synthesis for ASIP's, "in *Proceedings of the International Conference on Computer-Aided Design*, pp. 388–392, 1995.

[7] M. T.-C. Lee, and V. Tiwari, "A Memory Allocation Technique for Low-Energy Embedded DSP Software," in *Proceedings of the Symposium on Low Power Electronics*, pp. 44–45, 1995.

[8] F. Catthoor, F. Franssen, S. Wuytack, L. Nachtergaele, and H. DeMan, "Global Communication and Memory Optimizing Transformations for Low Power Signal Processing Systems," in *Proceedings, IEEE Workshop on VLSI Signal Processing*, pp. 178–187, 1994.

[9] A. Hasegawa, I. Kawasaki, K. Yamada, S. Yoshioka, S. Kawasaki, and P. Biswas, "SH3: High Code Density, Low Power," *IEEE Micro.*, pp. 11–19, 1995.

[10] T. Sato, Y. Ootaguro, M. Nagamatsu, and H. Tago, "Evaluation of Architecture-Level Power Estimation for CMOS RISC Processors," in *Proceedings of the Symposium on Low Power Electronics*, pp. 44–45, 1995.

[11] M. T.-C. Lee, V. Tiwari, S. Malik, and M. Fujita, "Power Analysis and Minimization Techniques for Embedded DSP Software," *IEEE Trans. Very Large Scale Integration (VLSI) Syst.*, vol. 5, no. 1, pp. 123–135, 1997.

[12] T. Chou and K. Roy, "Accurate Estimation of Power Dissipation in CMOS sequential circuits," *IEEE Trans. Very Large Scale Integration (VLSI) Syst.*, 1996.

[13] F. N. Najm, "A Survey of Power Estimation Techniques in VLSI Circuits," *IEEE Trans. Very Large Scale Integration (VLSI) Syst.*, vol. 2, no. 4, pp. 446–455, 1994.

[14] A. Chandrakasan, R. Allmon, A. Stratakos, and R. Brodersen, "Design of Portable Systems," In *IEEE Custom Integrated Circuits Conference*, pp. 259–266, 1994.

[15] A. Chandrakasan et al., "Optimizing Power Using Transformations," *IEEE Trans. Computer-Aided Design Integrated Circuits Syst.*, vol. 14 no. 1, pp. 12–31, 1995.

[16] M. Weiser, B. Welch, A. Demers, and S. Shenker, "Scheduling for Reduced CPU Energy," in *Proceedings, First USENIX Symposium on Operating System Design and Implementation (OSDI)*, pp. 13–23, 1994.

[17] D. B. Lidsky and J. M. Rabaey, "Low Power Design of Memory Intensive Functions. Case Study: Vector Quantization," in *IEEE Workshop on VLSI Signal Processing, Proceedings*, pp. 378–387, 1994.

[18] J. W. Davidson and S. Jinturkar, "Memory Access Coalescing: A Technique for Eliminating Redundant Memory Accesses," *ACM SIGPLAN Notices*, vol. 29, no. 6, 1994.

[19] V. Tiwari and M. T.-C. Lee, "Power Analysis of a 32-Bit Embedded Microcontroller, in *Proceedings of the Asia and South Pacific Design Automation Conference*, pp. 141–148, 1995.

[20] S. Liao, S. Devadas, K. Keutzer, S. Tijang, and A. Wang, "Code Optimization Techniques for Embedded DSP Microprocessors," in *Proceedings of the Design Automation Conference*, pp. 599–604, 1995.

[21] T. Wilson and G. Grewal, "A Global Mode Instruction Minimization Technique for Embedded DSPs," in *Proceedings, The Sixth Great Lakes Symposium on VLSI*, pp. 18–21, 1996.

[22] S. Ellis, "The Low Power Intel486SL Microprocessor," in *IEEE COMPCON*, pp. 88–95, 1993.

[23] B. W. Suessmith and G. P. II, "PowerPC 603 Microprocessor Power Management," *Commun. ACM*, vol. 37, no. 6, pp. 43–46, 1994.

[24] S. Gary, C. Dietz, J. Eno, G. Gerosa, S. Park, and H. Sanchez, "The PowerPC (TM) 603 Microprocessor: A Low-Power Design for Portable Applications," in *Digest of Papers, Spring COMPCON 94*, pp. 307–315, 1994.

[25] M. B. Srivastava, A. P. Chandrakasan, and R. W. Brodersen, "Predictive System Shutdown and Other Architectural Techniques for Energy Efficient Programmable Computation," *IEEE Trans. Very Large Scale Integration (VLSI) Syst.*, *vol. 4, no. 1, pp.* 42–54, 1996.

[26] R. Lauwereins, M. Engels, J. Peperstraete, and E. Steegmans, "GRAPE: A CASE Tool for Digital Signal Parallel Processing," *IEEE ASSP Mag.*, vol. 7, no. 2, pp. 32–43, 1990.

[27] P. N. Hilfinger, Silage reference manual, draft release 2.0, http://infopad.eecs.berkeley.edu/~ hyper/Hyper/Intro/Silage/reference.ps, University of California, Berkeley, 1993.

[28] M. Potkonjak and J. M. Rabaey, "Power Minimization in DSP Application Specific Systems Using Algorithm Selection," in *Proceedings, International Conference on Acoustics, Speech, and Signal Processing*, pp. 2639–2642, 1995.

[29] D. Genin, J. DeMoortel, D. Desmet, and E. V. deVelde, "System Design, Optimization and Intelligent Code Generation for Standard Digital Signal Processors," in *Proceedings of the International Symposium on Circuits and Systems*, pp. 565–569, 1989.

[30] J. Sato, A. Y. Alomary, Y. Honma, T. Nakata, A. Shiomi, N. Hikichi, and M. Imai, "PEAS-I: A Hardware/Software Codesign System for ASIP Development," *IEICE. Trans. Fund. Electron. Commun. Computer Sci.*, vol. E77A, no. 3, pp. 483–490, 1994.

[31] N. N. Binh, M. Imai, A. Shiomi, and N. Hikichi, "An Instruction Set Optimization Algorithm for Pipelined ASIP's," *IEICE Trans. Fund. Electron. Commun. Computer Sci.*, vol. E78A, no. 12, pp. 1707–1714, 1995.

[32] I.-J. Huang, and A. M. Despain, "Synthesis of Instruction Sets for Pipelined Microprocessors," in *Proceedings, Design Automation Conference*, pp. 5–11, 1994.

[33] J. M. Rabaey, "Reconfigurable Processing: The Solution to Low-Power Programmable DSP," in *Proceedings, International Conference on Acoustics, Speech, and Signal Processing*, pp. 275–278, 1997.

INDEX

Printed and bound by CPI Group (UK) Ltd, Croydon, CR0 4YY

27/10/2024

14580254-0003